In the Light of Logic

Logic and Computation in Philosophy

Series Editors
Wilfred Sieg (Editor-in-Chief)
Clark Glymour
Teddy Seidenfeld

This series will offer research monographs, collections of essays, and rigorous textbooks on the foundations of cognitive science emphasizing broadly conceptual studies, rather than empirical investigations. The series will contain works of the highest standards that apply theoretical analyses in logic, computation theory, probability, and philosophy to issues in the study of cognition. The books in the series will address questions that cross disciplinary lines and will interest students and researchers in logic, mathematics, computer science statistics, and philosophy.

Mathematics and Mind
Edited by Alexander George

The Logic of Reliable Inquiry
Kevin T. Kelly

In the Light of Logic
Solomon Feferman

IN THE LIGHT OF LOGIC

Solomon Feferman

New York Oxford

Oxford University Press

1998

Oxford University Press

Oxford New York
Athens Auckland Bangkok Bogotá Buenos Aires Calcutta
Cape Town Chennai Dar es Salaam Delhi Florence Hong Kong Istanbul
Karachi Kuala Lumpur Madrid Melbourne Mexico City Mumbai
Nairobi Paris São Paulo Singapore Taipei Tokyo Toronto Warsaw

and associated companies in
Berlin Ibadan

Published by Oxford University Press, Inc.
198 Madison Avenue, New York, New York 10016

Library of Congress Cataloging-in-Publication Data
Feferman, Solomon.
In the light of logic / Solomon Feferman.
p. cm. — (Logic and computation in philosophy)
Includes bibliographical references and index.
ISBN 0-19-508030-0
1. Logic, Symbolic and mathematical. I. Title. II. Series.
QA9.2.F44 1998
511.3—dc21 97-51336

1 3 5 7 9 8 6 4 2
Printed in the United States of America
on acid-free paper

For Anita
Light of my life

Preface

Mathematics is extraordinarily distinctive in both its concepts and its methods. From the point of view of classical logic it is a rigid, precise, timeless, deductive "infallible" system, whereas modern *anti*-logical accounts stress its imprecision, fallibility, and dependency on time and culture. My own views fall between these two extremes. I believe that the anti-logical picture fails to explain what is truly distinctive about mathematics; as I see it, that can only be captured by eliciting its conceptual and logical structures. On the other hand, it is a fact that throughout its history, mathematics has been subject to fits of vagueness, uncertainty, puzzlement and, on occasion, sheer contradiction; but it is just these very problems that have made it necessary for there to be concerns about the foundations of mathematics. In the past, mathematicians dealt with such questions on a case-by-case basis in their respective disciplines, while in this century—as the nature of the problems began to cross many fields—they have largely become the province of mathematical logic.

This volume consists of a selection of my essays of an expository, historical, and philosophical character which in the main are devoted to the light logic throws on problems in the foundations of mathematics. I began writing them in the late 1970s; other pieces written over the same period which did not fit in directly with the plan chosen for this volume have been reserved for a future occasion. The present essays are grouped thematically rather than chronologically, and to some extent according to degree of accessibility to the reader. In particular, the first chapter was presented as a talk for a general audience. Beyond that, in order to give substance to the case which is made here for the essential role of logic in getting at the nature of mathematics, it is necessary to explain a number of technical concepts and results from metamathematics, that is, the logical study of formal axiomatic systems. While no knowledge of that subject on the part of the reader is presumed, a modicum of familiarity with logic would be helpful, especially in the later chapters of the volume. But I hope the general reader will find at least some of each chapter rewarding, and that those sufficiently engaged with the material will persist in reading all the way through, since there is an arc of thought which ties the problems brought out in the first part to results described later. As further assistance, an annotated list of references is included at the conclusion of chapter 1 which can be pursued in various directions, and again at various levels of technicality, in order to enlarge the reader's understanding of the role of logic in addressing foundational problems.

Because these essays were written as stand-alone pieces of edification and persuasion for different audiences on different occasions, there is, inevitably, a certain amount of overlap and repetition. At the same time, each chapter has a distinct purpose, and takes its place in the successive parts of the volume as follows. Part I consists of two chapters on foundational problems raised by the work of David Hilbert and Georg Cantor. These mainly concern the role and proper treatment of the mathematical transfinite, whose modern development at the hands of Cantor in his theory of sets in the latter part of the nineteenth century brought new problematic concepts and principles to the fore. That is followed in part II, first by a defense of the primacy of the logical analysis of mathematics for an account of its essential nature, and then by two chapters presenting general surveys of the manifold ways in which logic succeeds in contributing to the solution of foundational problems. (The first of those two chapters is a "slimmed-down" version of the second. They are both included in order to give the reader a choice according to taste and specific interests.) Part III is devoted to the work and thought of Kurt Gödel, whose results in the 1930s—on the incompleteness of formal systems containing arithmetic and on the consistency with axiomatic set theory of Cantor's continuum hypothesis—have been of utmost significance for all further work in our subject, and whose latter-day staunch platonism is a lightning rod for the modern philosophy of mathematics.

Part IV of this volume concentrates on a part of metamathematics called proof theory, which is a subject initiated by Hilbert in an effort to justify all of mathematics in terms of completely finitary principles. Hilbert's program is almost universally regarded as having failed in consequence of Gödel's incompleteness results, but proof theory has continued to be employed in a relativized form of the program by showing how certain formal systems embodying what may be regarded as problematic concepts and/or principles may be reduced to other systems which have a more evident conceptual basis. Then the concluding part V returns to the question broached in part I as to the extent to which the Cantorian transfinite is actually necessary for mathematical practice. I explain here how certain proof-theoretical results allow one to reduce systems in which substantially all of scientifically applicable mathematics can be directly formalized, to systems which rest on completely arithmetical principles. These results advance considerably a program initiated by Hermann Weyl in 1918 for a development of classical mathematical analysis on what is called a predicativist basis, that is, in which all sets are considered to be introduced by definition beginning with the natural numbers, rather than regarded as preexistent entities. The successes of this program put into question arguments advanced by Willard Van Orman Quine and Hilary Putnam, among others, for the justification of substantial portions of impredicative set theory on the grounds of its indispensability to natural science.

While no individual essay here is of a sustained philosophical character, much of the work in this volume is philosophically motivated, and many of the pieces explicitly address philosophical questions. It will soon be clear to the reader that I am a convinced antiplatonist in mathematics. Briefly, according to the platonist philosophy, the objects of mathematics such as numbers, sets, functions, and spaces are supposed to exist independently of human thoughts and constructions, and statements concerning these abstract entities are supposed to have a truth value independent of our ability to determine them. Though this accords with the mental practice of the working mathematician, I find the viewpoint philosophically preposterous; despite that, I have not tried to engage here the considerable philosophical literature, both pro and con, which exists on platonism in mathematics. Rather, my aim is to see what logic has to tell us in this respect. There are two sides to that: first, to examine critically those results claimed to buttress the platonist position, and second, to explore the viability of alternative philosophies for the foundations of mathematics. In particular, as the previous paragraph suggests, I have much to say about the unexpected reach in mathematical practice of predicative mathematics. This is a semiconstructive philosophy, going back to ideas of Henri Poincaré in the early part of the twentieth century, whose point and programmatic development beginning with that of Weyl (mentioned above) are not nearly as well known as strictly constructive programs such as Brouwer's intuitionism and those of more modern schools of that character. I have relatively little to say about constructivity, partly because I do not see the necessity, insisted upon by Brouwer and his followers, to restrict to constructive reasoning in order to obtain constructive results, and partly because those ideas are well represented and accessible at a variety of levels elsewhere in the literature. It should not be concluded from this, or from the fact that I have spent many years working on different aspects of predicativity, that I consider it the be-all and end-all in nonplatonistic foundations. Rather, it should be looked upon as the philosophy of how we get off the ground and sustain flight mathematically without assuming more than the basic conception of the structure of natural numbers to begin with. There are less clear-cut conceptions which can lead us higher into the mathematical stratosphere, for example, that of various kinds of sets generated by infinitary closure conditions. That such conceptions are less clear-cut than the natural number system is no reason not to use them, but one should look to see *where* it is necessary to use them and what we can say about what it is we know when we *do* use them.

I am indebted to Wilfried Sieg for urging me to publish this collection of essays in the present series, *Logic and Computation in Philosophy*, and to Oxford University Press editors Angela Blackburn, Robert Miller, and Cynthia Read for helping to shepherd it through. The volume itself would not exist without the sustained work of Kathy Richards, who reset all the original articles in a uniform format using the LaTeX computerized typesetting system, and who bore with me patiently through endless rounds of proofreading and changes of plans. Thomas Hofweber helped me compile and check for accuracy the references combined from the separate pieces. A substantial part of my own work in organizing this volume for publication was carried out while I held a fellowship at the Center for Advanced Study in the Behavioral Sciences at Stanford, California, during the academic year 1995–1996. I am most grateful to the Center and its director, Neil Smelser, and associate director, Robert Scott, for having provided that opportunity for work and study under the most satisfying conditions, as well as to the Andrew W. Mellon Foundation and the National Science Foundation for their support during my tenure there.

The reader will find specific acknowledgments to various of my colleagues in one or another of the essays below. It is to my wife, Anita, to whom I turned whenever I needed a willing ear and a special sensitivity to felicities of language; my thanks, as ever, to her.

Stanford, California S.F.
July 1998

Note to the Reader

Each essay in this volume comes with a special introductory footnote explaining its provenance and containing the requisite permission to republish. The essays themselves are all reprinted in a uniform format with minor additions and corrections. In some cases there are essential additions to the text, and these are set off by square brackets. There are also some new footnotes, indicated by an asterisk (*), to distinguish them from the previously numbered footnotes. Only chapter 13 has new material provided as a postscript.

A comprehensive reference list of symbols used at one point or another in the book is to be found following the final chapter, for the convenience of the reader who may want an occasional reminder of their meaning.

Contents

V COUNTABLY REDUCIBLE MATHEMATICS

Part I

FOUNDATIONAL PROBLEMS

1

Deciding the Undecidable: Wrestling with Hilbert's Problems

In the year 1900, the German mathematician David Hilbert gave a dramatic address in Paris, at the meeting of the Second International Congress of Mathematicians—an address which was to have lasting fame and importance. Hilbert was at that point a rapidly rising star, if not superstar, in mathematics, and before long he was to be ranked with Henri Poincaré as one of the two greatest and most influential mathematicians of the era. Like Poincaré, Hilbert worked in an exceptional variety of areas. He had already made fundamental contributions to algebra, number theory, geometry, and analysis. After 1900 he would expand his researches further in analysis, then move on to mathematical physics and finally turn to mathematical logic. In his work, Hilbert demonstrated an unusual combination of direct intuition and concern for absolute rigor. With exceptional technical power at his command, he would tackle outstanding problems, usually with great originality of approach.

The title of Hilbert's lecture in Paris was simply, "Mathematical problems." In it he emphasized the importance of taking on challenging problems for maintaining the progress and vitality of mathematics. And with this, he expressed a remarkable conviction in the solvability of all mathematical problems, which he even called an axiom. To quote from his lecture:

> Is the axiom of the solvability of every problem a peculiar characteristic of mathematical thought alone, or is it possibly a general law inherent in the nature of the mind, that all ques-

"Deciding the undecidable: Wrestling with Hilbert's problems" is the text of a previously unpublished lecture for a general audience delivered at Stanford University on 13 May 1994. Some minor revisions have been made to bring it up-to-date and to make it more suitable for its appearance here. However, its character as a lecture text has been retained, so there are few footnotes and no source references. I have added a list of selected references at the end of the chapter, which can be pursued for further information.

> tions which it asks must be answerable? . . . This conviction of the solvability of every mathematical problem is a powerful incentive to the worker. We hear within us the perpetual call: There is the problem. Seek its solution. You can find it by pure reason, for in mathematics there is no *ignoramibus*.[1]

To be sure, one problem after another had been vanquished in the past by mathematicians, though sometimes only after considerable effort and only over a period of many years. And Hilbert's own experience was that he could eventually solve any problem he turned to. But it was rather daring to assert that there are *no* limits to the power of human thought, at least in mathematics. And it is just this that has been put in question by some of the results in logic that I want to tell you about today.

Among mathematicians, what Hilbert's Paris lecture is mainly famous for is the list he proposed of twenty-three problems at the then leading edge of research. These practically ran the gamut of the fields of mathematics of his day, from the pure to the applied, and from the most general to the most specific. Naturally, the choice of these problems was to some extent subjective, restricted by Hilbert's own knowledge and interests, broad as those were. But the work on them led to an extraordinary amount of important mathematics in the twentieth century, and individual mathematicians would become famous for having solved one or another of Hilbert's problems. In 1975, a conference was held under the title *Mathematical Developments Arising from Hilbert Problems*, which summarized the advances made on each of them to date. In many cases, the solutions obtained thus far led to still further problems which were being pursued vigorously, though no longer with the cachet of having Hilbert's name attached.

The solution of three of Hilbert's problems were to involve mathematical logic and the foundations of mathematics in an essential way; they are the ones numbered 1, 2, and 10 in his list, but for reasons that you'll see, I want to discuss them in reverse order.[2]

Problem 10 called for an *algorithm* to determine of any given *Diophantine equation* whether or not it has any integer solutions. I'll explain to you what is meant by these specialized terms later. For the time being, just think of an algorithm as some sort of mathematical recipe or computational procedure. Algorithms are ubiquitous; we use them automatically in our daily arithmetical calculations. And algorithms are built into the myriad kinds of electronic devices all around us and in the software we use to put our computers through their special paces. Diophantine equations are equations expressed entirely in terms of integers and operations on integers, whose unknowns are also to be solved for integers. And by

---

[1] For a fuller extract from Hilbert's lecture, see the appendix to this chapter.

[2] Hilbert's statements of these problems are reproduced in full in the appendix to this chapter.

integers, of course, we mean the whole numbers 1, 2, 3, ... extended to include 0 and the negative integers −1, −2, −3, ... The integers are one of the basic number systems of mathematics; others that I'll also have to say something about are the *rational number system*, which are simply ratios of integers, in other words, fractions, and the *real number system*, which are the numbers used for measuring arbitrary lengths to any degree of precision. The most famous Diophantine equation is that addressed in the so-called Fermat's Last Theorem.

Contrary to Hilbert's expectations, Problem 10 was eventually solved in the negative. This was accomplished in 1970 by a young Russian mathematician, Yuri Matiyasevich, who built on earlier work in the 1950s and 1960s by the American logicians Martin Davis, Hilary Putnam, and Julia Robinson. (Incidentally, Robinson was just finishing her Ph.D. work in Berkeley with Alfred Tarski, one of the leading logicians of our time, when I began graduate studies there, and Tarski was to become my teacher as well. Because of her contributions in the following years to logic and recursive function theory, Robinson was eventually elected to the mathematical section of the National Academy of Sciences and later became president of the American Mathematical Society; moreover, she was the first woman to be honored in each of these ways.) The result of the Davis-Putnam-Robinson-Matiyasevich work, as we describe it nowadays, is that *the general problem of the existence of integer solutions of Diophantine equations is algorithmically undecidable.* Now the word "undecidable" here is being used in a very special technical sense that I'll explain to you later; there is also a second technical meaning of the term which we'll come to shortly. Neither of these has to do with the kinds of minor and major *indecisions* that we face in our daily lives—for example, whether to repaint the bathroom "Whisper White" or "Decorator White," or what to do about Bosnia.

To return to the undecidability result that was just stated: it's quite definite—there's no question about it—but that's by no means the end of the story. For, number theorists have been working on decision procedures for special classes of Diophantine equations, and there's a sizeable gap between what's known to be decidable by their work, and what's established to be undecidable by the work of Davis, Putnam, Robinson, and Matiyasevich. Moreover, Hilbert's Tenth Problem is just one of a host of decision problems that have been settled one way or the other, and even where the result is positive there remain significant open questions; these turn out to lie on the borderline of logic, mathematics, and computer science.

Next comes Hilbert's Second Problem, which called for a proof of consistency of the arithmetical axioms. Now, in 1900 Hilbert was a bit vague in stating just which axioms he had in mind in this problem. But when he took up logic full scale in the 1920s, he made quite specific what axioms were to be considered. Moreover, in order not to beg the question, he placed strong restrictions on the methods to be applied in consistency proofs of these and other axiom systems for mathematics: namely, these methods were to be completely *finitary* in character. The proposal to obtain finitary consistency proofs of axiom systems for mathematics came to be called *Hilbert's Program* for the foundations of mathematics; to avoid confusion with his list of problems, I will refer to this as *Hilbert's Consistency Program.* Hilbert himself initiated specific work in the 1920s on his formulation of Problem 2. Here again, contrary to Hilbert's expectations, there was a negative solution, namely, through the stunning results of the young Austrian logician Kurt Gödel, whose incompleteness theorems of 1931 have become among the most famous in mathematical logic. The title of Gödel's paper was "On formally undecidable propositions of *Principia Mathematica* and related systems." *Principia Mathematica* was the landmark work of Whitehead and Russell whose aim was to give a formal axiomatic basis for all of mathematics. Gödel showed that for any such system (and even much more elementary ones), there will always be individual propositions in its language that are undecidable by the system, that is, which can neither be proved nor disproved from its axioms provided it is consistent. Even more, Gödel showed that the consistency of such a system can't be proved within the system itself, and so finitary methods can't suffice. Though these results apparently knocked down Hilbert's Consistency Program, several questions remain which reach to the very foundations of our subject: Is incompleteness an essential barrier to the process of discovery in mathematics, or is there some way that it can be overcome? In answer to that, Gödel himself proposed one way forward, in fact, a way that connects up with Hilbert's First Problem, as we'll see. Later, the English logician and proto-computer scientist Alan Turing proposed a quite different way to overcome incompleteness that will be explained at greater length. A second important question is whether Hilbert's Consistency Program is still viable in any way, either by restricting its scope or by somehow enlarging the methods of proof to be admitted. There's been considerable research in recent years on both aspects of this question, and toward the conclusion of my lecture I'll tell you something about the current state of progress in that respect.

Hilbert's First Problem is in a way the most technical of the three, which is why I leave it for last. I will also have to content myself with giving an indication of its character and significance and what's been done about it.* Here Hilbert called for a proof of Cantor's conjecture on the cardinal number of the continuum of real numbers, the so-called *Continuum Hypothesis*. I'll not even try to explain the details of this here. Let me just say that this is part of a subject called *set theory* and that all presently generally accepted facts in set theory have been derived from principles which have been codified in a specific system of axioms for this subject, called the *Zermelo-Fraenkel axioms* for set theory, including (what's called) the *Axiom of Choice*. So naturally, one would also seek to decide the Continuum Hypothesis on the basis of these axioms, that is, either prove or

*[For more on this problem, see chapter 2.]

disprove it from them. The latter possibility was shown to be excluded by Gödel's second outstanding result, in 1938: he showed that *the Zermelo-Fraenkel axioms cannot disprove the Continuum Hypothesis.* There the matter rested until 1963 when Paul Cohen, of our own [Stanford] mathematics department, obtained the major result that *these same axioms cannot prove the Continuum Hypothesis.* In other words, *Cantor's conjecture is undecidable on the basis of currently accepted principles for set theory,* provided, of course, that the Zermelo-Fraenkel axioms are consistent. There was also a second part to Hilbert's First Problem in that he called for the construction of a specific well-ordering of the continuum. For those of you who know the meaning of this concept, let me just take a moment to explain how matters turned out in that respect, too. As you know, the Axiom of Choice, which I'm here counting among the Zermelo-Fraenkel axioms, implies that for any set there *exists* a well-ordering of that set, but it doesn't tell you how to construct one. In fact, I was able to show using Cohen's methods that it is consistent with the Zermelo-Fraenkel axioms plus the Continuum Hypothesis, that there is *no* definable well-ordering of the continuum—again contrary to Hilbert's expectations.

But now to return to Cantor's continuum problem itself, the first question to ask following the undecidability results of Gödel and Cohen is whether that situation could change by adding further axioms for set theory in some reasonable way. As it turns out, their results apply to *all* plausible (and even not so plausible) such extensions that have been considered so far. There are sharply divergent views as to whether it is still hopeful to obtain a reasonable extension of the Zermelo-Fraenkel axioms which will settle the Continuum Hypothesis. And, as in any subject, we have both optimists and pessimists: the direction one leans may be a matter of basic differences of temperament, but in this case I believe it really comes down to basic differences in one's philosophy of mathematics, namely, as to whether one thinks mathematics in general, and set theory in particular, is about some independently existing abstract, "platonic" reality, or whether it is somehow the objective part of human conceptions and constructions. The platonists say the Continuum Hypothesis must have a definite answer and so for them it is still hopeful to find that out by some means or other. I, for one, am a pessimist or, better, antiplatonist about the Continuum Hypothesis: I think that the problem is an inherently vague or indefinite one, as are the propositions of higher set theory more generally. On the other hand, I'm an optimist of sorts concerning Hilbert's Second Problem, so it's not just a matter of temperament. Anyhow, if you agree with me that the Continuum Hypothesis does not constitute a *genuine* definite mathematical problem, then its undecidability relative to any given axioms ceases to be an issue with which to struggle; it simply evaporates as a problem. Because of the limitations of time, and the technicality of this subject, I'll have to leave the discussion there, and won't try to say any more about Hilbert's First Problem in this lecture.

In each of these cases, the outcome of what Hilbert apparently took to be a fairly definite problem for which he expected a positive solution not only led in the opposite direction, but also to the very foundations of our subject. More broadly speaking, these results connect with the question whether there are any essential limits to the power of human reasoning. I don't pretend to have an answer to this question. My purpose here is just to try to explain what the attacks on the Hilbert problems have led to thus far, and why and how the struggle with them is still continuing.

So now, let's get down to work by returning to Hilbert's Tenth Problem, which called for an algorithm for determining of any given Diophantine equation whether or not it has any integer solutions. The idea of an algorithm (as I said earlier) is that of a step-by-step procedure or sequence of rules to go from the data of any specific problem of a certain type to its solution. The data could be a number, or an expression, or a sequence of numbers and symbols, and so on. The word "algorithm" is derived from the name of the ninth-century Persian mathematician Mohammed Al-Khowârizmi, who wrote a book of rules for adding, subtracting, multiplying, and dividing numbers in our familiar base ten notation. However, algorithms themselves are as old as mathematics; one of the most famous is called Euclid's algorithm for computing the greatest common divisor of two integers, which is much faster than the more obvious algorithm of factoring each number completely in order to combine all the common prime factors.

The word "Diophantine" stems from the third-century A.D. Greek mathematician Diophantus, who was the first to study solutions of equations in integers, more or less systematically. The subject of Diophantine equations lapsed for a long time after that but was revived in the seventeenth century by Pierre de Fermat, and this subject has received steady attention by number theorists ever since. Though Fermat made many important and lasting discoveries, which have been verified in one way or another, the so-called Fermat Last Theorem has challenged mathematicians almost to the present day. While the specific proposition stated in the theorem is simple enough to explain, I don't want to distract attention from our main line to spell it out, but I do want to tell something about the circumstances because they provide an interesting cautionary tale. What Fermat had done was write in the margin of his copy of the *Arithmetica* by Diophantus that he had found a truly marvelous demonstration of this proposition, which the margin was too narrow to contain. Well, did he or didn't he? In all other cases, we know from Fermat's correspondence and other evidence that he had done what he claimed to have done. But in this case there is no such further evidence, and we suspect he was simply mistaken—though we have good reason to believe that he saw how to handle certain specific cases. At any rate, many mathematicians worked on Fermat's Last Theorem with only limited partial success in the following 300 years, up to the present. In the summer of 1993 the Princeton mathematician Andrew Wiles announced in a lecture for specialists that he had finally found a proof of the Fermat Last

Theorem. This proof was so advanced and difficult technically that only a few experts in the world could work through it to help verify its correctness. As it happens, in the process of going over his 200-page manuscript, Wiles himself found a gap in his proof, and so it was up in the air whether the work was complete. It took another year before he was able, in collaboration with Richard Taylor, to get around this stumbling block, and now the proof seems to be accepted by the appropriate experts. The history of all this shows that problems about even very simple Diophantine equations are notoriously difficult. Surely, Hilbert was aware of this even back in 1900, and his incautious optimism in the face of that in his statement of the Tenth Problem is surprising.

In the 1920s, other decision problems in algebra and logic emerged which looked equally difficult, and the feeling began to grow that for some of these, *no possible algorithm could work*, or—as we say nowadays—that the problem is *algorithmically undecidable*. Now, if someone comes along with a proposed algorithm to settle a given decision problem in a positive way, one can check to see that it does the required work (or at least try to do so), without inquiring into the general nature of what constitutes an algorithm. But if it is to be shown that the problem is undecidable, one has to have a precise explanation of what algorithms can compute in general. Analogous situations had arisen earlier in mathematics, for example, to show that there is no possible construction by straightedge and compass which will trisect any given angle, or that there is no possible explicit formula for finding the roots of any polynomial equation (in one unknown) of degree 5 or higher (we do have such formulas for degrees 1, 2, 3, and 4). In each of these cases, one first needed a precise characterization of everything that can be constructed or defined in the specified way, before showing that certain problems *cannot* be solved by such constructions or definitions.

Similarly, in order to establish undecidability results, one first needed to have a precise characterization of what in general can be computed by an algorithm. Several very different looking answers to this were offered by logicians—including Kurt Gödel, Alonzo Church, and Alan Turing—in the mid 1930s, but they were eventually all shown to be equivalent. The most familiar explanation is that due to Turing, who described what can be done by an ideal computer if no restrictions are placed on how much time or memory space is required to carry out a given computation. These are called *Turing Machines* nowadays, and Turing's conception is the foundation of theoretical computer science, at least of *what can be computed in principle*. Both Church and Turing applied their characterization of what is algorithmically computable, to show that *the decision problem for* [what is called] *first-order logic is undecidable*, that is, that no possible algorithm can be used to determine of any given proposition formulated in the symbolism of that logic whether or not it is universally valid. In the years following that first work of Church and Turing, many other logical and mathematical problems were shown to be undecidable. But it took a long time to settle Hilbert's Tenth Problem. Significant progress was made on this in the 1950s and '60s by Davis, Putnam, and Robinson, but it was not until 1970 that Matiyasevich was able to take the final crucial step to show that there is no possible decision procedure which will determine of an arbitrary Diophantine equation whether or not it has any integer solutions. Thus, at least in the terms that Hilbert posed the Tenth Problem, the answer is definitely negative.

So what's unsettled here? Well, the best that the logicians working on this have been able to show is that there is no decision procedure for equations in nine or more unknowns. But number theorists have been working on comparatively special classes of Diophantine equations, and the best *they* have done is to obtain decision procedures for certain classes of equations in two unknowns. So, there is here a sizable gap between the positive cases and the negative cases, and it is really up in the air how this will be filled, if at all.

The vast body of higher mathematics deals with problems where it's not even appropriate to ask whether they fall under a mechanical decision procedure, that is, whether they are algorithmically decidable or undecidable. Whatever one's expectations, the problems that I've mentioned about logical validity and about Diophantine equations *were* cases where it was appropriate to ask that kind of question but where, as we have seen, the answers turned out to be negative. These results tend to confirm our everyday impressions about the inherent difficulty of such general problems. It is thus surprising to see how far logicians were able to establish the *decidability* of various nontrivial classes of algebraic problems. For example, in the 1930s, before he immigrated to the United States, Alfred Tarski established a famous decision procedure for the algebra of real numbers. This allows one to determine, among other things, of any finite system of equations and inequalities in any number of unknowns whether or not it has some solution in real numbers. Tarski's procedure has both theoretical and practical applications. However, when questions of actual feasibility of computation came to the fore with the advent of high-speed electronic computers in the 1950s, such decision procedures began to be reexamined. So here we face the new question of *computability in practice* instead of computability in principle.

In the mid 1970s, it was shown by Michael Fischer and Michael Rabin that *no* algorithm for the algebra of real numbers can work faster than exponential rate in general. That is, given a problem expressed with n symbols, it will, in general, take on the order of 2^n steps (that is, $2 \times 2 \times 2 \times \ldots, n$ times) to settle it. For $n = 50$, taking 2^{50} steps is beyond the limits of actual computers—it would require about 35 years full time, and for $n = 60$ or beyond, it would take over 30,000 years full time. Fischer and Rabin also obtained similar results for other cases where the decision problem had been settled positively. Remarkably, the decision problem for the first-order theory of integers under addition alone requires 2^{2^n} steps for an input of length n, which becomes prohibitive for $n = 6$ or beyond, and for multiplication it goes even one exponential level higher. All these might be considered examples of *undeciding the decidable.*

There are some specific problems concerning multiplication which are of great practical importance, since modern cryptographic systems are based on them. These are the problems of determining of an integer whether or not it is a prime number, and the related problem of factoring an integer into prime parts. It is believed that the factorization problem is not, in general, computationally feasible for large integers, though there is no proof of that. On the other hand, there are relatively quick methods of determining primality. And—if one is willing to give up absolute certainty—there are even much faster methods that work within seconds using nondeterministic procedures, by making successive random guesses; one of these is due to Michael Rabin, and another is due to Robert Solovay and Volker Strassen. So, if you start looking at the question, what can be decided within probability, say, 99%, and within practical time limits, *all* problems such as those indicated above have to be reexamined.

You may have noticed that I hedged one statement above, namely, that if someone comes to you with a proposed algorithm, you can check whether it does the required work—to which I added: "or, at least, try to do so." The point is that there is *no* mechanical method to verify that an algorithm does what it is supposed to do, or even that it always terminates with an answer. Again, our picture of this has changed considerably with the advent of high-speed computers. Nowadays, algorithms are fed to machines in the form of programs, and many programs in actual practice are very long and complicated, sometimes requiring thousands of lines. It is thus of great practical importance to have good design principles for programs that allow one to break them into manageable, understandable parts and to have usable methods to prove their termination and correctness. This is really a problem in logic, and in recent years the theory of proofs originally developed for Hilbert's Second Problem has turned out to be one of the major tools to deal with these problems in computer science.

My own involvement in applications of proof theory to questions of termination and correctness of computer programs has been relatively recent, starting around 1989. But I've been making good use in this of a general approach to systems of reasoning for various forms of constructive mathematics that I initiated in the mid 1970s. It is very satisfying to see the central roles that have been played by ideas and methods (via all sorts of approaches) from proof theory and constructivity in the enormous ferment that has been taking place between logic and computer science, especially during the last decade.

Before getting into Hilbert's Second Problem, I have to say something in very broad terms about its historical background going back to antiquity, and the historical tension between the process of discovery and invention in mathematics on the one hand and its systematization and verification on the other. Greek mathematics was dominated by geometry; the best known names from that period are those of Euclid and Archimedes. In Euclid's

Elements, circa 300 B.C., geometry was developed axiomatically, and for many years this exposition was considered the ideal of what mathematics, if not all logical argument, is or should be like. Not much is known about Euclid personally, but what *is* known is that he was not himself one of the main creators of Greek geometry, but rather a systematizer of what was achieved by his day. Bertrand Russell said that Euclid's *Elements* is certainly one of the greatest books ever written; a recent commentator added that it is also one of the dullest. From Euclid you get no idea about how mathematics is actually discovered, how one arrives at the constructions, in many cases ingenious, that lead from the data to the conclusion; one can only go through his proofs step by step to see that they are indeed correct. What has come down to us from Archimedes is much closer in spirit and practice to modern mathematics, and in fact, some of his methods anticipate the modern calculus and can't be reduced to Euclidean geometry.

After the Dark Ages in Europe, mathematics was revived in the early Renaissance, first with the need for a more practical arithmetic for commerce, then with the development of algebra in the fifteenth and sixteenth centuries. The results of Greek mathematics were transmitted to the West through Arabian sources. In the seventeenth century, geometry and algebra were married in the so-called analytical geometry of Fermat and Descartes. Then mathematics exploded at the beginning of the eighteenth century through the creation of the calculus by Newton and Leibniz, and especially with Newton's applications of calculus and differential equations to physics. Here we have the wholesale use of infinite methods in mathematics, and the use of troublesome concepts such as infinitesimals and infinite sums. In the eighteenth century and into the nineteenth century mathematics was mostly developed in a free-wheeling way. The processes of discovery and invention outran careful verification. That only began to receive systematic attention in the nineteenth century, first with the foundations of the calculus and analysis and then going still deeper with the axiomatic foundations of the number systems basic to mathematics. Richard Dedekind gave axioms for the real number system with a kind of reduction of that to the rational numbers, and those are in turn easily reduced to the positive integers; finally both Dedekind and Giuseppe Peano gave axioms for the system of positive integers. In the latter part of the nineteenth century, the very basics of logical reasoning in mathematics were analyzed in a new symbolic form by Gottlob Frege, and that work was combined with Peano's symbolism by Bertrand Russell in the early twentieth century in his attempt to give a development of all of mathematics in completely symbolic logical form.

Even Euclid's geometry turned out to be lacking full rigor; without realizing it, Euclid had made implicit use of certain hypotheses that were not included in his axioms. Moreover, one of the most startling discoveries of the nineteenth century was the realization of the existence of non-Euclidean geometries, coming out of the independence of the parallel postulate from the other postulates. One of Hilbert's major pieces of work before 1900 was the development of Euclidean and non-Euclidean geometry in a completely rigorous way. One part of his work shifted attention to the question of *independence* of this or that axiom from the others. That's really a question of *consistency*: statement A is independent from a set of statements S if the negation of A is consistent with S.[3] One way to prove consistency of a set of statements is to produce a model for it, and that model has to be defined in terms that we already recognize to be consistent. Hilbert was here evolving what we nowadays call a *metamathematical* point of view: we treat mathematics formally as what can be carried out in an axiom system; then we investigate questions of independence, consistency, and completeness for these axioms. For Hilbert, a mathematical concept "exists" if it is determined by a system of axioms. And to be thus determined, the system must be both *consistent* and *complete*; that is, it should suffice to prove or disprove every proposition in the subject matter under consideration. What Hilbert did in verifying the consistency of various combinations of the geometrical axioms was to construct models by means of analytic geometry, that is, built up using the real number system. So, when in 1900 Hilbert raised his second problem, he was calling next for a proof of consistency of axioms for the real numbers. In addition, he raised there the following much more ambitious question. In the latter part of the nineteenth century, Georg Cantor's theory of sets had introduced revolutionary and surprisingly powerful concepts and methods to mathematics, through a full embrace of the mathematical infinite. Now, that had been shown to lead to inconsistencies when the basic set-theoretic principles were assumed unrestrictedly. In Problem 2, Hilbert also called for a consistency proof of a somehow restricted theory of sets, though he did not say anything about how that was to be regarded axiomatically. In fact, the first axioms for set theory were not introduced until a few years later by Ernst Zermelo, one of Hilbert's protégés.

In the first two decades of the twentieth century, Hilbert concentrated in his own work on problems in analysis and mathematical physics, but he continued to lecture regularly at Göttingen on the foundations of mathematics. Then, in the 1920s, when Hilbert turned to logic and metamathematics full scale (with the assistance of Wilhelm Ackermann and Paul Bernays), he proposed much more definite consistency problems to be solved than he had in 1900. But overarching these questions was what he considered to be the general problem of the infinite in mathematics. Even Peano's axioms for the integers involved, according to his view, an implicit appeal

[3] There is another usage of "independence" in logic whose meaning is close to this but different; namely, statement A is said to be independent of S if neither A *nor* its negation is provable from S. In the terminology used in the discussion of Hilbert's Problem 1 above, that is the same as saying that A is an undecidable proposition on the basis of S. And, in terms of consistency, that is equivalent to A and its negation being consistent with S. The two distinct senses of "independence" are a source of possible confusion when reading the literature, especially when it is not specified which is being used.

to the infinite, and one would also have to demonstrate their consistency and completeness. But if all infinitary methods would have to be justified on a *prior* basis, that could only admit finitary methods, that is, reasoning about finite combinations of objects confined entirely to such combinations. In Hilbert's view, with mathematics represented in axiomatic systems, and every proposition and proof represented in such a system as a finite sequence of basic symbols, one could hope to prove consistency of such a system by completely finitary methods, by showing that no possible proof could result in a contradiction. This would be done through a theory of proofs and their transformations which Hilbert initiated. Again, he was fully optimistic for his program to secure all of mathematics by finitary consistency proofs, first for number theory with Peano's axioms, then for the real number system and analysis, and finally for set theory.

Hilbert's program thus received a big shock in 1931 when Kurt Gödel showed that Peano's axioms and, moreover, any effectively described extension of those axioms—including analysis and set theory—are *incomplete* if consistent.[4] Moreover, the consistency of such a system cannot be proved within itself, so any such proof would have to use something from outside the system. In particular, any system which incorporates all finitary methods cannot be proved consistent by those methods.

How does all this look to us now? It's generally agreed that Gödel's results definitively undermine Hilbert's program, at least as originally conceived. Incidentally, Hilbert himself never admitted that, so strong were his convictions that his general program was the only way to go: he said that Gödel's results showed it would be more difficult to carry his program through than he originally thought. His co-workers, though, saw that some modification would be necessary, and that in fact is what happened in the subsequent development of proof theory.

But nowadays, the whole preoccupation with consistency proofs has receded as a matter of central concern. On the one hand, hardly anyone doubts that the axiom systems of number theory, analysis, and set theory in current use are consistent. The reason is that—whether or not one is a platonist—they accord with basic informal, reasonably coherent, conceptions we have of what these systems are supposed to be about, under which their axioms are recognized to be correct. Moreover, in the past, all inconsistencies resulted from very obvious defects of formulation, and those defects have been eliminated. To be sure, there could still be hidden defects we're not aware of, but none has emerged even after a massive amount of detailed investigation of these systems.

Also, Hilbert's idea that mathematical concepts "exist" *only* through axiom systems for them is accepted by very few. For, given that the sys-

[4]To be more precise, Gödel required a slightly stronger condition, called omega-consistency, for his result. Later, J. Barkley Rosser showed by a modification of Gödel's argument that simple consistency is sufficient for incompleteness. In any case, Gödel's theorem on unprovability of consistency did not require the stronger hypothesis.

tems we use are necessarily incomplete (granted their consistency), no such system can be said to fully determine its subject matter. So we are led back to philosophical questions about the nature of mathematical concepts and how we come to accept and have our knowledge about them, questions that are just the sort that Hilbert hoped to avoid by his consistency and completeness programs.

While consistency may no longer preoccupy us, Gödel's incompleteness results are still a live matter. For, proofs are the only means we have for final verification of mathematical truths. And proofs must proceed from what is evidently true or already known to be true by incontrovertible steps to what is to be established. And when we analyze this picture, we are led to the representation of proofs in formal axiomatic systems. Every system in use for other than very specialized parts of mathematics contains Peano's Axioms. So, by Gödel's results, it is incomplete (assuming it is consistent). From this line of argument, it appears that there will be sensible mathematical propositions which we can never prove or disprove, and that—contrary to Hilbert's general view—*ignoramus et ignorabimus*.[5] Is this conclusion really justified by Gödel's incompleteness results?

To begin with, let's look more specifically at how Gödel proved those results themselves. What he did was to translate properties of symbols and finite sequences of symbols, including formal propositions and proofs, into properties of numbers systematically assigned to those symbols and finite sequences of symbols. Then, for each of the axiom systems S to which his results apply, he was able to construct a number-theoretical proposition A that "says of itself" (under this translation) that it is not provable in S. Next he showed that if S is consistent, then the proposition A constructed in that way is indeed not provable in S, but since that's exactly what A asserts of itself, it follows that A is true, though we can't establish its truth *within* S. Now, if S only proves true statements, as we would want, then certainly the negation of A is also not provable in S: that is, A is undecidable by S. In fact, a much weaker hypothesis on S stated by Gödel is sufficient for this second part, and by a variant argument due to Barkley Rosser, it is sufficient for both parts to simply assume that S is consistent.[6]

Like Heisenberg's Uncertainty Principle, the Big Bang, and black holes, Gödel's incompleteness theorems and their self-referential aspects have attracted much popular attention. Paradoxical self-reference goes back to antiquity, and we still have no generally accepted way of accounting for the so-called Liar Paradox, "I am lying," or "This statement is false," which if true is false, and if false is true. What Gödel did was similar, using provability in S in place of truth, but unlike the liar statement, it is non-paradoxical, since we *do* have a proof of A, though not in S; that's the escape hatch. One of the most widely read popularizations of Gödel's the-

[5] The Latin phrase, translated as "we do not know and shall never know," is credited to Emil du Bois Reymond.

[6] Cf. ftn. 4.

orem is to be found in the book *Gödel, Escher, Bach: An eternal golden braid* by Douglas Hofstadter (1979), in which he made vivid but loose play with analogous uses of self-reference or self-return or "strange looping" in the art of M. C. Escher and the music of J. S. Bach, as well as in computer programs which refer back to or "call" themselves in the process of computation. This last use of self-reference is more aptly analogous to Gödel's than the others. Since Hofstadter's ideas are all over the place, in my opinion that's *not* where to go for a straight story on Gödel's theorem, but it is entertaining and stimulating and not *too* misleading. Other examples of popular bounce-offs that some of you may have seen are the books by Raymond Smullyan, *What is the name of this book?* (1978) and *Forever undecided: A puzzle guide to Gödel* (1987). And, science-fiction aficionados tell me of several recent works featuring Gödel's theorem in their plots. But I'll bet that few, if any, of you are aware of what I think is the most unusual spin-off from the Gödel incompleteness theorems, namely, the Second Violin Concerto by Hans Werner Henze, which incorporates a poem by Hans Magnus Enzensberger, entitled "Hommage à Gödel."[7]

But to come back to the Gödel example of an undecidable sentence A, it is disappointing in the following way: as soon as we follow Gödel's argument, we are ready to accept A; that is, it becomes *decided*, though by an argument which goes beyond those represented in the system S. A second source of disappointment is that A itself has no prior mathematical interest, unlike famous unsettled problems in number theory such as the Twin Primes Conjecture or Goldbach's Conjecture.

Now logicians, starting with work by Jeff Paris and Leo Harrington, have been chipping away at this second concern by coming up with propositions which use only ordinary mathematical notions—that is, are not obtained by translating logical symbolism into number theory—and which are independent of Peano Arithmetic and many still stronger systems S. Unlike Gödel's A, which is constructed in the same way for each system S, these are more and more complicated the stronger one takes S. But here's the kicker: just like Gödel's statement A, we always know how to decide these more mathematical looking propositions, because they're provably equivalent to a form of consistency of S (unless there is a question about that very consistency).

Returning to Gödel's own proposition A undecidable in S, he also show-linebreak ed that if C is the number-theoretical translation of the statement that S is consistent, then C implies A in S. Hence, also C is not provable in S if S is consistent. Now C is of prior *metamathematical* interest, since it is just this which was the object of Hilbert's Consistency Program, to be established by finitary means. And it is not a "cooked up" statement, like A.

[7]At this point in the lecture, I played a passage from a taped performance of this composition which was one of its high points vis à vis Gödel's theorem. To my knowledge, no recording of this piece was commercially available at the time of writing.

In Gödel's view, the "true reason" for his incompleteness theorems lies in the fact that beyond any system S, we must accept new axioms concerning arbitrary subsets of the universe of objects with which S deals. If S is itself a system of set theory, these new axioms are called "axioms of higher infinity," since the new sets obtained will be infinitely larger in a suitable sense than the sets which can be shown to exist in S. It is indeed the case that by adding such new axioms one is able to establish the consistency of S. Of course, we then obtain a new system S′—which is again incomplete—and then the process of adding new axioms must be repeated, so it must be iterated indefinitely. All this accords with Gödel's underlying belief in the platonic reality of set theory and that the kind of informal reasoning which led us to accept the Zermelo-Fraenkel axioms, as true of this reality, can be continued to expand these axioms indefinitely to settle hitherto undecided propositions.

If one does not subscribe to Gödel's philosophy, there are much more philosophically conservative ways of overcoming incompleteness to be considered. The first of these was due to Alan Turing in 1939, following his work on the general theory of computation. Namely, starting with any given system S_1 all of whose number-theoretical consequences are true, (for example, Peano's axioms), one simply adds the statement C_1 of consistency of S_1 as a new (and, in fact, true) axiom to S_1 in order to obtain S_2, then similarly adds C_2 to S_2 to obtain S_3, and so on. However, doing this an unbounded finite number of times is not the end of the story, because the "limit" system S_ω, which is S_1 together with C_1, C_2, C_3, etc., is still incomplete, so we can start the process all over again. As we say, the procedure of adding consistency statements can be iterated effectively into the transfinite. At any stage S in this iteration process, the consistency statement C for S is added as a new axiom to S in order to obtain a system S′ at the next stage, and at limit stages one simply combines all axioms previously obtained to form a new system.

Turing's main result about this procedure is a completeness theorem for purely universal arithmetical statements, that is, for statements of the form that all positive integers have an effectively checkable property. (The consistency statements C and also many number-theoretical statements such as Goldbach's Conjecture are of this form.) Turing's result is that any such statement which is true is provable at some stage in the transfinite iteration process. There is, however, a catch to his result, namely, that it is very dependent on *how* the passage into the transfinite is effected, and there is no obvious way in advance to prefer one such passage to another. This is a delicate matter which can't be explained without fairly technical notions, and so I won't try to say anymore about it here. But basically, the catch is that *we have traded incompleteness for uncertainty about which path to follow;* in other words, you don't get something for nothing.

Formal consistency statements are just one kind of statement of trust in the correctness of a system; more general such statements are called *reflection principles*, because they result from reflecting on what has led one to accept that system in the first place. In the early 1960s, I obtained a considerable strengthening of Turing's completeness result, and showed that for the transfinite iteration of adding a reflection principle as new axioms at each stage, one is able to prove *every* true arithmetical statement at some stage in this transfinite progression of systems. However, this completeness theorem suffers the same catch as Turing's earlier result, since it is dependent on which paths are taken into the transfinite, the choice of which is not justified in advance.

Well, is there anything that can be rescued from all this that doesn't require such a trade-off? The answer is yes, if we're willing to settle for less, namely, if we impose a kind of "boot-strap" or autonomous generation condition that requires justifying which path into the transfinite is to be taken before moving forward on that path. I've called what we obtain by imposing this condition an *autonomous transfinite progression of axiom systems*, and began investigating the properties of this notion some years ago. I've shown more recently that everything which is proved in such a progression for the case of Peano's axioms as S_1 lies in what I call the *reflective closure* of S_1, which is defined directly without the apparatus of transfinite ordinals.

Now, in general, while the reflective closure of a system S_1 overcomes the incompleteness of S_1 to a considerable extent, it too is still incomplete. Its significance, as I've argued especially in the case of Peano's axioms for the initial system S_1, is instead that it contains everything you *ought* to have accepted if you accepted the concepts and principles in S_1. In order to go beyond the reflective closure of S_1, one must use essentially new concepts or essentially new kinds of arguments. This accords with the historical necessity in mathematics to introduce, periodically, essentially new concepts and principles. However, more work needs to be done (and is waiting to be done) on what is obtained by reflective closure applied to other initial axiom systems, in order to test this general view of its significance.

What, finally, is to be said about the viability of Hilbert's Program? The part of logic that was developed to carry this out is called *proof theory*. In order to use this for the ultimate foundations of mathematics in which all vestiges of assumptions about the infinite would be eliminated, Hilbert had insisted on carrying out proof theory by strictly finitary means. Gödel's incompleteness theorems showed that if proof theory is to be applied to stronger and stronger systems, one will have to give up that restriction in one way or another. Indeed, once that was abandoned, proof theory made enormous strides forward, and has led to a number of technical results of very high order of importance, ones in which I happen to have been among the leaders in promoting. At the same time, I've been very concerned that they not be pursued simply for their own sake; this requires a new rationale and conceptual framework to take the place of the original Hilbert Program. In my view, a fruitful way of looking at what has been accomplished by this work is as a *relativized* form of Hilbert's program, in which we develop a global picture of *What rests on what?* in mathematics, that is, what higher-order and/or nonconstructive concepts and principles can be reduced to *prima facie* more basic ones.*

But there has also been unexpected progress in recent years in the opposite direction, namely, to see what can be done in systems to which strictly finitary proof theory applies. Here, the surprising result is the demonstration that an enormous amount of scientifically applicable mathematics, even using very modern methods, can be carried out in proof-theoretically very weak systems or, as I have put it elsewhere, that *a little bit goes a long way.*** This is of philosophical interest because it shows that infinitary concepts are not essential to the mathematization of science, all appearances to the contrary. And this also puts into question the view that higher mathematics is justified by science or is somehow embodied in the world, rather than that it is the conceptual edifice raised by mankind in order to make sense of the world. If, then, the search for absolute truth in mathematics is replaced by the progressive clarification and improvement of human understanding, incompleteness ceases to be a bug-a-boo.

If Hilbert were to reappear at a meeting of the International Congress of Mathematicians in the year 2000, he would have to take cognizance of all the work I've told you about with respect to these three problems. No doubt he would be shaken in his original naive formulations and his optimism concerning them. But I venture to say that he would come up with a new and more sophisticated version of his basic belief in the solvability of all mathematical problems, something like *there are no genuine absolutely undecidable problems.* The hard part for him would be to say what constitutes a *genuine* mathematical problem, and by what means it is to be decided. But since he isn't coming back, it will be up to us and those who follow to try to answer these questions.

Appendix: Hilbert's Problems 1, 2, and 10

The circumstances surrounding Hilbert's address to the meeting of the International Congress of Mathematicians in Paris in 1900 are well described in the biography *Hilbert* by Constance Reid (1970, pp. 69–83). Due to limitations of time, Hilbert actually presented only ten out of his list of twenty-three problems there. The text of his lecture with the full list of problems was first published in German in 1900 and was followed by an authorized English translation under the title "Mathematical problems" (Hilbert 1902); a reprinting of the latter is in Browder (1976, pp. 1–34). Prior to the list of problems, the first part of that text (pp. 1–7 *op. cit.*) consists of a general discussion of the vital role played by mathematical

*[This is spelled out in chapter 10.]

**[That work is reproduced in chapter 14.]

problems in the advancement of mathematics, their varied sources, the necessity of rigorous demonstration in their solutions, and an acknowledgment of the fecundity of good notation and informal geometric reasoning. It concludes with Hilbert's statement of his conviction in the solvability of all definite mathematical problems; that section of his lecture is reproduced below, along with his statements (in full) of the problems numbered 1, 2, and 10.

Special note should be taken of Hilbert's recognition that in certain cases a problem may be "solved" by showing its impossibility of solution by prescribed methods or "hypotheses." Among these he mentions the parallel postulate and equations of the fifth degree. The former is independent of the remaining axioms of geometry; that is, in our terminology, it is an undecidable proposition on the basis of those axioms. The latter are not solvable by radicals (that is, their roots are not in general expressible by formulas built up from the coefficients by means of the arithmetic operations and nth roots); this is similar to the situation of algorithmically unsolvable (that is, undecidable) problems, discussed above in connection with the Tenth Problem.

* * *

Herewith, the extracts from Hilbert (1902):

> Occasionally it happens that we seek the solution [of a problem] under insufficient hypotheses or in an incorrect sense, and for this reason do not succeed. The problem then arises: to show the impossibility of the solution under the given hypotheses, or in the sense contemplated. Such proofs of impossibility were effected by the ancients, for instance when they showed that the ratio of the hypotenuse to the side of an isosceles right triangle is irrational. In later mathematics, the question as to the impossibility of certain solutions plays a preeminent part, and we perceive in this way that old and difficult problems, such as the proof of the axiom of parallels, the squaring of the circle, or the solution of equations of the fifth degree by radicals have finally found fully satisfactory and rigorous solutions, although in another sense than that originally intended. It is probably this important fact along with other philosophical reasons that gives rise to the conviction (which every mathematician shares, but which no one has as yet supported by a proof) that every definite mathematical problem must necessarily be susceptible of an exact settlement, either in the form of an actual answer to the question asked, or by the proof of the impossibility of its solution and therewith the necesssary failure of all attempts. Take any definite unsolved problem, such as the question as to the irrationality of the Euler-Mascheroni constant C, or the

existence of an infinite number of prime numbers of the form $2^n + 1$. However unapproachable these problems may seem to us and however helpless we stand before them, we have, nevertheless, the firm conviction that their solution must follow by a finite number of purely logical processes.

Is this axiom of the solvability of every problem a peculiarity characteristic of mathematical thought alone, or is it possibly a general law inherent in the nature of the mind, that all questions which it asks must be answerable? For in other sciences also one meets old problems which have been settled in a manner most satisfactory and most useful to science by the proof of their impossibility. I instance the problem of perpetual motion. After seeking in vain for the construction of a perpetual motion machine, the relations were investigated which must subsist between the forces of nature if such a machine is to be impossible; and this inverted question led to the discovery of the law of the conservation of energy, which, again, explained the impossibility of perpetual motion in the sense originally intended.

This conviction of the solvability of every mathematical problem is a powerful incentive to the worker. We hear within us the perpetual call: There is the problem. Seek its solution. You can find it by pure reason, for in mathematics there is no *ignorabimus*. . . .

1. Cantor's Problem of the Cardinal Number of the Continuum

Two systems, that is two assemblages of ordinary real numbers or points, are said to be (according to Cantor) equivalent or of equal *cardinal number*, if they can be brought into a relation to one another such that to every number of the one assemblage corresponds one and only one definite number of the other. The investigations of Cantor on such assemblages of points suggest a very plausible theorem, which nevertheless, in spite of the most strenuous efforts, no one has succeeded in proving. This is the theorem:

Every system of infinitely many real numbers, that is every assemblage of numbers (or points), is either equivalent to the assemblage of natural integers, 1, 2, 3, . . . or to the assemblage of all real numbers and therefore to the continuum, that is, to the points of a line; *as regards equivalence there are, therefore, only two assemblages of numbers, the countable assemblage and the continuum.*

From this theorem it would follow at once that the continuum has the next cardinal number beyond that of the countable assemblage; the proof of this theorem would, therefore, form a new bridge between the countable assemblage and the continuum.

Let me mention another very remarkable statement of Cantor's which stands in the closest connection with the theorem mentioned and which, perhaps, offers the key to its proof. Any system of real numbers is said to be ordered, if for every two numbers of the system it is determined which one is the earlier and which is the later, and if at the same time this determination is of such a kind that, if a is before b and b is before c, then a always comes before c. The natural arrangement of numbers of a system is defined to be that in which the smaller precedes the larger. But there are, as is easily seen, infinitely many other ways in which the numbers of a system may be arranged.

If we think of a definite arrangement of numbers and select from them a particular system of these numbers, a so-called partial system or assemblage, this partial system will also prove to be ordered. Now Cantor considers a particular kind of ordered assemblage which he designates as a well ordered assemblage and which is characterized in this way, that not only in the assemblage itself but also in every partial assemblage there exists a first number. The system of integers 1, 2, 3, ... in their natural order is evidently a well ordered assemblage. On the other hand, the system of all real numbers, that is the continuum in its natural order, is evidently not well ordered. For, if we think of the points of a segment of a straight line, with its initial point excluded, as our partial assemblage, it will have no first element.

The question now arises whether the totality of all numbers may not be arranged in another manner so that every partial assemblage may have a first element, that is whether the continuum cannot be considered as a well ordered assemblage—a question which Cantor thinks must be answered in the affirmative. It appears to me most desirable to obtain a direct proof of this remarkable statement of Cantor's, perhaps by actually giving an arrangement of numbers such that in every partial system a first number can be pointed out.

2. The Compatibility of the Arithmetical Axioms

When we are engaged in investigating the foundations of a science, we must set up a system of axioms which contains an exact and complete description of the relations subsisting between the elementary ideas of that science. The axioms so set up are at the same time the definitions of those elementary ideas; and no statement within the realm of the science whose foundation we are testing is held to be correct unless it can be derived from those axioms by means of a finite number of logical steps. Upon closer consideration the question arises: *Whether, in any way, certain statements of single axioms depend upon one another, and whether the axioms may not therefore contain certain parts in common, which must be isolated if one wishes to arrive at a system of axioms that shall be altogether independent of one another.*

But above all I wish to designate the following as the most important among the numerous questions which can be asked with regard to the axioms: *To prove that they are not contradictory, that is, that a finite number of logical steps based upon them can never lead to contradictory results.*

In geometry, the proof of the compatibility of the axioms can be effected by constructing a suitable field of numbers, such that analogous relations between the numbers of this field correspond to the geometrical axioms. Any contradiction in the deductions from the geometrical axioms must thereupon be recognizable in the arithmetic of this field of numbers. In this way, the desired proof for the compatibility of the geometrical axioms is made to depend upon the theorem of the compatibility of the arithmetical axioms.

On the other hand a direct method is needed for the proof of the compatibility of the arithmetical axioms. The axioms of arithmetic are essentially nothing else than the known rules of calculation, with the addition of the axiom of continuity. I recently collected them* and in so doing replaced the axiom of continuity by two simpler axioms, namely, the well-known axiom of Archimedes, and a new axiom essentially as follows: that numbers form a system of things which is capable of no further extension, as long as all the other axioms hold (axiom of completeness). I am convinced that it must be possible to find a direct proof for the compatibility of the arithmetical axioms, by means of a careful study and suitable modification of the known methods of reasoning in the theory of irrational numbers.

To show the significance of the problem from another point of view, I add the following observation: If contradictory attributes be assigned to a concept, I say, that *mathematically the concept does not exist.* So, for example, a real number whose square is -1 does not exist mathematically. But if it can be proved that the attributes assigned to the concept can never lead to a contradiction by the application of a finite number of logical processes, I say that the mathematical existence of the

---

* *Jahresbericht der Deutschen Mathematiker-Vereinigung* 8 (1900), p. 180. [Hilbert footnote.]

concept (for example, of a number or a function which satisfies certain conditions) is thereby proved. In the case before us, where we are concerned with the axioms of real numbers in arithmetic, the proof of the compatibility of the axioms is at the same time the proof of the mathematical existence of the complete system of real numbers or of the continuum. Indeed, when the proof for the compatibility of the axioms shall be fully accomplished, the doubts which have been expressed occasionally as to the existence of the complete system of real numbers will become totally groundless. The totality of real numbers, that is the continuum according to the point of view just indicated, is not the totality of all possible series in decimal fractions, or of all possible laws according to which the elements of a fundamental sequence may proceed. It is rather a system of things whose mutual relations are governed by the axioms set up and for which all propositions, and only those, are true which can be derived from the axioms by a finite number of logical processes. In my opinion, the concept of the continuum is strictly logically tenable in this sense only. It seems to me, indeed, that this corresponds best also to what experience and intuition tell us. The concept of the continuum or even that of the system of all functions exists, then, in exactly the same sense as the system of integral, rational numbers, for example, or as Cantor's higher classes of numbers and cardinal numbers. For I am convinced that the existence of the latter, just as that of the continuum, can be proved in the sense I have described; unlike the system of *all* cardinal numbers or of *all* Cantor's alephs, for which, as may be shown, a system of axioms, compatible in my sense, cannot be set up. Either of these systems is, therefore, according to my terminology, mathematically non-existent.

10. Determination of the Solvability of a Diophantine Equation

Given a diophantine equation with any number of unknown quantities and with rational integral numerical coefficients: *To devise a process according to which it can be determined by a finite number of operations whether the equation is solvable in rational integers.*

Selected References

In keeping with the presentation of this essay as a delivered lecture, I have given no source references in the text. In place of that, here, for the benefit of the reader, are a variety of books, monographs, and articles from the most general to the most technical level which can be followed up for further information. They are arranged according to character and subject matter. Detailed publication information is given in the list of references at the end of this volume.

Biographical

John W. Dawson, Jr. (1996), *Logical Dilemmas: The Life and Work of Kurt Gödel.*

Andrew Hodges (1983), *Alan Turing. The Enigma.*

Constance Reid (1970), *Hilbert.*

Assorted Nontechnical References

Keith Devlin (1983), *Mathematics: The New Golden Age* [see especially chapters 2, 6, 8, and 11].

Douglas Hofstadter (1979), *Gödel, Escher, Bach: An Eternal Golden Braid.*

Ernest Nagel and James R. Newman (1960), *Gödel's Proof.*

Raymond M. Smullyan (1987), *Forever Undecided: A Puzzle Guide to Gödel.*

Assorted Technical References and Source Collections

Jon Barwise (ed.) (1977), *Handbook of Mathematical Logic.*

Martin Davis (ed.) (1965), *The Undecidable: Basic Papers on Undecidable Propositions, Unsolvable Problems and Computable Functions.*

Kurt Gödel (1986/1990/1995), *Collected Works. Vol. I. Publications 1929–1936*; *Vol. II. Publications 1938–1974*; *Vol. III. Unpublished Essays and Lectures.*

Jean van Heijenoort (ed.) (1967), *From Frege to Gödel: A Source Book in Mathematical Logic.*

Hilbert's Problems

Felix E. Browder (ed.) (1976), *Mathematical Developments Arising from Hilbert Problems.*

David Hilbert (1902), Mathematical problems. [English translation of lecture delivered before the International Congress of Mathematicians at Paris in 1900].

Hilbert's First Problem and Cantor's Continuum Hypothesis

Paul J. Cohen (1966), *Set Theory and the Continuum Hypothesis.*

Kurt Gödel (1940), *The Consistency of the Axiom of Choice and of the Generalized Continuum Hypothesis with the Axioms of Set Theory.* [Reprinted in Gödel 1990].

———(1947, 1964), What is Cantor's continuum problem? [Reprinted in Gödel 1990].

Solomon Feferman (1965), Some applications of the notions of forcing and generic sets.

———(1987a), Infinity in mathematics: Is Cantor necessary? [Reprinted as chapters 2 and 12 in this volume].

Donald A. Martin (1976), Hilbert's first problem: The continuum hypothesis. [In Browder 1976].

Hilbert's Second Problem, Hilbert's Consistency Program, and the Incompleteness Theorems

Kurt Gödel (1931), Über formal unentscheidbare Sätze der *Principia Mathematica* und verwandter systeme I. [Reprinted, with English translation, as On formally undecidable propositions of *Principia Mathematica* and related systems I, in Gödel 1986].

David Hilbert and Paul Bernays (1939), *Grundlagen der Mathematik,* Vol. II.

Georg Kreisel (1976), What have we learned from Hilbert's second problem? [In Browder 1976].

Jeff Paris and Leo Harrington (1977), A mathematical incompleteness in Peano Arithmetic. [In Barwise 1977].

Craig Smorynski (1977), The incompleteness theorems. [In Barwise 1977].

Raymond M. Smullyan (1992), *Gödel's Incompleteness Theorems.*

[See also references in sections on Overcoming Incompleteness and Proof Theory below.]

Hilbert's Tenth Problem on Diophantine Equations

Martin Davis, Yuri Matiyasevich, and Julia Robinson (1976), Hilbert's tenth problem. Diophantine equations: positive aspects of a negative solution. [In Browder 1976].

Yuri Matiyasevich (1993), *Hilbert's Tenth Problem.*

Julia Robinson (1996), *The Collected Works of Julia Robinson.*

Algorithms and Decision Problems

Nigel Cutland (1980), *Computability. An Introduction to Recursive Function Theory.*

Martin Davis (1977), Unsolvable problems. [In Barwise 1977].
David Harel (1987), *Algorithmics. The Spirit of Computing.*
Michael O. Rabin (1977), Decidable theories. [In Barwise 1977].

Logics of Programs

Solomon Feferman (1992), Logics for termination and correctness of functional programs.
Patrick Cousot (1990), Methods and logics for proving programs.
Raymond Turner (1991), *Constructive Foundations for Functional Languages.*

Overcoming Incompleteness

Solomon Feferman (1962), Transfinite recursive progressions of axiomatic theories.
————(1988b), Turing in the land of $O(z)$.
————(1991), Reflecting on incompleteness.
————(1996), Gödel's program for new axioms: Why, where, how and what?
Alan Turing (1939), Systems of logic based on ordinals. [Reprinted in Davis 1965].

Proof Theory

Solomon Feferman (1987), Proof theory: a personal report. [Appendix to Takeuti 1987].
————(1988), Hilbert's program relativized: Proof-theoretical and foundational reductions.
————(1993), What rests on what? The proof-theoretic analysis of mathematics. [Reprinted in chapter 10 in this volumc].
————(1993a), Why a little bit goes a long way. Logical foundations of scientifically applicable mathematics.[Reprinted in chapter 14 in this volume].
Gaisi Takeuti (1987), *Proof Theory*, 2nd edition.

2

Infinity in Mathematics: Is Cantor Necessary?

Dedicated to the memory of my friend
and colleague Jean van Heijenoort

Infinity is up on trial . . .
(Bob Dylan, *Visions of Johanna*)

Introduction

Since the rise of abstract mathematics in Greek times, mathematicians have had to grapple with the problems of infinity in many guises. When mathematics became an integral part of physical science it could be used to formulate precise answers to age-old questions: Is space infinite? Did time have a beginning? Will it have an end? Modern cosmological theories now marshal considerable physical evidence to support the finitude of space and time. But whether or not (or how) infinity is manifested in the physical universe, mathematics requires at its base the use of various infinite arithmetical and geometrical structures. Without these no coherent system of mathematics is possible, and since mathematics is essential for the formulation of physical theories, there is also no science without these uses of the infinite at the base.

Beginning in the 1870s, Georg Cantor came to realize that one must distinguish different *orders* or *sizes of infinity* of the underlying sets of objects in these basic mathematical structures. As he continued to work out

---

"Infinity in mathematics: Is Cantor necessary?" first appeared in G. Toraldo di Francia (ed.), *L'infinito nella scienza* (*Infinity in Science*), Istituto della Enciclopedia Italiana, Rome (1987), 151–209 (Feferman 1987a). It is reprinted here with the kind permission of the publisher, and appears in this volume in two parts. The present chapter constitutes the first (approximately) two thirds of the original article; the last third, which is more technical, appears as chapter 12. Some minor corrections and additions to bring it up-to-date have been made.

the implications of his ideas, he was led to the introduction of a *series of transfinite cardinal numbers* for measuring these sizes. Many mathematicians of Cantor's time were disturbed by his work, partly due to the novelty of his concepts and partly due to the uncertain grounds on which his computations and arguments with the scale of cardinals rested. But some reacted in direct opposition to Cantor's work, for his reintroduction of the "actual infinite" into mathematics (ironically, after that seemed to have been finally eliminated from analysis by the previous foundational work of the nineteenth century). One of Cantor's most vigorous and severe critics was his former teacher Leopold Kronecker, who would admit only "potentially infinite" sets to mathematics and, indeed, only those reducible to the natural number sequence 0, 1, 2, 3, ...

Worries about the Cantorian approach were compounded when, around the turn of the century, *paradoxes* appeared in the theory of sets by taking its ideas to what appeared to be their logical conclusion. The most famous and simplest of these contradictions was due to Bertrand Russell, but earlier ones had already been discovered for Cantor's transfinite numbers by Cesare Burali-Forti[1] and even by Cantor himself. While these paradoxes did not worry Cantor, the vague distinctions between the transfinite and the "absolute" infinite that he made in order to avoid them could not at first be made precise. But the contradiction plagued Russell and he attempted many solutions to escape them; that is, he sought precise systematic grounds for accepting substantial portions of Cantor's theory while excluding the paradoxes. The means at which Russell finally arrived is called the *theory of types*, and though it proved to be very cumbersome as a framework for set theory, it did restore a measure of confidence in Cantor's work. Independently, Ernst Zermelo introduced an *axiomatic theory of sets*, which featured a simple device for limiting the size of sets so as to avoid the paradoxes while providing a very flexible and ready means for the precise development of a considerable portion of Cantor's theory of higher infinities. Extensions of Zermelo's axiom system, which are described below, allow one to develop Cantorian theory in full while avoiding all known paradoxical constructions.

Parallel to the work by logicians providing an axiomatic foundation for higher set theory, Cantor's ideas were put to use more and more in mathematics, so that these days they are largely taken for granted and permeate the whole of its fabric. But there are still a number of thinkers on the subject who, in continuation of Kronecker's attack, object to the panoply of transfinite set theory in mathematics. The reasons for doing so are no longer the paradoxes, which have apparently been blocked in an effective way by means of the axiomatic theories devised by Zermelo and his succes-

[1] Actually, Burali-Forti's paradox was only implicitly contained in his work. Incidentally, Russell's paradox was found independently by Zermelo in 1902. While the work was not published, Zermelo claimed this earlier discovery in his 1908 paper; that has subsequently been confirmed by a variety of evidence in Rang and Thomas (1981).

sors. Rather, the objections to the Cantorian ideas reside in fundamentally differing views as to the nature of mathematics and the objects (numbers, points, sets, functions, ...) with which it deals. In particular, these opposing points of view reject the assumption of the actual infinite (at least in its nondenumerable forms). Following this up, alternative schemes for the foundations of mathematics have been pursued with the aim to demonstrate that everyday mathematics can be accounted for in a direct and straightforward way on philosophically acceptable non-Cantorian grounds. While genuine progress has been made along these lines, the successes obtained are not widely known and the alternative approaches have attracted relatively few adherents among working mathematicians. The general impression is that non-Cantorian mathematics is too restrictive for the needs of mathematical practice, regardless of the merits of the guiding non-Cantorian philosophies.

Some logicians would now go farther to bolster this impression by giving it a theoretical underpinning; their aim is to demonstrate that Cantor's higher infinities are in fact *necessary* for mathematics, even for its most finitary parts. The results that have been obtained in this direction do, at first sight, appear to justify such a reading. Nevertheless, it is argued below [now in chapter 12] that *the necessary use of higher set theory in the mathematics of the finite has yet to be established.* Furthermore, a case can be made that *higher set theory is dispensable in scientifically applicable mathematics*, that is, in that part of everyday mathematics which finds its applications in the other sciences. Put in other terms: the *actual infinite is not required for the mathematics of the physical world.* The reasons for this depend on other recent developments in mathematical logic, the description of which is the final aim of this essay [now chapter 12; cf. also chapter 14].

In order to explain the objections to Cantor's ideas in mathematics that lead one to search for viable alternatives, one must first provide some understanding of their nature and use. This will be done here more or less historically though necessarily in outline; we start with a rather innocent looking problem about the existence of certain special kinds of numbers.[2]

From Transcendental Numbers to Transfinite Numbers

There are two basic number systems (having ancient origins), the set **N** of *natural numbers* $0, 1, 2, 3, \ldots$ used for counting, and the set **R** of *real numbers* use for measuring. The latter represent positions of points on a two-way infinite straight line, relative to any initial point 0 (the "origin") and any unit of measurement 1. **R** is pictured as follows:

---

[2]This is not where Cantor himself started, though he came to it soon enough. For a good detailed introduction to the history of the development of Cantor's ideas, see Grattan-Guinness (1980), chapters 5, 6 (by J. W. Dauben and R. Bunn, resp.). For a deeper pursuit I would recommend most highly the books of Moore (1982) and Hallett (1984). [Cf. also Lavine (1994).]

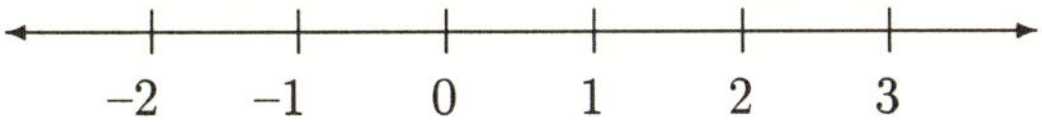

N can thus be identified with a subset of **R**. Other subsets of use are $\mathbf{Z} = \{\ldots, -3, -2, -1, 0, 1, 2, 3, \ldots\}$ (the set of all *integers*) and $\mathbf{Q} = \{n/m : n, m \in \mathbf{Z} \text{ and } m \neq 0\}$, the set of *rational numbers*, consisting of all *quotients* or *ratios* of integers with nonzero denominators. We here write "$x \in S$" for the relation of membership of an object x to a set S, and "$\{x : P(x)\}$" for the set of all objects x satisfying a determinate property $P(x)$. If x does not belong to S we write "$x \notin S$." Finite sets S can be denoted directly by a listing of their elements as $S = \{a_0, a_1, a_2, \ldots, a_n\}$. This notation is extended to certain infinite sets such as $\mathbf{N} = \{0, 1, 2, \ldots, n, \ldots\}$ and **Z** (as above), when we have a complete survey of their elements.

If S_1 and S_2 are sets, then S_1 is a *subset* of S_2, in symbols $S_1 \subseteq S_2$, if for every $x \in S_1$ we have $x \in S_2$; $S_2 - S_1 = \{x : x \in S_2 \text{ and } x \notin S_1\}$ then denotes the *difference* of these two sets. $S_2 - S_1$ might be empty, when $S_1 = S_2$; the *empty set* is denoted by $\emptyset$.

The set **Q** is densely dispersed throughout **R** and cannot be distinguished from **R** by a simple picture as above. A basic realization from Greek geometry was the existence of *irrational magnitudes*, that is, elements of $\mathbf{R} - \mathbf{Q}$. For, Pythagoras' law giving the hypotenuse c of a right triangle in terms of its legs a, b by $c^2 = a^2 + b^2$, or equivalently $c = \sqrt{a^2 + b^2}$, leads directly to quantities such as $\sqrt{2} = \sqrt{1^2 + 1^2}$ and $\sqrt{5} = \sqrt{1^2 + 2^2}$, which are easily proved to be irrational. Other kinds of irrational numbers also arise naturally in geometry, for example, $\sqrt[3]{2}$ in the classical problem of the duplication of the cube (that is, construction of a cube with double the volume of a given one), and $\pi(= 3.14159\ldots)$, the ratio of the circumference of a circle to its diameter. However, the proofs of the irrationality of numbers like π only came much later.

Other irrational numbers arise from the solution of algebraic equations; for example, one solution of $x^4 - 7 = 0$ is $x = \sqrt[4]{7}$ and of $x^6 - x^3 - 1 = 0$ is $x = \sqrt[3]{(1 + \sqrt{5})/2}$. Some equations, such as $x^2 + 1 = 0$, have no solutions in real numbers, although we can treat their solutions in the extension of the real number system by the *imaginary numbers* such as $\sqrt{-1}$; however, those will not concern us here. A real number is called *algebraic* if it is the solution x of a nontrivial polynomial equation $p(x) = 0$ with integer coefficients; that is, $p(x) = k_n x^n + k_{n-1} x^{n-1} + \ldots + k_1 x + k_0$ with $n > 0$ and $k_n \neq 0$, and each $k_i \in \mathbf{Z}$. We use "**A**" to denote the set of all (real) algebraic numbers. Thus $\mathbf{Q} \subseteq \mathbf{A}$ and **A** contains all the irrational numbers shown above, except possibly π. However, it is natural to suspect that $\pi \notin \mathbf{A}$, since there is no known polynomial equation of which it is a root. This was in fact first conjectured by Legendre in the eighteenth century, but a proof did not come until a century later.

A number is called *transcendental* if it is in the set $\mathbf{T} = \mathbf{R} - \mathbf{A}$. Another specific number which, like π, is ubiquitous in mathematics and was conjectured to be transcendental is the base e $(= 2.71828\ldots)$ of "natural" logarithms. The first proof that there *exist* any transcendental numbers *at all*, that is, that $\mathbf{T} \neq \emptyset$, was given by Liouville in 1844. His method was to find a property $P(x)$ which applies to all algebraic numbers x and which says something (technical) about how well x can be approximated by rational numbers. Liouville then cooked up a real number ℓ which does not satisfy the property P, so ℓ must be transcendental. Infinitely many other transcendental numbers can also be produced in this way, but Liouville's method did not help answer the specific questions as to whether e and π are transcendental. Those results were not obtained until somewhat later, by Hermite (in 1873) and Lindemann (in 1882), respectively, using rather special methods. In the meantime, Cantor published in 1874 a new and extremely simple but striking argument to prove the existence of transcendental numbers. Cantor's *result* in this respect was no better than Liouville's, but the methodology of his proof turned out to be one of the main starting points for his novel conception of infinity in mathematics.

First of all, Cantor defined two sets S_1 and S_2 to be *equinumerous* if their elements can be placed in one-to-one correspondence with each other; symbolically this is indicated by $S_1 \sim S_2$. A set S is *finite* if it is equivalent to some initial segment of $\mathbf{N}$, possibly empty, that is, $S \sim \{0, \ldots, n-1\}$, where $n \geq 0$. A set S is called *denumerable* if S is finite or $S \sim \mathbf{N}$. Every nonempty denumerable set S can be listed as $S = \{a_0, a_1, \ldots, a_n, \ldots\}$, possibly with repetitions, and conversely. From this follows directly several basic facts:

(1) a denumerable union of denumerable sets is denumerable;

(2) the set $\mathbf{Q}$ is denumerable;

(3) the set $\mathbf{A}$ is denumerable.

In (1) we are considering sets $S_0, S_1, \ldots, S_n, \ldots$, each of which is denumerable and may be assumed nonempty; these are listed as

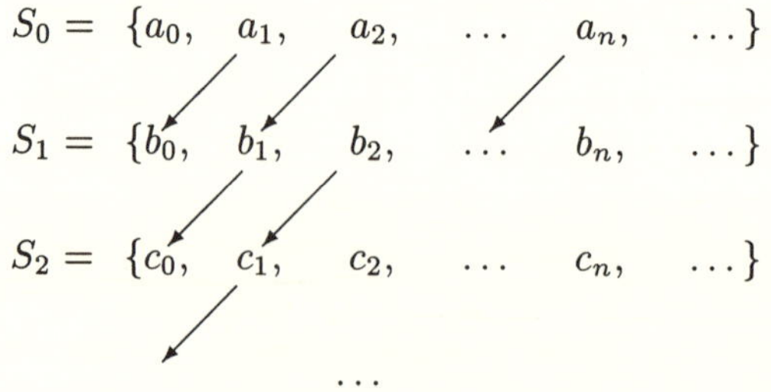

The arrows have been added to show that the union S can be listed following the indicated arrows as

$$S = \{a_0, a_1, b_0, a_2, b_1, c_0, \ldots\}.$$

Now, for (2), take S_n to be the set of all multiples m/n for $n \neq 0$ and m in $\mathbf{Z}$; each S_n is denumerable and their union is $\mathbf{Q}$, so (2) follows. To prove (3), one shows first of all that all equations of the form $k_m x^m + \ldots + k_1 x + k_0 = 0$ with integer coefficients and $k_m \neq 0$ can be enumerated. If $p_n(x) = 0$ is the nth such equation, take S_n to be the set of its solutions. This set is finite (possibly empty), hence certainly denumerable. But the set $\mathbf{A}$ of algebraic numbers is the union of these S_n's, so it also is denumerable; that is, (3) holds.

Now, in contrast, Cantor showed that

(4) $\mathbf{R}$ is non-denumerable.

In other words there is no way to list $\mathbf{R}$ as a simple sequence of real numbers, $\{x_0, x_1, x_2 \ldots\}$. Cantor's first proof of (4) made use of special properties of $\mathbf{R}$. Later he gave a simpler proof that could be generalized to other sets; this used his famous *diagonal argument.* It is sufficient to show that the set of reals x between 0 and 1 cannot be enumerated. Indeed, given any enumeration $\{x_0, x_1, x_2 \ldots\}$ of a subset of $\mathbf{R}$, we shall construct a number x that is not in the enumeration. First write each member of $\{x_0, x_1, x_2 \ldots\}$ as an infinite decimal:

$$\begin{aligned} x_0 &= 0.a_1 a_2 a_3 \ldots a_n \ldots \\ x_1 &= 0.b_1 b_2 b_3 \ldots b_n \ldots \\ x_2 &= 0.c_1 c_2 c_3 \ldots c_n \ldots \\ &\ldots \end{aligned}$$

Now form $x = 0.k_1 k_2 k_3 \ldots k_n \ldots$ not in $\{x_0, x_1, x_2 \ldots\}$ by choosing $k_1 \neq a_1, k_2 \neq b_2, k_3 \neq c_3$, etc. For example, take $k_1 = 0$ if $a_1 \neq 0$ and $k_1 = 1$ if $a_1 = 0$, etc. This proves (4); it then follows immediately from (3) that $\mathbf{A} \neq \mathbf{R}$. Hence $\mathbf{T} = \mathbf{R} - \mathbf{A}$ must be nonempty, and thus the existence of transcendentals is newly established. In fact, $\mathbf{T}$ must also be nondenumerable, for otherwise by (1) and (3) we would have $\mathbf{R}$ denumerable. This is a stronger conclusion than Liouville's, which was merely that $\mathbf{T}$ is infinite.

Clearly there are two senses of the *size* of a set involved here. If S_1 is a proper subset of S_2, then it is smaller than S_2 in the sense of containing fewer elements. But it may well be possible for S_1 to be a proper subset of S_2 and still have S_1 and S_2 being of the same size in the sense that $S_1 \sim S_2$. For example, $\mathbf{N} = \{0, 1, 2, \ldots, n, \ldots\}$ is equinumerous with the set $\mathbf{E} = \{0, 2, 4, \ldots, 2n, \ldots\}$ of even integers, though $\mathbf{E}$ is a proper subset of $\mathbf{N}$. Similarly, $\mathbf{Q} \sim \mathbf{A}$ by the above, though $\mathbf{Q}$ is a proper subset of

A. Intuitively, any infinite set S is of the same size as some proper subset, while finite sets are just those which are not equinumerous with any proper subset.

So if S_1 is a subset of S_2 and we do not know whether S_1 is a proper subset of S_2, one way to establish that is to show that S_1 and S_2 are *not* equinumerous, in symbols, $S_1 \not\sim S_2$; however, to do so would require a special argument, since by the preceding $S_1 \sim S_2$ is well possible for proper subsets. In the case of $S_1 = \mathbf{A}$ and $S_2 = \mathbf{R}$, that is accomplished by the results (3) and (4) above. Cantor's argument is novel not only in the concepts (set, equinumerosity, denumerability, nondenumerability) and results (1)–(4) involved but also in a basic point of methodology: Cantor finds a property $P^\star$ of *sets* and shows two sets S_1, S_2 to be distinct by showing that one of them has the property $P^\star$ while the other does not. Moreover, the *existence of elements of a special kind* follows by a purely logical argument: if S_1 is a subset of S_2 and $P^\star(S_1)$ holds while $P^\star(S_2)$ does not hold, then S_1 must be a *proper* subset of S_2; that is, *there exists an element x of $S_2 - S_1$*.

Since two finite sets $\{a_1, a_2, \ldots, a_n\}$ and $\{b_1, b_2, \ldots, b_m\}$ (of distinct elements) are equinumerous just in case they have the same number of elements ($n = m$), Cantor was led to say in general that two sets S_1 and S_2 have the *same number of elements* if $S_1 \sim S_2$. One may regard the *number of elements in S* as an abstraction from its specific nature which isolates just what it has in common with all equinumerous sets. Thus, for example, $\{1,2,3\} \sim \{1,3,2\} \sim \{\sqrt{2}, \sqrt[3]{5}, \pi\}$ all have the same number 3. For Cantor this was a process of double abstraction; the first level $\bar{S}$ abstracts away all that is distinctive about the elements of S except how they are placed in a certain order, and the second level $\bar{\bar{S}}$ abstracts away the order as well; $\bar{\bar{S}}$ is called the *cardinal number of S*. Here instead we shall write *card(S)* for $\bar{\bar{S}}$. For example $card(\{\sqrt{2}, \sqrt[3]{5}, \pi\}) = 3, card(\{5\}) = 1$, and $card(\emptyset) = 0$. To identify the cardinal number of infinite sets we need new names. Cantor used the Hebrew letter $\aleph$ (aleph) with subscripts to name various infinite cardinal numbers, beginning with $card(\mathbf{N}) = \aleph_0$. Thus also $card(\mathbf{Z}) = card(\mathbf{Q}) = card(\mathbf{A}) = \aleph_0$, but $card(\mathbf{R}) \neq \aleph_0$. To name the cardinal number of the continuum $\mathbf{R}$, a new symbol $\mathbf{c}$ is introduced by definition as $\mathbf{c} = card(\mathbf{R})$.

There is another way of naming $card(\mathbf{R})$ that comes from the extension of the arithmetic operations of addition, multiplication, and exponentiation to infinite cardinals, as follows. Suppose $n = card(S_1)$ and $m = card(S_2)$; we can assume, without loss of generality, that S_1 and S_2 are disjoint. Then define $n + m = card(S_1 + S_2)$, where $S_1 + S_2$ is the union of S_1 and S_2. Next, define $n \times m = card(S_1 \times S_2)$ where $S_1 \times S_2$ is the set of all possible ways (x, y) of combining an element x of S_1 with an element y of S_2. Finally, define $n^m = card(S_1^{S_2})$ where $S_1^{S_2} = \underbrace{\ldots S_1 \times \ldots \times S_1 \times \ldots}_{S_2}$ consists of all possible combinations of elements of S_1, one for each position in S_2. For S_1, S_2 finite these definitions of $+$, $\times$, and exponentiation agree with the usual ones, but for S_1 or S_2 infinite we obtain new results. As examples of calculations with the latter, one has

$$n + \aleph_0 = \aleph_0 + \aleph_0 = \aleph_0$$

for any finite n. For this reflects the relations

$$\{0, 1, \ldots, n-1\} + \{n, n+1, \ldots\} \sim \{0, 2, 4, \ldots\} + \{1, 3, 5, 7, \ldots\} \sim \{0, 1, 2, \ldots\}.$$

Thus unlike the case for finite m where $n + m$ is greater than m (for $n \neq 0$), we have $m + m = m$ for $m = \aleph_0$. Similarly, for the same m we have $m \times m = m$; this corresponds to the fact that a denumerable union of denumerable sets is denumerable.

In order to represent the cardinal number $\mathbf{c}$ of $\mathbf{R}$ in terms of these operations, we return to the relationship of $\mathbf{R}$ with $\mathbf{Q}$. Each element x of $\mathbf{R}$ is approximated by sequences or sets of rationals in various ways. One way is to associate with x the subset $\mathbf{Q}_x$ of $\mathbf{Q}$ consisting of all rationals r with $r < x$; then x is uniquely determined by $\mathbf{Q}_x$. Let $\mathcal{P}(\mathbf{Q})$ be *the set of all subsets of* $\mathbf{Q}$, in symbols $\mathcal{P}(\mathbf{Q}) = \{S : S \subseteq \mathbf{Q}\}$; thus $card(\mathbf{R}) \leq card(\mathcal{P}(\mathbf{Q}))$.[3] Now for any set S the set $\mathcal{P}(S)$ of all subsets S' of S is equinumerous with $\{0,1\}^S$, the set of all possible ways of associating a 0 or 1 with each element i of S (the correspondence puts 1 if the element i belongs to S' and 0 otherwise). Hence $card(\mathcal{P}(S)) = 2^{card(S)}$ (and for this reason, $\mathcal{P}(S)$ is called the *power set of* S). In particular, $card(\mathcal{P}(\mathbf{Q})) = 2^{\aleph_0}$, so, by the above, $\mathbf{c} \leq 2^{\aleph_0}$. On the other hand, it is not difficult to produce $2^{\aleph_0}$ distinct members of $\mathbf{R}$, by looking for example at real numbers $x = 0.a_1a_2a_3, \ldots$ in binary expansion, that is, where each $a_i = 0$ or 1. The conclusion is thus that

(5) $\mathbf{c} = 2^{\aleph_0}$.

Obviously $\aleph_0 \leq 2^{\aleph_0}$ since $\mathbf{N}$ is a subset of $\mathbf{R}$; but since $\mathbf{R}$ is nondenumerable, we must have

(6) $\aleph_0 < 2^{\aleph_0}$.

Thus there are at least two transfinite numbers $\aleph_0$ and $\mathbf{c}$; can these be all?

A Plethora of Transfinite Numbers

Cantor was thus led directly to the following questions:

I. Are there any infinite cardinal numbers besides $\aleph_0$ and $\mathbf{c}$?

[3] For $n = card(S_1), m = card(S_2)$ define $n \leq m$ to hold if $S_1 \sim S_1'$ for some $S_1' \subseteq S_2$ and $n < m$ if $n \leq m$ but $n \neq m$.

II. In particular, are there any cardinal numbers n strictly between $\aleph_0$ and $\mathbf{c}$?

Cantor first computed the cardinal numbers of other familiar structures besides $\mathbf{N}$ and $\mathbf{R}$. For example, the Euclidean plane when considered from the point of view of analytic (coordinate) geometry may be represented as $\mathbf{R} \times \mathbf{R}$, and three-dimensional space as $(\mathbf{R} \times \mathbf{R}) \times \mathbf{R}$. But Cantor succeeded in showing (to his own surprise) that

$$(1) \qquad \mathbf{c} \times \mathbf{c} = (\mathbf{c} \times \mathbf{c}) \times \mathbf{c} = \mathbf{c}.$$

The fact $\mathbf{c} \times \mathbf{c} = \mathbf{c}$ is analogous to the fact $\aleph_0 \times \aleph_0 = \aleph_0$ noted above. Initially, Cantor indexCantor, Georg proved this by considering arbitrary pairs (x, y) of elements x of $\mathbf{R}$ and y of $\mathbf{R}$. Expanding the decimal parts of each as $0.a_1a_2a_3\dots$ and $0.b_1b_2b_3\dots$, one obtains a corresponding new decimal number by alternating their terms: $0.a_1b_1a_2b_2a_3b_3\dots$. That $(\mathbf{c} \times \mathbf{c}) \times \mathbf{c} = \mathbf{c}$ then follows immediately from $\mathbf{c} \times \mathbf{c} = \mathbf{c}$. For a while, Cantor found no other infinite cardinal numbers besides $\aleph_0$ and $\mathbf{c}$. But eventually he proved that $\mathbf{c} < 2^{\mathbf{c}}$ and in fact that for any cardinal number n,

$$(2) \qquad n < 2^n.$$

The proof of this makes use again of his diagonal argument, via the representation of 2^n as $card(\{0, 1\}^S)$ for $n = card(S)$.

Now starting with $n = \aleph_0$ we can form the sequence of larger and larger infinite cardinals

$$(3) \qquad n < 2^n < 2^{2^n} < \dots .$$

But we can go still further beyond these by forming the infinite sum $n + 2^n + 2^{2^n} + \dots$ (suitably defined), so that

$$(4) \qquad n < 2^n < 2^{2^n} < \dots < (n + 2^n + 2^{2^n} + \dots).$$

Then one can continue still higher by exponentiating and, at limits, taking sums.

One has here the beginnings of what appears to be a *scale* for representing infinite cardinal numbers. But in order to obtain a *complete scale* or *system of representations*, every cardinal number must appear somewhere in the list. This brings us back to question II above: is there any cardinal number n between $\aleph_0$ and $2^{\aleph_0}$? Cantor conjectured that there is *not*, and he expended considerable effort over many years trying to prove just that. This conjecture has come to be called the *Continuum Hypothesis*, abbreviated CH. If CH is true, then $\mathbf{c}$ is the first cardinal larger than $\aleph_0$, but independently of whether or not CH is true, it is natural to ask whether there is a first such cardinal. Cantor argued that must be so by invoking what has come to be called the *Well-Ordering Principle* (WO). A set S is said to be *well-ordered* by a relation $<$ of ordering between its elements if every nonempty subset S' of S has a first element. Thus, for example, the rearrangements of $\mathbf{N}$

$$\text{(5)}\qquad \begin{array}{ll} \text{(i)} & 1,2,3,\ldots,0, \\ \text{(ii)} & 0,2,4,\ldots,1,3,5,\ldots \\ \text{(iii)} & \ldots,5,3,1,0,2,4\ldots \end{array}$$

determine three different orderings $<_1, <_2$, and $<_3$ of $\mathbf{N}$. Under the first two of these, $\mathbf{N}$ is well-ordered (as it is under its standard order $0, 1, 2\ldots$), but for the third, $<_3$ is not a well-ordering of $\mathbf{N}$ since the subset $\{\ldots 5, 3, 1\}$ does not contain a first element. Note that $<_3$ is the same as the natural ordering of the set $\mathbf{Z}$ of all integers.

Now Cantor claimed WO is true by the following kind of argument[4]: if S is any nonempty set, choose an element x_0 of S arbitrarily. Either $S = \{x_0\}$ or $S - \{x_0\}$ is nonempty; in the latter case choose an element x_1 of $S - \{x_0\}$ arbitrarily. Now either $S = \{x_0, x_1\}$ or $S - \{x_0, x_1\}$ is nonempty; in the latter case we proceed by choosing an element x_2 of $S - \{x_0, x_1\}$. If S is finite, then at some stage n, $S = \{x_0, x_1, \ldots, x_n\}$; otherwise $\{x_0, x_1, \ldots, x_n \ldots\}$ is an infinite subset of S. If $S = \{x_0, x_1, \ldots, x_n \ldots\}$ we are through; otherwise choose an element x_ω of $S - \{x_0, x_1, \ldots, x_n, \ldots\}$. We proceed on, as necessary, to choose distinct elements $x_{\omega+1}, x_{\omega+2} \ldots$ of S; eventually we must exhaust all the elements of S. Then we arrive at

$$\text{(6)}\qquad S = \{x_0, x_1, \ldots, x_n, \ldots, x_\omega, x_{\omega+1}, \ldots, x_\alpha, \ldots\}$$

where at each stage in the procedure of generating this transfinite sequence, there is a *first* element beyond all those already chosen. This determines a well-ordering of S by the order of generation,

$$\text{(7)}\qquad x_0 < x_1 < \ldots < x_n < \ldots < x_\omega < x_{\omega+1} < \ldots < x_\alpha < \ldots .$$

The problem with this argument is that it assumes there is a method for making an unlimited number of successive arbitrary choices, so that for each subset $S' = \{x_0, x_1, \ldots, x_\beta, \ldots\}_{\beta<\alpha}$ of S, if $S' \neq S$, the method chooses an element x_α of $S - S'$. If one takes S to be the set $\mathbf{N}$, there is no problem since the standard ordering of $\mathbf{N}$ already provides a well-ordering. But if one takes S to be the set $\mathbf{R}$, there in no known method to make the required choices. The assumption of the existence of such a sequence of choices may thus be considered to be unjustified, and that was one cause of opposition to Cantor's theory.

[4]Actually, Cantor initially asserted WO to be a "law of thought" when he first stated it in 1883. Around 1895 he began to recognize the need for an argument and indicated such in letters to Hilbert and Dedekind between 1896 and 1898; see Moore (1982), pp. 51–53.

If one assumes the Well-Ordering Principle WO, then a complete scale for the infinite cardinal numbers follows as a consequence. For if $n = card(S_1)$ and $m = card(S_2)$ and $n < m$, we have $S_1 \sim S_1'$, where S_1' is a subset of S_2. Well-order S_1', and then extend that to a well-ordering of S_2. Thus S_1 is equinumerous with an initial segment of the well-ordering of S_2. But since $S_1 \not\sim S_2$, there is a first initial segment S_2' of S_2 with $S_1 \not\sim S_2'$. Let $p = card(S_2')$; then p is larger than the cardinal number n of S_1 and, by the way it is defined, there is no cardinal number between n and p; furthermore, $p \leq m$. Hence, beyond any cardinal number n there is a next larger one, n^+, and for any m with $n < m$ we have $n^+ \leq m$.

Cantor now defined his complete scale of infinite cardinals (which he labeled using the Hebrew aleph with subscripts), as follows:

$$(8) \qquad \aleph_0 < \aleph_1 < \aleph_2 < \ldots < \aleph_n < \ldots < \aleph_\omega < \aleph_{\omega+1} < \ldots < \aleph_\alpha \ldots,$$

where at each successor stage $\aleph_{\alpha+1}$ is taken to be $(\aleph_\alpha)^+$ (the first cardinal number greater than $\aleph_\alpha$), and at limit stages $\aleph_\alpha$ is taken to be the first cardinal number larger than all preceding $\aleph_\gamma$ for $\gamma < \alpha$. The Continuum Hypothesis can now be reexamined in terms of the relationship of $\mathbf{c}$ to $\aleph_0$: since $\aleph_0 < \mathbf{c}$ and $\aleph_1$ is the first cardinal larger than $\aleph_0$, we have $\aleph_0 < \aleph_1 \leq \mathbf{c}$. If there is any cardinal number between $\aleph_0$ and $\mathbf{c}$, then $\aleph_1$ would certainly be such a number. Hence the question as to whether CH is true is equivalent to the question whether $\mathbf{c} = \aleph_1$; in other words,

$$(9) \qquad \text{is } 2^{\aleph_0} = \aleph_1?$$

To summarize: Cantor achieved his complete representation (8) of infinite cardinal numbers at the cost of assuming a problematic principle (WO), and even with this he was left with the fundamental open question (9) about the location of $\mathbf{c}$ in the resulting scale.

Justifying Cantor: Enter Zermelo

At the (Second) International Congress of Mathematicians, held in Paris in 1900, David Hilbert gave a famous lecture entitled "Mathematische Probleme," in which he presented twenty-three major unsolved problems.[5] First in Hilbert's list, which began with problems in the foundations of mathematics, was *Cantor's problem of the cardinal number of the continuum.* In his discussion of this, Hilbert states CH in the form that for every subset S of $\mathbf{R}$, either S is denumerable or $S \sim \mathbf{R}$, that is, that there are no cardinal numbers n between $\aleph_0$ and $2^{\aleph_0}$. Hilbert found CH plausible, but he also raised a question in this connection about Cantor's principle WO and, in

[5] Hilbert's article on his talk was published in English translation in 1902 and is reproduced in Browder (1976), which also contains separate articles detailing the developments arising from the various problems. Hilbert is here quoted in translation. [See also the Appendix to chapter 1 in this volume.]

particular, its consequence that **R** can be well-ordered. Concerning the latter Hilbert said, "It appears to me most desirable to obtain a direct proof of this remarkable statement, perhaps by actually giving an arrangement of [real] numbers such that in every partial system a first number can be pointed out." (Hilbert (1902), in Browder (1976), p. 9) Clearly, Hilbert was not convinced by the argument for WO that Cantor had communicated to him in 1896 or 1897—presumably in its use of a transfinite succession of arbitrary choices—since he is here calling for an *explicit* construction of a well-ordering of **R**.[6]

Hilbert was not the only one to recognize the significance of CH or to question the grounds for accepting the WO principle, despite his favorability to Cantor's ideas. Since Kronecker, Cantor had received criticism from many sides, though he had also gained adherents to his new mathematics. But as one of the leading mathematicians of the time, Hilbert's selection of the Continuum Hypothesis and the Well-Ordering Principle to be the first in his entire list of questions gave added prominence to both the importance of Cantor's ideas and their problematic aspects.

Ernst Zermelo was one of the first to take up Hilbert's challenge. Zermelo had initially worked on topics in analysis and mathematical physics, but in 1899 he became a Privatdozent at Göttingen, where he came under Hilbert's influence and began pursuing set theory. Zermelo's first main contribution to the subject was achieved in 1904 with his proof of the Well-Ordering Theorem on the basis of a new principle, which has come to be called the *Axiom of Choice*, or AC in abbreviated form.[7] Zermelo's original statement of AC is to the effect that if S is any nonempty set, then there is a function f which is defined on the collection C of all nonempty subsets X of S such that for each X in C we have $f(X)$ in X. In other words, f provides for the *simultaneous choice* from each nonempty subset X of S of a distinguished element (namely, $f(X)$) of X. By proving WO from AC, Zermelo was able to reduce the construction of a transfinite sequence of successive choices (which appear to proceed through time) needed in Cantor's attempt to prove WO, to the assumption of a single simultaneous collection of choices. AC itself is an immediate consequence of WO, for if S is well-ordered by a relation $<$, one can define $f(X)$ to be the least element of X under $<$ for each nonempty subset X of S. Since WO and AC are thus equivalent, Zermelo's proof can only be considered to have achieved progress if AC is evident in a way that WO is not.

In the years directly following Zermelo's 1904 publication, his proof provoked considerable controversy, centering on both the assumption of AC

[6] Modern logical work on set theory has shown that Hilbert's hope is impossible to realize; the precise sense of this is explained in a later section.

[7] Moore (1982) gives the most comprehensive history available concerning the background, development, and importance for set-theoretical mathematics of Zermelo's Axiom, as well as the extensive controversies to which it has given rise. It was subsequently proved that many important set-theoretical propositions are equivalent to AC. One which is closest to it is the *Multiplicative Axiom*, formulated independently by Russell, which emerged from work of Whitehead and Russell.

as well as his use of other set-theoretical concepts and principles. (Some critics mistakenly believed that a form of the Burali-Forti paradox was involved in Zermelo's argument.) The main criticism of AC was that a function f is asserted to *exist* without any means being provided to define it uniquely. In this, AC violates a view of mathematics according to which existence of an object satisfying a certain condition can only be asserted when such an object has been *explicitly constructed* or *defined*. In contrast, Zermelo's principle asserts the outright existence of an object, f, independent of any means of definition.

Later, Zermelo formulated another principle AC′, which may be considered intuitively clearer than AC, but which can be shown equivalent to it, namely, that if C is any collection of nonempty *disjoint* sets,[8] then there is a function f defined on C such that for each X in C we have $f(X)$ in X. AC′ can be pictured as follows:

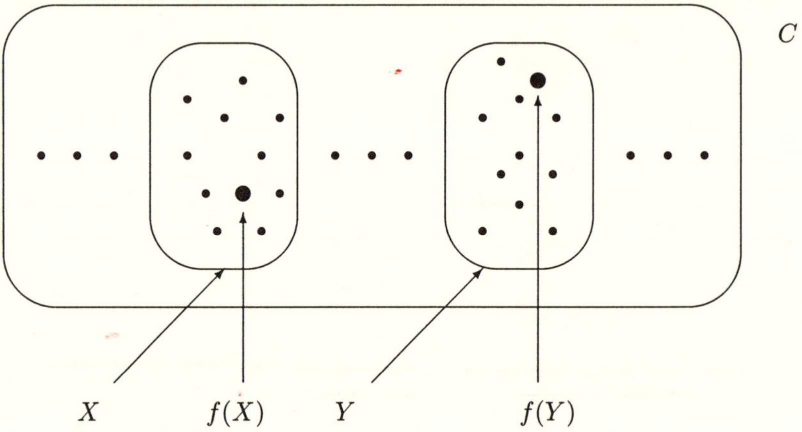

However, the basic objection to AC on the grounds that it asserts existence without explicit definability applies equally well to AC′. On the other hand, if one conceives sets to be *arbitrary collections* of entities existing independently of human constructions and definitions, then AC′ is immediately intuitively evident. Thus the question whether to accept or reject AC comes down to a fundamental difference of viewpoints as to the nature of mathematics.[9]

In order to disentangle the various objections to his proof of the Well-Ordering Theorem and shore up his defense of it, Zermelo arrived in 1908

---

[8]This means that if X, Y are in C and $X \neq Y$, then the intersection of X and Y is empty.

[9]AC was by no means the first example in mathematics of the assertion of existence of an object without any means being known to construct it. However, it was the first in which this difference is so blatantly revealed.

at an explicit, systematic presentation of basic notions and principles for sets, in the form of an *axiomatic system of set theory*. Two of Zermelo's axioms assert the outright existence of certain sets (namely, the empty set and an infinite set representing the natural numbers **N**), while further axioms state that if certain sets exist then other sets related to them also exist. The latter include the nonconstructive existence statement AC but also include axioms for the usual constructions in set theory such as those of pairing, union, intersection, difference, product, set-of-all subsets, etc. A principal general means of construction is guaranteed by Zermelo's *Axiom of Separation* according to which if S is any set and P is any *definite property* of elements of S, then the set $S' = \{x \in S : P(x)\}$ (that is, $\{x : x \in S \text{ and } P(x)\}$) exists. By this means Zermelo finessed the paradoxes which had arisen from the unrestricted formation of sets $\{x : P(x)\}$. The latter had led to what Cantor called *inconsistent* or *absolutely infinite* sets such as the set C of all cardinal numbers, and the set V of all sets. While not all sets $\{x : P(x)\}$ are necessarily inconsistent in the logical sense, some among them are, as for example Russell's set $R = \{x : x \notin x\}$ consisting of all sets which are not members of themselves (the paradox results when one asks whether $R \in R$ or not). Absolutely infinite or inconsistent sets in Cantor's sense cannot be constructed in Zermelo's axiom system,[10] since there is *no universal set* V from which to separate $\{x : P(x)\}$.

Aside from the existence axioms of Zermelo's system, there is one further axiom which reflects the underlying philosophical view as to the nature of mathematics on which the system rests. This is the *Axiom of Extensionality*, according to which if S_1 and S_2 are sets having exactly the same members, then $S_1 = S_2$; in other words, a set is determined entirely by its members and is to be regarded independently of any specific means of determining just what those members are.[11]

Zermelo showed how the proof of WO from AC could be carried out on the basis of his other axioms. Furthermore, the Well-Ordering Theorem has the following three basic consequences, as Zermelo had already shown in his 1904 article: (i) for any cardinal numbers n and m, exactly one of the three possibilities: $n < m, n = m, m < n$, must hold; (ii) every infinite cardinal n is of the form $\aleph_\alpha$ for some (ordinal) α; and (iii) for any infinite cardinal number m, we have $m = m + m = m \times m$. The first of these is called the *Trichotomy Principle* and says that the cardinal numbers are linearly ordered; in other words, any two of them can be compared as in a scale. The second result gives the complete system of representation for infinite cardinals in terms of the "alephs." The latter are defined in such a way that each $\aleph_\alpha$ represents a well-ordered set; hence (ii) implies WO

[10]Unless the system is logically inconsistent for some other, less obvious reason; then every statement would be derivable according to the usual laws of logic.

[11]For more detail on Zermelo's axiom system and its development see Moore (1982), pp. 153 ff., and, more extensively, Fraenkel, Bar-Hillel, and Levy (1973). Translations of Zermelo's 1904 and 1908 papers referred to above are in van Heijenoort (1967).

and thus AC. It was shown later (by Hartogs) that (i) implies AC as well. Still later Alfred Tarski proved that the statement $m = m \times m$ for all infinite cardinals m also implies AC. Hence even this special fact from the arithmetic of infinite cardinal numbers is as basic as the general principles (i) and (ii) concerning their scale of representation.

Zermelo's axiomatization of set theory did not succeed in satisfying his critics as he hoped. Indeed, it supplied even more fuel for criticism, particularly concerning Zermelo's use of the vague notion of *definite property* in his formulation of the Axiom of Separation. It was some years before this was given a satisfactory explanation by means of formal languages which had been developed for symbolic logic. This was done first by Hermann Weyl and then in a simpler, improved way by Thoralf Skolem (in 1922). In Skolem's account, "definite properties" are replaced by (*well-formed*) *formulas* in a language of the *first-order predicate calculus*, whose basic predicates are $\in$ and $=$. The basic formulas are thus of the form $x \in y$ and $x = y$; formulas in general are built up using symbols for propositional connectives including negation ($\neg$), disjunction ($\vee$), conjunction ($\wedge$), implication ($\rightarrow$), and by existential and universal quantification, $(\exists x)(\dots)$ ("there exists $x, \dots$") and $(\forall x)(\dots)$ ("for all $x, \dots$"). The reasoning employed in making derivations from the axioms is that of first-order logic, that is, the logic of propositional connectives and quantifiers with respect to variables ranging over the domain of discourse. This formal version of *Zermelo Set Theory*, without the Axiom of Choice, is often simply denoted Z. The full system with AC added is denoted Z + AC, or ZC in abbreviated form.

Though basic facts about the arithmetic and ordering of cardinal numbers can be derived in ZC, it turns out that it is insufficient to prove the existence of the limit of the increasing sequence of cardinals $\aleph_0, 2^{\aleph_0}, 2^{2^{\aleph_0}}, \dots,$ or equivalently of the set $M = \{\aleph_0, 2^{\aleph_0}, 2^{2^{\aleph_0}}, \dots\}$. We have here a correspondence R between the elements of a set whose existence is already given, namely $\mathbf{N}$, and those of a set M whose existence should be provable but is not; moreover, this correspondence is definable in the first-order language of set theory. In order to overcome this limitation in Zermelo's system, Abraham Fraenkel introduced (in 1922) a new axiom called *Replacement*. This says that if we have a first-order definable relation $R(x, y)$ which associates with each element x of an already existing set A a unique element y, then the set B of all the associated y's exists; intuitively each element x of A is replaced by the unique y such that $R(x, y)$ holds. The system Z with Fraenkel's Axiom of Replacement added is called *Zermelo-FraenkelSet Theory* and is denoted ZF.[12] Again, the adjunction of AC to ZF is indi-

[12]Actually, the need for the new axiom was realized independently by Skolem, and his formulation of it was even superior to Fraenkel's. Thus, properly speaking, the system should be called *Zermelo-Skolem Set Theory* and denoted ZS, or to be more generous, *Zermelo-Fraenkel-Skolem Set Theory* (ZFS). However, we are stuck with ZF as the standard designation these days. Translations of Skolem's and Fraenkel's 1922

cated by ZF + AC or simply ZFC. All of Cantor's theory of infinite sets and numbers can be formalized in ZFC, that is, can be developed in a logically exact manner within that system.

While the logicians were improving and extending Zermelo's axiom system, mathematicians were employing its basic principles more and more and with increasing assurance in the areas of algebra, analysis, and topology. It would take us too far afield here to outline this development, and we must refer the reader to other sources.[13] Though concern about AC never died down, set-theoretical methods and ideas were widely applied in mathematical practice by 1940 and eventually took over in textbook expositions of many parts of mathematics following the Second World War. Simultaneously, a growing mild acquaintanceship with logic gave mathematicians a better (though not deep) understanding of what Zermelo, Skolem, and Fraenkel had accomplished; in this superficial sense, then, Zermelo finally received his due.

The Foundational Issue (Part 1): Cantorian Set Theory as Mathematical Platonism

Zermelo's axiomatization and its extensions diminished one area of concern about Cantorian set theory, namely, that it led to paradoxical constructions; at least there is no obvious way to carry those out in systems such as ZC and ZFC. But the objections based on more fundamental grounds as to the nature of mathematics remained, and fueled a vigorous and continuing controversy between the opponents and defenders of higher set theory.

The history of mathematics has been punctuated repeatedly by the introduction of problematic concepts (for example, imaginary numbers, infinitesimals, and points at infinity), problematic principles (for example, the parallel postulate), and problematic methods (for example, Fourier series expansions of arbitrary functions, and the operational calculus). These have been dealt with in each case by either complete or partial (perhaps modified) acceptance or outright rejection. Each instance raised anew the question: *What is it justified to say and do in mathematics?* In the past, such questions were dealt with by mathematicians sensitive to the issue of justification on a case-by-case or "local" basis. Moreover, this tradition of mathematics taking good care of its own house has continued to operate right up to the present.[14]

---

papers are to be found in van Heijenoort (1967).

[13]See, to begin with, Moore (1982), chapters 3 and 4. Of interest for the sociology of the subject is that set-theoretical mathematics was given a great boost in Poland after the First World War, largely under the leadership of Wacław Sierpiński.

[14]I have analyzed various ways of dealing with such local foundational problems in Feferman (1985), using both classical and modern cases as illustrations. [An updated version of that paper is reproduced as chapter 5 in this volume; cf. also chapter 4 for an abbreviated version.]

The advent of higher set theory gave rise to a complex question of justification of an altogether more fundamental and "global" nature, which again engaged mathematicians first and foremost, but now brought it to the doors of philosophy. Even the most basic modes of reasoning employed in mathematics had to be reconsidered, and here the technical developments of modern logic provided essential new tools for that purpose.

What were the aspects of Cantorian and Zermeloan set theory found to be problematic by its opponents? Here there was no united view, but the following features were of recurrent concern:

(i) *Sets as independent existents.* Sets are conceived to be objects having an existence independent of human thoughts and constructions. Though abstract, they are supposed to be part of an external, objective reality.

(ii) *Actual infinity.* Infinite sets are supposed to exist as completed objects; the most basic of these is the totality of natural numbers.

(iii) *Arbitrary (sub)sets.* For each set, any arbitrary combination of its elements is supposed to exist as a well-determined set in its own right.

(iv) *Power sets.* For each set S, the totality $\mathcal{P}(S)$ of arbitrary subsets of S is supposed to exist as another set.

(v) *The axiom of choice.* For any set S and any subset C of $\mathcal{P}(S)$ consisting of disjoint nonempty sets, there is a choice set, existing as an arbitrary combination of elements, one from each member of C.

(vi) *Relations and functions as sets.* A relation R between elements of a set S_1 and those of a set S_2 is supposed simply to be an arbitrary set of pairs in $S_1 \times S_2$. Functions from S_1 to S_2 are supposed simply to be many-one relations.

(vii) *Objectivity of truth and classical logic.* Each proposition P about sets has a definite truth value (true or false), independent of any means we may have to verify it. The *Law of Excluded Middle* (LEM), that P or $\neg P$ holds, is thus accepted for any such P.

The following remarks expand on these features:

ad(i). Views of mathematical objects as independently existing abstract entities are generally called a form of *platonism.* In its particular set-theoretical manifestation, this reveals itself most obviously in such principles as the Axiom of Extensionality and the Axiom of Choice.

ad(ii). The platonist position *per se* does not necessarily require the existence of infinite sets (or, for that matter, of sets at all), but of course that is essential to Cantorian set theory.

ad(iii). Though the Axiom of Separation—according to which for any set S and well-determined property $P(x)$ (of elements of S) the set $\{x \in S : P(x)\}$ exists—is fundamental and widely applied in set theory, this must not be taken to mean only definable subsets of a set need be assumed to exist.

ad(iv). The formation of power sets leads to the higher infinities by the increase in cardinality of $\mathcal{P}(S)$ over that of S, for any set S. In addition, the existence of power sets justifies *impredicative definition*, whereby a subset S' of S can be singled out by reference (by quantification over $\mathcal{P}(S)$ to arbitrary subsets X of S (for example, as the intersection of all subsets X of S satisfying a given condition $Q(X)$).

ad(v). As already explained at length in the preceding sections, the Axiom of Choice is essential to establish a linear (in fact, well-ordered) scale of infinite cardinals.

ad(vi). The reduction of functions to sets goes far beyond mathematicians' previous conception of functions as given by laws. Since the idea of one-to-one correspondence is explained in terms of one-to-one functions, this is also essential for the theory of equinumerosity.

ad(vii). An immediate consequence of LEM and the other laws of logic is the method of *proof by contradiction*: if the assumption that P is false leads to a contradiction, then P is true; formally,

$$[(\neg P) \rightarrow (Q \wedge \neg Q)] \rightarrow P.$$

Furthermore, the usual laws of quantification coupled with LEM yield

$$(\exists x)P(x) \leftrightarrow \neg(\forall x)\neg P(x).$$

In other words, to establish an existence result $(\exists x)P(x)$, it is sufficient to assume $(\forall x)\neg P(x)$ and derive a contradiction. In general this will not show how to produce a solution x to satisfy $P(x)$.

According to the platonist picture of set theory, then, statements such as the Continuum Hypothesis CH have a definite truth value, which it would be our aim as mathematicians to determine with all means at our disposal.

The Foundational Issue (Part 2): Brouwer's Way Off, Hilbert's Way Out, and Weyl's Way to Get By

The possible responses to those who found the platonist philosophy of mathematics unacceptable as a justification for higher set theory were either to reject that theory in whole or in part, or attempt some alternative form of justification. Here we review three basically different responses, due respectively to L. E. J. Brouwer, David Hilbert, and Hermann Weyl. To carry out their ideas, Brouwer and Hilbert created substantial foundational programs; these drew much attention but (as we shall see) proved far from successful. Weyl's initiative was more limited in its scope but achieved its aims within that; his basic work and approach is a forerunner of the results to be reported at the end of the essay [in chapter 12].

Brouwer's Way Off

Brouwer's program was the most radical and the most original of the three considered here. He advanced this with passionate conviction and persistence (despite its reception with general incomprehension and/or rejection) from his doctoral dissertation in 1907 through his last paper in 1955.[15] Brouwer strongly criticized platonism and formalism (of which, see below). In his 1908 paper "The unreliability of logical principles," he argued that the Law of Excluded Middle for statements about infinite sets is obtained by an unjustified extension from finite sets, where it is correct. For example, when P is a testable property, we can verify $(\exists x)P(x) \vee (\forall x)\neg P(x)$ for "x" ranging over a finite S, by inspecting each element of S in turn, but this method is evidently inapplicable to infinite S. For Brouwer, questions of truth are restricted to statements that can be verified or disproved. Thus $P \vee \neg P$ cannot be declared true until we have verified one or the other of its disjuncts. Similarly $(\exists x)P(x)$ cannot be recognized as true until we have found an instance a which makes $P(a)$ true. On the other hand, to verify $(\forall x)P(x)$ in an infinite S we cannot check each element of S, and other methods must be found according to the nature of S. Thus, for example, the *Principle of Mathematical Induction*, that $P(0) \wedge (\forall x)[P(x) \rightarrow P(x+1)]$ implies $(\forall x)P(x)$, provides a method of drawing conclusions of the form $(\forall x)P(x)$ in the set $\mathbf{N}$ of natural numbers. Brouwer also rejected the completed infinite, and he called collections S such as $\mathbf{R}$ and Cantor's second number class Ω (the countable ordinals) *denumerably unfinished*: beyond any well-determined denumerable subset S' of such S we can associate an element of $S - S'$. For Brouwer, then, the statement CH, which states the equinumerosity of $\mathbf{R}$ and Ω, has no definite meaning and the question of its truth has no interest for him.

Brouwer thus took a completely *constructivist* stance in his critique of platonism, continuing the way advanced by Kronecker. But Brouwer gave his constructivism a particularly subjectivist stamp which he labeled *intuitionism*, emphasizing the origin of mathematical notions in the human intellect. Moreover, beginning in 1918, Brouwer went on to provide a systematic constructive redevelopment of mathematics for the first time, going far beyond anything actually done by Kronecker.

---

[15]Brouwer was Dutch and did his dissertation in Amsterdam. All his published papers on philosophy and the foundations of mathematics are to be found in volume 1 of his *Collected Works* (1975); the papers originally in Dutch are there translated into English. Incidentally, the second volume in his *Collected Works* consists of papers in nonconstructive mathematics, mainly topology, to which Brouwer made fundamental contributions during the period 1908–1913.

The theory of real numbers provided the first obstacle to a straightforward constructivist redevelopment. With real numbers identified as convergent sequences of rational numbers, all sorts of classical results would apparently fail to have reasonable constructive versions if one restricted attention only to sequences determined by effective laws. Brouwer introduced instead a novel conception, that of *free choice sequences* (f.c.s.), which might be determined in nonlawlike ways (for example, by random throws of a die), and of which one would have only finite partial information at any stage. Then with the real numbers viewed as convergent f.c.s. of rationals, a function f from **R** to **R** can be determined using only a finite amount of such information at any given argument. Brouwer used this line of reasoning to conclude that any function from **R** to **R** must be *continuous*, in direct contradiction to the classical existence of discontinuous functions. With this step Brouwer struck off into increasingly alien territory, and he found few to follow him even among those sympathetic to the constructive position.[16]

Hilbert's Way Out

When Hilbert addressed the International Congress of Mathematicians in 1900, he was nearing the age of forty and was already considered to be one of the world's greatest mathematicians. Hilbert had by that time made his mark in algebra, number theory, geometry, and analysis. He would go on to make further substantial contributions in analysis, mathematical physics, and the foundations of mathematics. At Göttingen, where he was a professor, Hilbert had many first-rate colleagues and students who often helped with the detailed development of his ideas. In particular, Hilbert's program for the foundation of mathematics was taken up by von Neumann, Ackermann, Bernays, Gentzen, and others, beginning in the 1920s. This program was shaped initially by both Hilbert's general tastes and interests as well as his specific experience with axiomatic geometry. Hilbert was noted for his clarity, rigor, and a penchant for systematically organizing subjects; at the same time he had the knack of putting these pedagogical tendencies to work to develop powerful methods for the solution of concrete problems. His work in algebra and algebraic number theory was part of the growing tendency toward abstract methods in the nineteenth century, which then came to dominate twentieth-century mathematics. One advantage of such methods is their generality—any mathematical structure meeting the basic principles must satisfy all the conclusions drawn from them.

In his work on axiomatic geometry, Hilbert returned to the axiom system that had come down from Euclid, for which he gave a superior development meeting modern standards of rigor. In addition, Hilbert moved on

[16]Nowadays, Brouwer's theories of f.c.s. are much better understood and are demonstrably coherent. Dummett (1977) gives an excellent introduction to the development of mathematics based on Brouwer's intuitionistic ideas. [Cf. also Troelstra and van Dalen (1988) and van Stigt (1990).]

to what we would now call "metageometry," through the study of questions such as the *independence* of certain axioms from the remaining ones. This he achieved by a series of constructions of unusual *models* which satisfy all the axioms but the one to be shown independent. Also, Hilbert demonstrated the *consistency* of his axioms by the methods of analytic geometry, which interpret the statements in the Cartesian plane **R** × **R** and space **R** × **R** × **R**. In other words, geometry is shown consistent *relative* to a theory of real numbers (and for weaker combinations of axioms, relative to subsets of **R** such as the algebraic numbers). In the second problem of his 1900 address (Hilbert 1900, 1902), the compatibility of the arithmetical axioms, Hilbert proceeded to call for a proof of consistency for a system of axioms for **R**, which he recognized would, in some sense, have to be *absolute*.

Hilbert's statement of Problem 2 already reveals some of his key positions, though they would be extended and elaborated later. He says there that the foundations of any science must be provided by setting up an *exact* and *complete* system of axioms. "The axioms so set up are at the same time the definitions of [the] elementary ideas [of that science]; and no statement within the realm of the science whose foundation we are testing is held to be correct unless it can be derived from those axioms by means of a finite number of logical steps" (Hilbert (1902), in Browder (1976), p. 10). After explaining the relative consistency proofs for geometry, he says: "On the other hand a direct method is needed for the proof of the compatibility of the arithmetical axioms" (*ibid.*) (that is, for a theory of real numbers). Hilbert goes on to posit that a mathematical concept *exists* if, and only if, it can be shown to be consistent (noncontradictory); thus, for him "the proof of the compatibility of the axioms [for real numbers] is at the same time the proof of the mathematical existence of the complete system of real numbers" (*ibid.*, p. 11). The real numbers are not to be regarded as all possible convergent sequences of rational numbers, but rather as a structure determined by or governed by certain axioms. Finally, in his statement of Problem 2, Hilbert expresses the conviction that the existence of Cantor's higher number classes will be demonstrated by a consistency proof "just as that of the continuum." Evidently, at that time Hilbert thought the consistency proof of both the theory of real numbers and set theory would be straightforward. But within a few years (Hilbert 1904) he was presenting a less sanguine view: the paradoxes of set theory seemed to him to indicate that the problem of establishing the consistency of set theory presented greater difficulties than he had anticipated. Furthermore, according to Paul Bernays, "although he strongly opposed Leopold Kronecker's tendency to restrict mathematical methods, he nevertheless admitted that Kronecker's criticism of the usual way of dealing with the infinite was partly justified."[17]

[17]Bernays (1967), p. 500. For more information about Hilbert's shifting views concerning foundations see Sieg (1984), pp. 166 and 170ff. [cf. also Hallett (1995)].

In the following decades Hilbert was preoccupied with the theory of integral equations and mathematical physics, and he did not return to problems of foundations until 1917.* In the meantime, Zermelo had laid the axiomatic grounds for Cantorian (platonist) set theory while Brouwer had launched his attacks on mathematical platonism and formalism. Concerning the latter, Brouwer stressed that *consistency* was not enough to justify the use of mathematical principles; what was necessary was to assure their *correctness*.

Hilbert's mature program for the foundations of mathematics via *finitist proof theory* was announced in his 1917 address, "Axiomatic thought" (Hilbert 1918); it was then elaborated in a succession of almost yearly publications through 1931. Briefly, the idea is that a given body of mathematics (such as number theory, analysis, or set theory) is to be treated as formally represented in an axiomatic theory T. Each such T is to be specified by precisely described rules for generating its well-formed formulas (statements) from a finite stock of basic symbols. Certain of these formulas are then selected as axioms (both logical ones and axioms concerning the specific subject matter of T), and rules for drawing inferences from the axioms are specified. For a formal axiomatic theory T presented in this way, the set of *provable formulas* is defined to consist of just those formulas for which there exists a proof (or derivation) in the system, that is, a finite sequence ending with the formula, each term of which is an axiom or is obtained from preceding terms by one of the rules of inference. Then T is *consistent* just in case there is no contradiction $(P \wedge \neg P)$ which is provable in T. Hilbert's *Beweistheorie*, or *theory of proofs*, was developed as a tool to analyze all possible derivations in formal axiomatic systems. With proofs represented as finite sequences of formulas, and formulas as finite sequences of basic symbols, whose structure in both cases is regulated by effective conditions, the question of consistency of T in no way assumes the actual infinite. Now Hilbert's idea was to use entirely *finitary* methods in establishing the consistency of formal systems which otherwise required for their justification the assumption of the actual infinite. Not only the theory of real numbers but already a formal system of elementary number theory would have to be shown consistent. Such a system would be a first-order version PA (Peano Arithmetic) of Peano's axioms for **N**, formulated by replacing the second-order axioms of induction by the corresponding first-order scheme: $P(0) \wedge \forall x(P(x) \rightarrow P(x+1)) \rightarrow \forall x P(x)$, for all formulas $P(x)$ in the language of PA. Hilbert wanted to justify the use of such a system including classical logic, which leads by LEM to statements like $(\exists x)P(x) \vee (\forall x)\neg P(x)$; in this, he indirectly acknowledged Brouwer's criticism of the assumption of the actual infinite.

In addition to his general program, Hilbert proposed some specific proof-theoretic techniques to carry it out. These methods were shown

*[It has been brought to my attention by W. Sieg that this is not quite correct: Hilbert continued to lecture throughout that period on the foundations of mathematics; cf. Hallett (1995).]

to work in relatively simple cases for theories T much weaker than PA. The success of the program would depend next on extending them to a finitary proof of the consistency of PA, and so on, up to the consistency of set-theoretical systems like ZFC. As we shall see, these hopes were to be dashed by Gödel's incompleteness theorems of 1931.

Hilbert's 1926 paper "On the infinite"[18] is a very readable exposition of the finitist program for the foundations of mathematics and is a typical example of Hilbert's style of heroic optimism, which nowadays may be considered bombastic: "the definitive clarification of the *nature of the infinite* has become necessary, not merely for the special interests of the individual sciences, but rather for the *honor of the human understanding* itself" (Hilbert (1926), in van Heijenoort (1967), pp. 370–371).

The first portion of Hilbert's 1926 paper reviews the general "problem of the infinite" in mathematics, from which he turns to the particular problems raised by Cantor's theory of transfinite numbers. Here "the infinite was enthroned and enjoyed the period of its greatest triumph" (*ibid.*, p. 375). But the paradoxes discovered by Russell and others brought one to an intolerable situation: is there no way to retain what Cantor achieved while escaping the paradoxes? Yes, by careful investigation and proof of the complete reliability everywhere of our inferences, "no one shall drive us from the paradise that Cantor created for us" (*ibid.*, p. 375).

The plan laid out in the mid-portion of Hilbert's paper is that of expressing mathematical propositions formally and representing mathematical inferences by derivations in precisely described formal systems. The formulas of these systems are divided into *finitary propositions* and *ideal propositions*. Finitist proof theory is to be used to show how the latter can be eliminated in terms of the former; this is to be achieved by finitary proofs of consistency. The procedure of elimination here is analogous to that used to justify the introduction of *ideal elements* in mathematics such as imaginary numbers, points at infinity, and the algebraic number ideals introduced by Kummer. Apropos of this last, Hilbert remarks that "it is strange that the modes of inference that Kronecker attacked so passionately are the exact counterpart of what . . . the same Kronecker admires so enthusiastically in Kummer's work and praised as the highest mathematical achievement" (*ibid.*, p. 379). Moving on to more definite proposals, a formal system of elementary number theory is presented which is equivalent to the first-order axioms PA of Peano Arithmetic indicated above. Hilbert claims that the problem of its consistency is "perfectly amenable to treatment"; moreover, what a "pleasant surprise that this gives us the solution also of a problem that became urgent long ago," (*ibid.*, p. 383) namely, the consistency of a system of axioms for the real numbers.

Finally, in the third part of his 1926 paper, Hilbert moves into high gear, leaving even his acolytes standing bewildered in the dust. For here

[18] We refer here to the English translation in van Heijenoort (1967), pp. 369–392.

he claims "to play a last trump" and to show how the continuum problem can be solved by use of proof theory. Hilbert's plan has intriguing aspects roughly related to later work on constructive hierarchies of number-theoretic functions as well as to Gödel's own definitive results on the consistency of CH (to be discussed below), but it has never been worked out in any coherent form.* In his bravado and eagerness to demonstrate the success of his program, Hilbert promised far more than he could deliver. Indeed, even the claims of finitist proofs of the consistency of elementary number theory and real number theory would not survive Gödel's incompleteness theorems of 1931. For one who trumpeted the cause of absolute reliability, how wrong could he be? Nevertheless, Hilbert's influence and program were decisive factors in what followed: as we shall see, the attack on the "problem of the infinite" was shifted to the arena of metamathematics and in that proof theory became one of the primary tools.

One final remark concerning Hilbert vis à vis the constructivists needs to be made. As Paul Bernays tells it, "there was a fundamental opposition in Hilbert's feelings about mathematics . . . namely his resistance to Kronecker's tendency to restrict mathematical methods . . . particularly, set theory . . . [and his] thought that Kronecker had probably been right. . . . It became his goal to do battle with Kronecker with his own weapon of finiteness."[19] But in his efforts to outfight the constructivists, Hilbert was hoist with his own petard.

Weyl's Way to Get By

Hermann Weyl was one of Hilbert's most illustrious students, and his work was almost as broad (and deep) as that of his teacher. After a period as Dozent in Göttingen following his doctoral work there, Weyl took a position in Zürich shortly before World War I. Hilbert made several efforts to bring him back, and Weyl finally returned as Hilbert's successor in 1930, only to leave for Princeton when the Nazis came into power in 1933. All these personal and professional connections made it difficult for Weyl to express his strong differences with Hilbert on foundational matters, but he did so tactfully on a number of occasions.[20]

Weyl evolved his first approach to the foundations of mathematics in the monograph *Das Kontinuum* (1918). In the introduction thereto he criticized axiomatic set theory as a "house built on sand" (though the objects of, and reasons for, his criticism are not made explicit). He proposed to replace this with a solid foundation, but not for all that had come to

*[Cf. the discussion by R. M. Solovay of one of Gödel's proofs of the consistency of CH in Gödel (1995), 114–127.]

[19] Quoted in Reid (1970), p. 173.

[20] Weyl's best known works on logic and the philosophy of his mathematics are his monograph (1918) and book (1949), both discussed below. But there are also many less familiar articles on these subjects scattered through his *Collected Works* (1968), spanning the period 1918–1955. [Chapter 13 in this volume provides a detailed examination of Weyl (1918); see also chapter 14.]

be accepted from set theory; the rest he gave up willingly, not seeing any other alternative. Weyl's main aim in this work was to secure mathematical analysis through a theory of the real number system (the continuum) that would make no basic assumptions beyond that of the structure of natural numbers **N**. Unlike Hilbert, Weyl did not attempt to reduce first-order reasoning about **N** to something supposedly more basic. In this respect Weyl agreed with Henri Poincaré that the natural number system and the associated principle of induction constitute an irreducible minimum of theoretical mathematics, and any effort to "justify" that would implicitly involve its assumption elsewhere (for example, in the metatheory). And, unlike Brouwer, Weyl accepted uncritically the use of classical logic at this stage (though at a later date he was to champion Brouwer's views). Formulated in modern terms, a system like PA was thus accepted by Weyl as basic. However, for a theory of real numbers one would have to provide a means to treat sets or sequences of natural numbers (using the reduction of **Q** to **N**) and then, for analysis, explain how to deal with functions of real numbers as functions from and to such sets. Weyl added axioms for existence of sets of natural numbers which are arithmetically definable; these are of the form $(\exists X \subseteq \mathbf{N})(\forall n \in \mathbf{N})[n \in X \leftrightarrow P(n)]$ where the formula $P(n)$ contains no quantified variables other than those which range over **N**. Using two sorts of variables, lowercase for elements of **N** and uppercase for subsets of **N**, these axioms take the form $(\exists X)(\forall n)[n \in X \leftrightarrow P(n)]$. Weyl deliberately excluded axioms of this type in which P contains quantified variables ranging over subsets of **N**; in particular, he thus excluded statements of the form $(\exists X)(\forall n)[n \in X \leftrightarrow (\forall Y)Q(n, Y)]$, even when Q is an arithmetical formula. Weyl's reason for doing so was that otherwise one would be caught in a *circulus vitiosus.* The matter at issue here requires a lengthy aside.

The *vicious circle principle*, first enunciated by Poincaré, was designed to block certain purported definitions, in which the object introduced is somehow defined in terms of itself. According to Poincaré, all mathematical objects beyond the natural numbers are to be introduced by explicit definitions. But a definition which refers to a presumed totality—of which the object being defined is itself to be a member—involves one in an apparent circle, since the object is then itself ultimately a constituent of it own definition. Such "definitions" are called *impredicative*, while proper definitions are called *predicative*; put in more positive terms, in the latter one refers only to totalities which are established prior to the object being defined. The formal definition of a set X of natural numbers as $X = \{n : (\forall Y)Q(n, Y)\}$, taken from the axiom $\exists X \forall n[n \in X \leftrightarrow (\forall Y)Q(n, Y)]$, is impredicative in Poincaré's sense because it involves the quantified variable "Y" ranging over arbitrary subsets of **N**, of which the object X being defined is one member. Thus in determining whether $(\forall Y)Q(n, Y)$ holds, we shall have to know in particular whether $Q(n, X)$ holds—but that can't be settled until X itself is determined.

Poincaré raised his ban on impredicative definition thinking doing so would exclude the paradoxes. However, that succeeds only by taking a very broad reading of what it means for a definition to refer to a totality. For example, in Russell's paradox, derived from $\exists a \forall x[x \in a \leftrightarrow x \notin x]$, or $a = \{x : x \notin x\}$, the formula $P(x) = (x \notin x)$ does not refer to the presumed totality of *all* sets, since it does not contain any *quantified variables* ranging over sets. On the other hand, it is true that in order to determine by its "definition" which members a has, we must already know the answer; in particular, $a \in a$ would be determined true only if we already knew that $a \notin a$. However, this vicious circle is not itself excluded by the principle enunciated by Poincaré in its prima facie reading. On the other hand, Poincaré's principle as it stands would certainly exclude impredicative definitions in analysis of the sort $X = \{n : P(n)\}$ where P contains quantified variables ranging over arbitrary subsets of $\mathbf{N}$; the objection to such definitions is not that they are paradoxical, but rather that they are implicitly circular and hence not proper.

Russell was one of the first to accept Poincaré's ban on impredicative definitions as applied to sets, and he turned to the construction of a formalism for generating predicative definitions. Roughly speaking, instead of talking about arbitrary sets, one may only talk in this formalism about *sets of level 1, sets of level 2*, etc. A set of level 1 is defined by a formula using quantified variables ranging over individuals only. This determines the totality of sets of level 1; then a set of level 2 is defined using just quantified variables ranging over individuals *and* over arbitrary sets of level 1. The problems with this kind of *ramification* (as Russell called it) is that in the resulting theory of real numbers (introduced as sets of rational numbers) one would have reals of level 1, reals of level 2, etc., but analysis with such distinctions would be unworkable. As an ad hoc device, Russell then introduced his *Axiom of Reducibility*, which says that every set of higher level is coextensive with one of lowest level. In effect, this "axiom" nullified the point of ramification and indirectly permitted the use of impredicative definitions in analysis. Russell recognized that this step did not jibe with the initial philosophical outlook stemming from Poincaré's definitionism, but saw no alternative to developing analysis. As it happens, his *theory of types*, which distinguished sets according as to whether they were sets of individuals (type 1), sets of such sets (type 2), etc., and which restricted the membership relation to objects of successive types (with individuals taken to be of type 0), already blocked the paradoxes without in any way forcing one to insist on predicative definitions. This was achieved by the later move (due to F. P. Ramsey) to *Simple Type Theory* as opposed to Russell's original *Ramified Type Theory* (which distinguished sets both as to types, and as to levels of definition).

Now if one accepts the platonist philosophy in set theory, the totality $\mathcal{P}(\mathbf{N})$ of subsets of $\mathbf{N}$ exists independently of how its objects may be defined, if at all. According to this view any formula $P(x)$ involving variables ranging over $\mathbf{N}$ and variables ranging over $\mathcal{P}(\mathbf{N})$, connected by arithmetical relations between individuals and relations of membership between individuals and sets, has a definite meaning, independent of whether we have any means to "determine" it; then the definition $X = \{x : P(x)\}$ simply serves to single out one member of $\mathcal{P}(\mathbf{N})$. According to the platonists, this is entirely analogous to singling out a natural number as the minimum one satisfying a certain (nonempty) property, formally $k = (min\, n)P(n)$, where we may have no way to determine k effectively, even when P only involves numerical quantification. But according to the predicativist (or "definitionist") there *is* a difference: the totality of natural numbers is granted as clear and definite and each of its members can be singled out by a prior representation, while there is no justification in the assumption of a totality of subsets of **N** independent of how these may be defined, since sets can be introduced only by definition on the basis of (successively) established totalities.

To return to Weyl, we can say that he positioned himself as follows in *Das Kontinuum*.[21] First, he rejected the platonist philosophy of mathematics as manifested in the impredicative existence principles of axiomatic set theory, though he accepted classical quantifier logic when applied to any established system of objects. Second, he accepted the predicativist viewpoint taking the system of natural numbers as its point of departure. Third, he recognized that, whatever their justification on predicative grounds, ramified theories would not give a viable account of analysis. Weyl's main step, then, was to see what could be accomplished in analysis if one worked just with sets of level 1, in other words, only with the principle of arithmetical definition. Here it is to be understood that such definitions may be *relative*; that is, if $P(x, Y_1, \ldots Y_n)$ contains variables $Y_1, \ldots, Y_n$ but no quantified set variables, then P serves to define $X = \{x : P(x, Y_1, \ldots, Y_n)\}$ relative to $Y_1, \ldots Y_n$. Given any specific definitions of $Y_1, \ldots, Y_n$ this will produce by substitution in P a specific definition of X. Finally, such means of relative arithmetical definition serve to explain which *functions* from sets to sets are to be admitted, namely, just those given as $F(Y_1, \ldots, Y_n) = \{x : P(x, Y_1, \ldots, Y_n)\}$ with P arithmetical.

In this way, Weyl was able to set up a formal axiomatic framework for *arithmetical analysis* in which he could define rational numbers in terms of (pairs of) natural numbers, then real numbers as certain sets (namely, lower Dedekind sections) of rational numbers, and finally functions of real numbers as functions from sets to sets in the way just explained. He then went on to examine which parts of classical analysis could be justified on these grounds. To begin with, the general *least upper bound (l.u.b.) principle for sets of real numbers* could not be accepted in its full generality. According to that principle, any set S of real numbers which is bounded above has a l.u.b. X. In Weyl's framework, S is given as consisting of lower Dedekind sections Y satisfying an arithmetical con-

[21] See also his elaboration of this position in Weyl (1919).

dition $P_1(Y)$; then its supposed l.u.b. X is the union $\bigcup Y[P_1(Y)]$; that is, $X = \{r \in \mathbf{Q} : \exists Y(P_1(Y) \wedge r \in Y)\}$. Translating rational numbers to natural numbers, this gives $X = \{n : (\exists Y)P(n, Y)\}$ for suitable arithmetical P; but this is an impredicative definition which is not justified in Weyl's system. On the other hand, the *l.u.b. principle for sequences of real numbers* does hold in his system. For, a sequence $\langle Y_k \rangle_{k \in \mathbf{N}}$ is given by an arithmetical condition $P_1(r, k)$, which holds just in case $r \in Y_k$. If the sequence is bounded above, its l.u.b. X is the union $\bigcup Y_k[k \in \mathbf{N}]$, that is, $X = \{r \in \mathbf{Q} : (\exists k)P_1(r, k)\}$, and this reduces to an arithmetical definition of the form $X = \{n : (\exists k)P(n, k)\}$ with P arithmetical.[22] Thus Weyl's task was reduced to seeing which parts of classical analysis rest simply on the l.u.b. principle for *sequences* of reals rather on the more general l.u.b. principle for *sets* of reals. Here, in fact, he found that the entire theory of *continuous functions* of reals goes through in a straightforward manner in arithmetical analysis, including for such (i) the intermediate value theorem, (ii) the attainment of maxima and minima on a closed interval, and (iii) uniform continuity on a closed interval.[23] From these it is a direct step to develop the theory of *differentiation and integration* for continuous functions. Here it is Riemann integration that can be treated in a straightforward predicative way. (Weyl remarks[24] that it is less simple to deal with the more modern theories of integration through theories of *measure*, but gives no indications; their treatment would be left for the future developments of predicative analysis [outlined in chapters 12–14 in this volume].)

There is no mention of Brouwer or intuitionism, and no restriction on the logic employed, in *Das Kontinuum*. However, within the next few years Weyl became more familiar with Brouwer's work and somewhat of a convert. He is quoted as saying, during some lectures on Brouwer's ideas in 1920: "I now give up my own attempt and join Brouwer."[25] Over the following years Weyl was to help champion Brouwerian intuitionism as against Hilbert's program in various publications, much to Hilbert's annoyance. However, in later years he became pessimistic about the prospects for the Brouwerian revolution. Moreover, it is not true to say that Weyl ever completely gave up his "own attempt," which he continued to mention over the years in articles on foundational matters, where his criticisms of mathematical platonism in set theory remained constant. A relatively mature expression of Weyl's view is provided by his 1949 book, which is

[22] Actually, Weyl establishes *Cauchy's convergence principle* for sequences of reals, which is equivalent to the l.u.b. principle for sequences. Note also that the l.u.b. principle for sets (sequences) is equivalent to the g.l.b. principle for the same, and with the given representation of real numbers this leads to a definition of the g.l.b. as an intersection rather than a union.

[23] As sketched in Weyl (1918), pp. 61–65. [Cf. also chapters 13 and 14 in this volume.]

[24] *Ibid.*, p. 65.

[25] See van Heijenoort (1967), p. 480; the original statement may be found in Weyl's *Gesammelte Abhandlungen* (1968), vol. 2, p. 158.

readily accessible.[26] The following quotations from this book reveal both his settled convictions as well as his unsettled state of mind about the eventual foundations of mathematics:

> The leap into the beyond occurs when the sequence of numbers that is never complete but remains open toward the infinite is made into a closed aggregate of objects existing in themselves. Giving the numbers the status of ideal objects becomes dangerous only when this is done. . . . The vindication of this transcendental point of view forms the central issue of the violent dispute . . . over the foundations of mathematics.[27]

> From an aggregate of individually exhibited objects we may by selection produce all possible subsets and thus make a survey of them one after another. But when one deals with an infinite set like **N**, then the existential absolutism for the subsets becomes still more objectionable than for the elements.[28]

There follows an explanation of the vicious circle principle, of levels of predicative definition, and of the resulting "dilemma" for analysis concerning the l.u.b. principle:

> Russell, in order to extricate himself from the affair, causes reason to commit harakiri, by postulating the above assertion [the Axiom of Reducibility] in spite of its lack of support by any evidence. . . . In a little book *Das Kontinuum*, published in 1918, I have tried to draw the honest consequence and constructed a field of real numbers of the first level, within which the most important operations of analysis can be carried out.[29]

> Mathematics with Brouwer gains its highest intuitive clarity. He succeeds in developing the beginnings of analysis in a natural manner, all the time preserving the contact with intuition much more closely than had been done before. It cannot be denied, however, that in advancing to higher and more general theories the inapplicability of the simple laws of classical logic eventually results in an almost unbearable awkwardness. And the mathematician watches with pain the larger part of his towering edifice which he believed to be built of concrete blocks dissolve into mist before his eyes.[30]

Finally, "Hilbert's mathematics may be a pretty game with formulas . . . but what bearing does it have on cognition, since its formulas admittedly

[26]That is a revised and considerably augmented English edition of an article Weyl wrote for the *Handbuch der Philosophie* in 1926.

[27]Weyl (1949), p. 38.

[28]*Ibid.*, p. 49.

[29]*Ibid.*, p. 50.

[30]*Ibid.*, p. 54.

have no material meaning by virtue of which they could express intuitive truths?"[31] In this connection, Weyl says that a consistency proof of arithmetic "would vindicate the standpoint taken by the author in *Das Kontinuum*, that one may safely treat the sequence of natural numbers as a closed sequence of objects."[32]

Incidentally, a famous wager was made in Zürich in 1918 between Weyl and George Pólya, concerning the future status of the following two propositions:

(1) Each bounded set of real numbers has a precise upper bound.
(2) Each infinite set of real numbers has a countable subset.

Weyl predicted that within twenty years either Pólya himself or a majority of leading mathematicians would admit that the concepts of number, set, and countability involved in (1) and (2) are completely vague, and that it is no use to ask whether these propositions are true or false, though any reasonably clear interpretation would make them false (unless the concepts involved were to acquire totally new meanings). The loser was to publish the conditions of the bet and the fact that he lost in the *Jahresberichten der Deutschen Mathematiker Vereinigung*; this never took place as such.[33]

Weyl's viewpoint in making this wager is often mistakenly identified as being that of Brouwer's intuitionism, though it was made at the time of publication of *Das Kontinuum*, prior to Weyl's taking up Brouwer's views. As we have seen, the l.u.b. principle was rejected in his 1918 publication on the grounds of the vicious circle principle and the rejection of impredicative definitions. And (2) requires some form of the Axiom of Choice applied to subsets of **R** for its proof, so was rejected by Weyl along with his rejection of the conception of **R** as a completed totality. As to the settlement of the wager, there is no question that Weyl lost under its stated conditions. Moreover, the *vast* majority of mathematicians (leading or otherwise) would say then, as they would say now, that Weyl's side of the bet was simply wrong-headed. But Weyl would not be shaken by this: "The motives are clear, but belief in this transcendental world taxes the strength of our faith hardly less than the doctrines of the early Fathers of the Church or of the scholastic philosophers of the Middle Ages."[34] So, the "Middle Ages" are

[31] Weyl (1949), p. 61.

[32] *Ibid.*, p. 60. For elaboration of Weyl's views, the reader should not overlook Appendix A of Weyl (1949), and see further his 1944 obituary article "David Hilbert and his mathematical work," reproduced in Reid (1970), pp. 245–283, as well as Weyl (1946). [Cf. also chapter 13 in this volume.]

[33] The document spelling out the bet was reproduced by George Pólya in a brief note (1972). According to Pólya, "The outcome of the bet became a subject of discussion between Weyl and me a few years after the final date, around the end of 1940. Weyl thought that he was 49% right and I, 51%; but he also asked me to waive the consequences specified in the bet, and I gladly agreed." Pólya showed the wager to many friends and colleagues and, with one exception, all thought that he had won.

[34] Weyl (1946), p. 6.

simply taking somewhat longer to wane than Weyl expected—but that's the way it goes with Middle Ages.

The Rise of Metamathematics and the Crumbling of Hilbert's Program

Hilbert's emphasis on the axiomatic approach to mathematics and on the metatheoretical questions concerning axiom systems constituted one of the principal sources for the development of metamathematics as a new, distinctive, and coherent subject. Here "meta" has taken the meaning "about"; that is, axiom systems are objects of study examined externally. Such a treatment fits very well with axiomatic mathematics in its traditional sense, as first exemplified in geometry, but also with modern axiomatizations of number theory, algebra, and topology; each of these deals with a restricted or "local" part of mathematics. The metamathematical questions that Hilbert raised about axiom systems concerned their consistency, completeness, categoricity, and independence (of the axioms, one from another). In his program for the foundations of mathematics via finitist proof theory, Hilbert imposed further requirements on how the investigation of such questions was to be carried out, but those restrictions are not essential to metamathematics and only influenced part of its development. What *is* essential is that axiom systems are to be described precisely in formal terms so that results concerning them may be established rigorously. The term "metamathematics" then is suitable for the study in each case of that part of mathematics formalized in a given axiom system. But this is already troublesome, insofar as such study is to be carried out by informal mathematical means which may themselves be formalized. This novel feature is exactly what was capitalized on in the first striking results of metamathematics, namely, the Löwenheim-Skolem theorem and Gödel's incompleteness theorems; these, furthermore, combined to undermine both Hilbert's conception of mathematical existence and his finitist program for establishing "existence" via consistency proofs.

The following is devoted primarily to the highly discomfiting (if not devastating) effects on Hilbert's program of several metamathematical results of a general character (including those just mentioned). At the same time, these results served to undermine the programs for the universal or "global" axiomatization of mathematics initiated by Frege and carried on in the work of Russell, Zermelo, Fraenkel, and others. In contrast to the axiom systems for restricted areas of mathematics of the sort described above, here one aimed at systems of such generality that "all" mathematics could be formalized within them. As such, the idea of investigating these systems from the outside was antithetical to the motives for their creation,[35] but as

[35] This point has been emphasized particularly by van Heijenoort; cf., for example, his 1967 essay reproduced in van Heijenoort (1986), pp. 13–14.

we shall see, the claims to universality could not withstand the onslaught of metamathematics.

The Löwenheim-Skolem Theorem

In 1915 Leopold Löwenheim established a theorem which is now stated as follows: if a formula A of the first-order predicate calculus has a model (is satisfiable in some domain), then it has a denumerable model. This was improved by Thoralf Skolem in 1920 (and again in 1922) both in the proofs and in the statement of results. Skolem's theorem tells us that if S is any set of first-order formulas which has a model, it has a denumerable model. The Löwenheim–Skolem theorem was derived later (1930) as a consequence of Gödel's completeness theorem, by which if S is consistent in first-order predicate logic, then S has a denumerable model. For it is trivial that if S has a model at all, then it must be consistent.[36]

At first sight, Gödel's completeness theorem would seem to support Hilbert's dictum that existence of a mathematical concept is the same as consistency of an axiom system for it. But Hilbert had in mind axiomatizations S like that of Peano for the natural numbers and of Dedekind for the reals (as well as certain of his own for geometry) which are *categorical*, that is, such that the models of the axiom system are *uniquely* determined up to isomorphism. This implies that if M, M' are any two models, then they are equinumerous; that is, $M \sim M'$. But the Löwenheim-Skolem theorem shows that no set of first-order axioms S for the real numbers, that is, for which **R** is a model, can be categorical since S has a denumerable model while **R** is nondenumerable.

A later result by Skolem showed that no set of first-order axioms for **N** can be categorical; he did this by constructing a *nonstandard model* N' satisfying exactly the same first-order statements as **N**.[37] A still more general theorem of Tarski showed that if a first-order set of statements S has a model M of any infinite cardinality m, then it has a model M' of any other infinite cardinality m'. So *no* set of first-order axioms S with an infinite model can be categorical.

Since Peano's original axioms for **N** and Dedekind's for **R** are categorical, they must have an essentially non-first-order component. In fact these are just, respectively, the axioms of induction for **N** and of completeness

[36]The various basic papers of Löwenheim, Skolem, and Gödel may be found translated in van Heijenoort (1967). [The paper of Skolem dated 1922 in that source was not published until 1923 and appears as Skolem (1923a) in our bibliography; the reason given by van Heijenoort for the earlier dating is that Skolem had lectured on this material in 1922.] See also Gödel (1986). [Chapter 6 in this volume provides a survey of Gödel's life and work.]

[37]In the language of current model theory, two structures M, M' which provide interpretations of the same language are called *elementary equivalent*, and one writes $M \equiv M'$, if they satisfy the same first-order statements in that language. Then by a nonstandard model of the natural numbers is meant an N' not isomorphic to **N** with $\mathbf{N} \equiv N'$. Similarly a nonstandard model of the reals is an R' such that $\mathbf{R} \equiv R'$ but R' is not isomorphic to **R**.

(or the l.u.b. principle) for **R** expressed set-theoretically. For example, in the case of Peano's axioms this takes the form

$$\forall X[0 \in X \wedge \forall n(n \in X \rightarrow (n+1) \in X) \rightarrow \forall n(n \in X)],$$

where the (set) variable "X" ranges over subsets of the domain. But for categoricity there is a further requirement, called that of *standard second-order logic*: in any interpretation of the axioms in a domain M, the set variables are to be interpreted as ranging over the set $\mathcal{P}(M)$ of arbitrary subsets of M.

Now this takes on a more puzzling aspect when we move to axioms for sets like those of Zermelo-Fraenkel, ZF (or even Zermelo's system Z). Since ZF is first-order, if it has a model M at all then it has a denumerable model M'. But any set A in M' must have an interpretation of its powerset $\mathcal{P}(A)$ in M', say $\mathcal{P}_{M'}(A)$. In ZF it is a theorem that if A is infinite then $\mathcal{P}(A)$ is nondenumerable. But in M', $\mathcal{P}_{M'}(A)$ has only denumerably many members. This peculiar situation is referred to as "Skolem's paradox" though it is not actually paradoxical; the puzzle is resolved by noting that *externally*, $A \sim \mathcal{P}_{M'}(A)$, since both sets are denumerably infinite. But *internally*, that is, within M', there is no function which can establish the one-to-one correspondence between A and the interpretation of $\mathcal{P}(A)$ in M'. In other words, the formal statement $\neg(A \sim \mathcal{P}(A))$ is still true in M'.

Now if we are to follow Hilbert's dictum of mathematical existence through in this context as well, the existence of set-theoretical concepts must be secured by the consistency of axiom systems like ZF. As we see, however, what would thereby be secured is not the intended concept but a variety of possible interpretations in models of different cardinality. The only way to secure the concept of set axiomatically is to add to the axioms of ZF a second-order "completeness" axiom referring to arbitrary subsets of the intended model M. This now puts us in a bind: one can't capture the notion of arbitrary set axiomatically without somehow already assuming the notion of arbitrary set to be understood.

There is still a final problem for Hilbert's conception of mathematical existence when combined with his finitist program. According to the latter, each axiomatic system must be prescribed finitarily: the syntax, axioms, and rules of inference are all to be determined by finitary effective procedures. For example, this is achieved in the axiom system PA by replacing Peano's second-order axiom of induction by the axiom scheme consisting of all formulas of the form

$$P(0) \wedge \forall n(P(n) \rightarrow P(n+1)) \rightarrow \forall n P(n),$$

where $P(n)$ is any formula in the first-order language of arithmetic using the basic symbols $(0, 1, +, \cdot,$ and $=)$. Now from a purely formal point of view, the statement of Peano's original axiom,

$$\forall X[0 \in X \wedge \forall n(n \in X \rightarrow (n+1) \in X) \rightarrow \forall n(n \in X)],$$

appears to be equally finitary. Indeed this is so, but being so does not suffice for categoricity; there is nothing in such a formal axiomatization within a second-order language that can force the system to be categorical. Only the *semantic* requirement that we are dealing with standard second-order logic, that is, that in any interpretation M, "X" ranges over $\mathcal{P}(M)$, will ensure categoricity. But that requirement is on its face nonfinitary; in fact, this is demonstrated as a consequence of the results with which we deal next.

Gödel's Incompleteness Theorems

In his famous paper "On formally undecidable propositions of *Principia Mathematica* and related systems," Gödel (1931) proved that for a wide class of effectively presented consistent extensions S of a formal system PM containing PA, there are elementary statements A such that neither A nor $\neg A$ is provable in S. For his proof, Gödel introduced a class of finitarily defined functions which are obtained from the zero and successor functions by explicit definition and inductive definition on **N**.[38] The class of functions so generated is now called the *primitive recursive functions*. All such are effectively computable; through later work of Church, Kleene, Turing, and Post,[39] one arrived at a definition of the most general class of effectively computable functions on **N**, which is now called the *recursive functions*.

Gödel's incompleteness results were achieved by first coding up syntactic objects (expressions, terms, formulas, proofs) as sequences (or sequences of sequences, etc.) of numbers and then representing such sequences $(m_1, \ldots, m_k)$ as single numbers n using the prime power representation $n = p_1^{m_1} \ldots p_k^{m_k}$ (this process is called *Gödel numbering*). Gödel then used this numbering to formally represent the syntax of the system PM within itself. More generally, he showed that if S is any extension of PM with a primitive recursive set of axioms (under its Gödel numbering) and S is (what he called) *ω-consistent*, then S is *incomplete*; that is, there are statements A such that neither A nor $\neg A$ is provable in S. Here ω-consistency is a mild technical extension of the usual concept of consistency. The hypothesis of ω-consistency was later replaced by that of simple consistency by an argument of Rosser,[40] who further strengthened Gödel's theorem so as to apply to any recursive system S extending PM. Gödel pointed out in 1931 that much weaker systems than PM served the same purpose. In fact, other later improvements gave the result that any recursive consistent extension S of PA is incomplete.[41]

---

[38] In the simplest case, of a function of one argument, inductive definition takes the form $F(0) = c, F(n+1) = H(n, F(n))$, where H is a previously defined function; more generally, one defines functions $F(n, m_1, \ldots, m_k)$ by $F(0, m_1, \ldots, m_k) = G(m_1, \ldots, m_k), F(n+1, m_1, \ldots m_k) = H(n, m_1, \ldots, m_k, F(n, m_1, \ldots, m_k))$ where G and H are previously defined functions. Nowadays, one uses "recursive" for "inductive" in such definitions of functions.

[39] See Davis (1965).

[40] See his 1937 paper reproduced in Davis (1965).

[41] See Tarski, Mostowski, and Robinson (1953) for the history and still more general

One can say more about the kinds of statements A which are not decided by such S; they can be chosen in the form $\forall n(F(n) = 0)$ where F is a primitive recursive function. For primitive recursive S, Gödel arrives at this by showing that the relation $Proofs_{\mathrm{S}}(m, n)$ which expresses that n is the number of a proof in S of the statement with number m, is primitive recursive; that is, it has a primitive recursive characteristic function. He then constructs by a diagonal procedure a statement A such that

$$(1) \qquad A \leftrightarrow \forall n \neg Proofs_{\mathrm{S}}(\ulcorner A \urcorner, n)$$

is provable in S, where $\ulcorner A \urcorner$ is the number of A; informally speaking, A expresses of itself that it is not provable in S. By the above we obtain a primitive recursive function F such that $F(n) = 0 \leftrightarrow \neg Proofs_{\mathrm{S}}(\ulcorner A \urcorner, n)$, so

$$(2) \qquad A \leftrightarrow \forall n(F(n) = 0).$$

For this choice of A, Gödel shows that

(3) (i) if S is simply consistent then A is not provable in S, and

(ii) if S is ω-consistent then $\neg A$ is not provable in S.

This is Gödel's first incompleteness theorem. (The construction of A must be modified for Rosser's improvement to simple consistency in (ii)).

The consistency of S can be expressed as the statement that no proof in S ends in a contradiction; this may also be expressed by a statement Con_{S} of the form $\forall n(G(n) = 0)$, with G primitive recursive. By ordinary logic, consistency is equivalent to the statement that *some* formula is not provable in S; hence the statement $\forall n \neg Proofs_{\mathrm{S}}(\ulcorner A \urcorner, n)$ implies Con_{S}. Now (3)(i) expresses informally that the following formal statement is true:

$$(4) \qquad Con_{\mathrm{S}} \rightarrow \forall n \neg Proofs_{\mathrm{S}}(\ulcorner A \urcorner, n).$$

Gödel succeeded in showing that not only is (4) true, but it is also provable in S—this, by formalizing the argument for (3)(i). Combining all of these facts it follows that

(5) if S is simply consistent then Con_{S} is not provable in S

(Gödel's second incompleteness theorem).

These results were both stunning and disappointing. Consider the latter aspect first: since, under the given hypothesis, the A constructed in (1) is not provable in S, then A is obviously true, for that is just what A expresses of itself. No statement of prior mathematical interest (such as the twin prime conjecture or the Riemann Hypothesis) has been shown

results. The reduction to PA essentially makes use of Gödel's result from his 1931 paper that all primitive recursive functions are arithmetically definable, that is, definable in the first-order language having $0, 1, +, \cdot, =$ as basic symbols.

to be independent of S. Rather, the statement produced has simply been cooked up for the occasion, again by a diagonal trick. Even more, since A is obviously true, there is no question which of $A, \neg A$ we would add to S as an axiom if we were to try to overcome *this* instance of incompleteness; if S is to reflect truth in $\mathbf{N}$ then A should be added. Of course then $S' = S + \{A\}$ is again subject to the incompleteness theorem, but that is another matter.

What was stunning about the incompleteness theorems was how they put Hilbert's entire conception of mathematical existence and his consistency program into question. First of all, if one gives up categoricity as the requirement on an axiom system S for it to uniquely specify the concept of a mathematical structure (such as that of the natural numbers), still one might hope that all truths in the supposed structure would be derivable in S, in other words, that S would be complete. Indeed, Hilbert had said as much in the statement of Problem 2 in his (1900) list of problems (quoted above), and he later specifically conjectured that a formal system for arithmetic is complete. This was shown false by Gödel's first incompleteness theorem.

Second, Hilbert had divided the propositions of a formal system into those which are *real* and those which are *ideal*, the latter being used simply as an aid to derive the former. Among the "real" propositions would be all statements of the form $\forall n(F(n) = 0)$ where F is a finitarily defined function. Since Hilbert granted that all primitive recursive functions are finitarily defined, Gödel's results from (2) and (3) above showed that no one consistent formal system S could even serve to derive all true "real" propositions. Thus, even the fallback position from completeness for all statements to completeness only for the class of "real" statements gives way.

Third, Hilbert's call for proofs of consistency of axioms systems to secure the concepts they were supposed to express fails to secure the correctness of some of the most elementary theorems in those systems. For if A is a true statement of the form $\forall n(F(n) = 0)$ with F primitive recursive which is not provable in S, then $S + \{\neg A\}$ is consistent and proves the false statement $\exists n(F(n) \neq 0)$.[42] Still, it is at least the case that consistency of S (of the sort to which the incompleteness theorems apply) ensures that if S proves a "real" statement B of the form $\forall n(G(n) = 0)$, with G primitive recursive, then B must be true; for, otherwise, there would exist a specific k such that $G(k) \neq 0$, hence also $\exists n(G(n) \neq 0)$ would be provable in S, contradicting $\forall n(G(n) = 0)$.[43]

Finally, there are the unsettling consequences that Gödel's second incompleteness theorems ((5) above) have for Hilbert's program of finitary

[42]This was pointed out by Gödel in his first public announcement (in 1930) of the incompleteness theorems; see Gödel (1986), pp. 196–204.

[43]This point has been stressed particularly by G. Kreisel. See his article (1976) on Hilbert's Second Problem, in Browder (1976).

consistency proofs. Since all mathematics is supposed to be formalizable, this holds in particular for finitary mathematics. Now if all finitary methods can be represented in a single formal system S, then either S is not consistent or its consistency cannot be proved by finitary methods. A possible alternative is that no one consistent formal system can serve to formalize *all* of finitary mathematics, though any given portion can be formalized in one system or another. However, it is difficult to conceive how single global systems like PM or ZF could fail to formalize all finitary arguments. In fact, all the finitary proofs of consistency of various systems that had been carried out by workers in the Hilbert school were already formalizable in PA, and this explains why their best efforts failed with PA itself. Nevertheless, it remained conceivable that the use of finitary methods involving novel and/or more difficult arguments could serve to handle PA and eventually go on to succeed with systems like PM and ZF. Gödel himself was cautious on this point at the conclusion of his 1931 paper, but eventually he came to the view that Hilbert's conception of finitism *is* limited by PA.[44]

A basic problem here is that the general concept of finitary proof is not sufficiently clear that one can draw definitive conclusions about its possible limitations. Every actual finitary proof which had been carried out could be recognized as such, and most proofs of current mathematics are evidently nonfinitary, but this experience has not been sufficient to determine the concept in any sharp way. Beginning with Gerhard Gentzen in 1936,[45] workers in proof theory extended the evidently finitary methods employed for various consistency proofs to include certain *transfinite* elements, such as principles of induction on various effectively presented systems of ordinal notations, while hewing to Hilbert's requirement to use only "real" statements $\forall n(F(n) = 0)$, with F primitive recursive. [Cf. chapter 10 below.]

Postscript on Second-Order Logic

Gödel's 1930 completeness theorem showed that an effectively given system PC_1 of axioms and rules of inference for the first-order predicate calculus serves to prove just those statements which are valid in all models. His 1931 results could be used to show that there is no corresponding completeness theorem for second-order predicate calculus PC_2, if by validity is meant validity in the standard sense, that is, that in each interpretation M the set variables required are to range over $\mathcal{P}(M)$. Suppose to the contrary that PC_2 has an effectively given system of axioms and rules which proves just those statements that are valid in this sense. Then Peano's axioms with induction in its set-theoretical form are categorical when we use with them the logic of PC_2; call this system PA_2. Now since PA_2 is true in the structure of natural numbers, every theorem A of PA_2 is a truth about $\mathbf{N}$. Conversely, every truth A about $\mathbf{N}$ is equally a truth about any model

[44]See the introduction to Gödel (1958); this paper is reproduced in translation, together with a later extended, revised version (1972) in Gödel (1990).

[45]See Gentzen (1969).

isomorphic to **N**, hence any model of PA_2; so, finally, A is a theorem of PA_2. Thus PA_2 is formally complete: for every statement A in its language, either PA_2 proves A or it proves $\neg A$. Since PA_2 contains PA, and is supposed to be effectively given, the conclusion violates Gödel's first incompleteness theorem. This fulfills the promise made in the previous subsection on the Löwenheim-Skolem theorem, that second-order logic cannot be axiomatized effectively.

Elimination of the Law of Excluded Middle

For Hilbert, the completed infinite already made its appearance, albeit implicitly, in the use of instances of the Law of Excluded Middle (LEM) of the form $\forall n(F(n) = 0) \vee \neg\forall n(F(n) = 0)$; these in turn lead to $\forall n(F(n) = 0) \vee \exists n(F(n) \neq 0)$ and the method of proof by contradiction for simple existential results. This is what made PA problematic for Hilbert and brought him to call for a proof of its consistency. In 1930, Arend Heyting set up a formal system of logic without LEM which was acceptable to the Brouwerians and has thus come to be called *intuitionistic logic*. One can associate with any axiomatic system S based on classical logic a corresponding system S^i based on intuitionistic logic while otherwise retaining the same mathematical axioms; obviously S^i is contained in S. The particular system PA^i is called *Heyting Arithmetic* and is denoted HA. By a straightforward argument in a paper of 1933 on the relationship between classical and intuitionistic arithmetic, Gödel showed that PA can be translated into HA in such a way as to preserve statements of the form $\forall n(F(n) = 0)$ with F primitive recursive. To be more precise, with each statement A of the language of PA (which is the same as that of HA) is associated a statement A' such that if PA proves A then HA proves A'. Moreover, for A of the form $\forall n(F(n) = 0), A' = A$.[46] Gödel's translation was subsequently extended to a wide variety of systems besides PA.[47]

Thus once more Gödel undermined Hilbert, who had stressed establishing the consistency of the *tertium non datur* (excluded third, or LEM) in the promotion of his program. Since the intuitionists said that they too reject the completed infinite, and since HA is intuitionistically acceptable, and since, finally, PA is reducible to HA by Gödel's result, what more would a finitary consistency proof of PA accomplish in the way of eliminating the "actual" infinite? Hilbert himself gave no answer to that question.

The Elusiveness of Cantor's Continuum Problem

With the general crumbling of Hilbert's program, his specific aim to use it in order to solve the continuum problem came, in the end, to nought.

[46]Gödel's result had a precursor in a similar one by Kolmogorov for the predicate calculus. The result for arithmetic was found independently by Gentzen and Bernays in papers withdrawn from publication when Gödel's work appeared; see Gödel (1986), p. 284.

[47]See the introductory note to his paper *ibidem*, especially pp. 284–285.

The alternatives then seemed to be either to follow Brouwer and Weyl and grant no definite meaning and no interest to the Continuum Hypothesis, or to take seriously the claims for the reality and cogency of set theory based on the platonistic vision of the set-theoretical universe. Apparently Gödel himself took this latter position early on, though he did not elaborate his view until much later, in the 1940s. One can find only a few scattered, brief indications thereof in his writings up to 1940.[48] One early sign of the direction in which his views pointed is in footnote 48a in Gödel's 1931 paper, evidently added as an afterthought:

> the true reason for the incompleteness inherent in all formal systems of mathematics is that the formation of ever higher types can be continued into the transfinite . . ., while in any formal system at most denumerably many of them are available. For it can be shown that the undecidable propositions constructed here become decidable whenever appropriate higher types are added. . . . An analogous situation prevails for the axiom system of set theory.[49]

The full extent of Gödel's platonism only began to emerge in his paper "Russell's mathematical logic" (1944). His attitude toward the Continuum Hypothesis was then spelled out more specifically in the article "What is Cantor's continuum problem?" (1947, and in revised and expanded form, 1964). There he stated in no uncertain terms his views that Cantor's notion of cardinal number is definite and unique and that the Continuum Hypothesis CH has a determinate truth value. His own conjecture was that CH is false, because of various "implausible" consequences it has. In any case, it was Gödel's conviction that it made sense to try to settle CH. At the same time, he acknowledged the failure thus far to come even remotely close to a solution. And finally, it was metamathematics that served to explain why it was proving so difficult to arrive at an answer. Indeed, once again, Gödel himself (had already) provided the first definitive results of that character, as follows.

Consistency of the Axiom of Choice and the Continuum Hypothesis

In a series of brief descriptions of results, 1938–1939, and finally at length in his 1940 monograph, Gödel proved the following result:

(1) if ZF is consistent then it remains consistent when we add to it the Axiom of Choice AC and the Generalized Continuum Hypothesis GCH, $\forall\alpha(2^{\aleph_\alpha} = \aleph_{\alpha+1})$, as new axioms.

---

[48]See the discussion of Gödel's philosophy of mathematics in Feferman (1986), pp. 28–32 [reproduced in chapter 6 in this volume].

[49]Gödel (1986), p. 181.

Gödel's method of proof was to introduce a notion of "constructible set" in the language of set theory and to show that when the universe of sets is restricted to the constructible sets then all the axioms of ZF together with AC and GCH become validated. Another way Gödel had of putting this (in his system of sets and classes of his 1940 monograph), using V to denote the class of all sets and L to denote the class of all constructible sets, is that

(2) (i) if ZF is consistent then ZF + (V = L) is consistent, and

(ii) ZF + (V = L) proves AC and GCH.

Here the equation V = L expresses the assumption that all sets are constructible, which is shown to be true in the universe of constructible sets (a seemingly obvious proposition yet one whose precise statement requires some technical work to establish, because the relativization of the notion of constructibility to L must be shown to be *absolute*, that is, unchanged thereby).

As we have seen, almost all the work with cardinal arithmetic requires the assumption of AC. Once its consistency with ZF is established via (2)(i), (ii), it makes sense to speak of the scale of alephs $\aleph_\alpha$ and then to consider the truth value of $2^{\aleph_\alpha} = \aleph_{\alpha+1}$, for any and all α. But one still needs the stronger hypothesis V = L to prove GCH itself.

What Gödel's consistency result showed was that one could not hope to *disprove* AC, and that if AC is assumed, one could not hope to *disprove* any instance of GCH using Zermelo-Fraenkel Set Theory. It was still conceivable that ZF could *prove* AC, or that ZFC (= ZF + AC) could *prove* some instance of GCH, in particular, CH itself. Gödel himself made efforts to establish these results, with only partial success, and only concerning the independence of AC. In fact, no progress on this problem was made in over twenty years, despite the general expectation that both AC and CH would be independent, respectively, of ZF and ZFC.[50]

Independence of the Axiom of Choice and the Continuum Hypothesis

In 1963, Paul Cohen obtained the following results:

(3)(i) if ZF is consistent then AC is independent of ZF; that is, it cannot be derived from ZF;

(ii) if ZF is consistent then CH is independent of ZFC;

(iii) if ZF is consistent then V = L is independent of ZFC + GCH.

Cohen's method of proof[51] involved a novel technique for building models of set theory, called the method of *forcing and generic sets*. Where

[50] According to Gödel's views in his article of 1947 and 1964, AC is true and CH is false, so he would certainly expect independence of CH from ZFC.

[51] See Cohen (1966) for an exposition.

Gödel had restricted a presumed model of set theory to obtain that of the constructible sets, Cohen extended such a model by adjunction of (a variety of) "generic" sets. For example, by adjoining sufficiently many generic subsets of **N**, he was able to construct a model of ZFC in which $2^{\aleph_0} = \aleph_2$, thus contradicting CH.

Subsequently, building on Cohen's work, it was shown by Easton that for each regular $\aleph_\alpha$, the power $2^{\aleph_\alpha}$ "can be anything it ought to be"; that is, one can arrange the simultaneous equations $2^{\aleph_\alpha} = \aleph_{F(\alpha)}$ in a suitable model of ZF, for any function $F(\alpha) = \beta$, from ordinals to ordinals satisfying a few simple restrictions.

Consistency and Independence of Definable Well-Orderings

Gödel's consistency result for (V = L) and proof that $2^{\aleph_0} = \aleph_1$ holds under that assumption showed

(4) it is consistent to assume with ZFC that there is a definable well-ordering of the continuum **R**; in fact, this is provable in ZF + (V = L).

Behind (4) lies the definition of L as $\bigcup_\alpha \mathrm{L}_\alpha$, where at each stage, L_α consists of the sets explicitly definable (in a predicative way) from the sequence of L_β for $\beta < \alpha$. (Gödel pointed out that this was an extension of Russell's predicative ramified hierarchy through all the transfinite ordinals.) Each constructible set is definable, and their definitions can be laid out in a transfinite sequence $\varphi_0, \ldots, \varphi_\alpha, \ldots$ without repetitions. Taking A_α to be the set in L defined by φ_α, this determines a definable well-ordering of all of L by $A_\beta < A_\alpha \leftrightarrow \beta < \alpha$. In particular, the restriction of this well-ordering to **R** (or to $\mathcal{P}(\mathbf{N})$) gives a definable well-ordering of **R** (resp. $\mathcal{P}(\mathbf{N})$) provably in ZF + (V = L).

Using Cohen's generic model of ZFC + GCH + (V $\neq$ L), I was able to show the following:[52]

(5) it is consistent with ZFC + GCH that there is no definable well-ordering of **R** (or $\mathcal{P}(\mathbf{N})$).

In other words, even if one assumes with ZF both the Axiom of Choice and the Generalized Continuum Hypothesis (and hence in particular $2^{\aleph_0} = \aleph_1$), one will not be able (provably) to arrive at any explicit definition of a well-ordering for the real numbers. It is in this sense that Hilbert's expressed hope in Problem 1 of his 1900 address cannot be realized.[53]

Of course Hilbert's hope could be satisfied if one granted the truth of V = L (or perhaps some similar axiom). By Cohen's work (3)(iii) the truth of V = L is not automatic if one accepts ZFC + GCH. In fact, most

[52] Feferman (1965).

[53] See footnote 6 above.

everyone who holds a platonist view of set theory denies the truth of V = L, for it is an axiom which says that every set is definable and, moreover, in a special way. But the restriction to definable objects is just opposite to the platonist position according to which the objects of set theory exist independent of any means of definitions. So, for the avowed platonist, there is nothing disturbing in (5).

New Axioms?

Gödel already projected in his 1947 paper that CH would be independent of ZFC, and that new axioms might be required to decide it; he developed these ideas further in the 1964 revision of that paper.

> For if the meanings of the primitive terms of set-theory . . . are accepted as sound, it follows that the set-theoretical concepts and theorems describe some well-determined reality, in which Cantor's conjecture must be either true or false. Hence its undecidability from the axioms being assumed today can only mean that these axioms do not contain a complete description of that reality. . . . [T]he axioms of set theory by no means form a system closed in itself but, quite the contrary, the very concept of set on which they are based suggests their extensions by new axioms which assert the existence of still further iterations of the operation "set of."[54]

The main kinds of new axioms thus suggested are called *axioms of infinity* or *large cardinal axioms.* The first of them asserts the existence of an *inaccessible cardinal* m. This is supposed to satisfy the properties (i) $\aleph_0 < m$, (ii) if $n < m$ then $2^n < m$, and (iii) if $card(A) < m$ and F is a function from A to cardinals less than m then $\sum_{x \in A} F(x) < m$. In other words, such m is a transfinite cardinal closed under exponentiation and summation over any smaller number of cardinals. It is not hard to show (in ZFC) that the assumption $(\exists m)$ "m is inaccessible" implies the existence of a model for ZFC, and hence the consistency of ZFC. Thus by Gödel's second incompleteness theorem, ZFC cannot prove $(\exists m)$ "m is inaccessible," if it is consistent.

The existence of inaccessible cardinals can be iterated further into the transfinite; that is, one may assume as a new and stronger axiom the statement that there is a strictly increasing sequence of inaccessible cardinals m_α, indexed by arbitrary ordinals α. Then one could go on to postulate the existence of inaccessible cardinals p such that whenever $\alpha < p$ then $m_\alpha < p$. The existence of such p cannot be proved under the assumption of the existence of the sequence of $m_\alpha s$. Now p is an inaccessible fixed point of the function $F(\alpha) = m_\alpha$, that is, $F(p) = p$, and it may be assumed that there is a function of ordinals, $F'(\alpha) = p_\alpha$, which enumerates

[54] Gödel (1964), p. 264 [or Gödel (1990), p. 260].

all such fixed points in increasing order. One is led in this way to notions of higher and higher levels of inaccessibility, which were first formulated in a systematic way by P. Mahlo in 1911; the cardinals of these types are thus called Mahlo cardinals. One is led correspondingly to stronger and stronger axioms of infinity; according to Gödel these "show clearly, not only that the axiomatic system of set theory as used today is incomplete, but also that it can be supplemented without arbitrariness by new axioms which only unfold the content of the concept of set."[55]

The study of large cardinal axioms has been carried on intensively since the 1960s, and by the introduction of new ideas vastly stronger axioms than those for the existence of Mahlo cardinals have been proposed: existence of "measurable" cardinals, "compact" cardinals, "supercompact" cardinals, etc. These involve extremely technical notions which can be understood only with somewhat advanced training in axiomatic set theory, so no attempt will be made to explain them here.[56]

There are two questions to be asked concerning the various existence statements for infinite cardinals that have been indicated here. First, what would lead one to accept them as axioms to be added to ZFC? Second, do they help decide the continuum problem? Concerning the first question, Gödel thought that these or other types of proposed new axioms need not "force themselves upon us" as being true in the same way as the axioms of ZFC, but that "a more profound understanding of the concepts underlying logic and mathematics would enable us to recognize [them] as implied by these concepts."[57] Here the final arbiter would be that of "mathematical intuition [which] need not be conceived of as a faculty giving an *immediate* knowledge of the objects concerned."[58] Gödel goes on to compare mathematical intuition with physical experience as a source of our ideas about underlying physical objects, but he is not specific as to how this intuition serves to decide between opposing theories of underlying mathematical objects in general, and between the existence or nonexistence of some huge cardinal in particular. While Gödel's remarks may be heartening to platonist set-theoreticians, they are too vague to be decisive in any particular case. Rather, their tendency seems to be that if one's mathematical intuition leads one to judge that a statement about sets is true, then it *is* true.

In a technical survey piece on large cardinal axioms, Kanamori and Magidor suggest the acceptance of such axioms on "theological" grounds, but have alternative arguments to engage those who are not already "true believers." For the latter, what is offered is an investigation of such statements on a purely formal level whose interest lies in the fascinating and

---

[55] *Loc. cit.*

[56] For a substantial introduction to the subject see Drake (1974). [Cf. also the more recent comprehensive work, Kanamori (1994).]

[57] Gödel (1964), p. 265.

[58] *Ibidem*, p. 271.

"aesthetic" intricacy of the net of consequences and interrelationships between them.[59]

Others have attempted to provide an overall rationale which would make large cardinal axioms the consequence of some very general principles. Main attention has been given to forms of the set-theoretical *reflection principle*, according to which any property of the universe of all sets must already be true of a level C_α in the cumulative hierarchy of sets (where $C_0 = \emptyset, C_{\alpha+1} = C_\alpha \bigcup \mathcal{P}(C_\alpha), C_\lambda = \bigcup_{\alpha<\lambda} C_\alpha$ for limit λ, and the universe is the union of all $C_\alpha s$). Initially formulated by Azriel Levy for first-order set-theoretical properties, this principle was extended to include second-order properties in an elegant way by Paul Bernays, who showed that its assumption leads to the existence of the Mahlo cardinals. This line was continued in Reinhardt's "Remarks on reflection principles, large cardinals and elementary embeddings," which provides a useful introduction to and motivation for the approach. Reinhardt considers this work "a first step in recognizing axioms . . . which will make [them] seem worth considering as axioms rather than merely as conjectures or speculations."[60] On the other hand, in his first footnote to this paper he says that Gödel finds Reinhardt's own proposed justification "rather unsatisfactory" and that preferably an alternative one should be given in terms of some concept of structural properties of sets.

It is apparent that those who accept large cardinal axioms have some rationale or other for doing so (perhaps rationalization would be the better word), but are well aware that the whole enterprise may be put in question. One of the most forceful statements in opposition is that of Paul Cohen (1971) in his statement of reasons for rejecting the platonist realist position and accepting a formalist position in the set-theoretical foundations of mathematics.

> A weakness which I believe any realist would have to accept is his inability to explain the source of the never-ending sequence of higher axioms, such as the higher axioms of infinity. Certainly even the staunchest realist must flinch when contemplating cardinals of a sufficiently inaccessible type. Also there are axioms such as that of the Measurable Cardinal which are more powerful than the most general Axiom of Infinity yet considered, but for which there seems absolutely no intuitively convincing evidence for either rejection or acceptance.[61]

He goes on to accuse "zealous set-theoreticians" of a certain "opportunism" in the acceptance and/or pursuit of these axioms. In opposition to that view we have, in turn, the comments of Kreisel (1971), who agrees

[59] Kanamori and Magidor (1978), pp. 103–105. [Cf. also Kanamori (1994).]

[60] Reinhardt (1974), p. 205. [Cf. also Feferman (1996) for a discussion of Gödel's program for new axioms, as well as Feferman (1998).]

[61] Cohen (1971), pp. 11–12.

that those who work on large cardinals may have a "vested interest" in the subject, but only such people are "competent to judge axioms about them."[62]

What of Gödel's original hope, that new axioms and, specifically, large cardinal axioms could help decide the Continuum Hypothesis? Here the situation was surveyed in the report by Donald A. Martin on Hilbert's First Problem,[63] with a conclusion that is easily summarized: CH is consistent with and independent from every large cardinal axiom A that has been proposed as at all plausible.[64]

Besides large cardinal axioms, little has been offered as a new axiom which would be plausible and decide CH. In unpublished work, Gödel proposed axioms about "scales of functions" which he stated imply $2^{\aleph_0} = \aleph_2$, but his proof turned out to be wrong, and he withdrew the paper containing it. Still later, he proposed other axioms about scales which imply $2^{\aleph_0} = \aleph_1$ (also unpublished[65]). Informed opinion about both of these efforts is that they are very inconclusive. Indeed, the whole of Gödel's thought about the definiteness of the continuum problem and the program to find new axioms to decide CH has so far come to nought, somewhat to the chagrin of his followers. In the words of Martin,

> I have been assuming a naive and uncritical attitude toward CH. While this *is* in fact my attitude, I by no means wish to dismiss the opposite viewpoint. Those who argue that the concept of set is not sufficiently clear to fix the truth-value of CH have a position which is at present difficult to assail. As long as no new axiom is found which decides CH, their case will continue to grow stronger, and our assertion that the meaning of CH is clear will sound more and more empty.[66]

Metamathematics, of which Gödel was the first real master and clear-eyed practitioner, has thus brought the platonist position in set theory, of which Gödel was the foremost exponent, to a very embarrassing position. The continuum problem—to locate $2^{\aleph_0}$ in the scale of cardinals $\aleph_\alpha$ and, more specifically, to decide whether or not it is $\aleph_1$—is the very first chal-

---

[62]Kreisel (1971). p. 194.

[63]Martin (1976).

[64]To be more precise, for each such A it has been shown that if ZFC + A is consistent then it remains consistent when we add either CH or its negation. The independence results make use of strong extensions of Cohen's forcing methods but even the consistency results require novel ideas, since Dana Scott (1961) showed that the existence of measurable cardinals implies $V \neq L$. There are large cardinal axioms that give some partial information about GCH. For example, Robert Solovay has shown that the existence of a compact cardinal implies $2^{\aleph_\alpha} = \aleph_{\alpha+1}$ for all sufficiently large cardinals of a special kind (cf. Martin 1976, p. 86).

[65]Cf. Gödel (1986), pp. 26–27. [Gödel's unpublished work on scales of functions is now to be found in Gödel (1995) as **1970a*, *b*, and *c*; cf. also the introduction to those items by Solovay, *ibid.*, p. 405ff.]

[66]Martin (1976), pp. 90–91.

lenging problem of Cantorian set theory. The fact that it has not been settled by any remotely plausible assumption leads me, for one, to agree with Weyl that it is an inherently indefinite problem which will never be "solved."[67] Of course, this conclusion will be difficult to accept for anyone who regards "ordinary" set theory (for example, ZFC) as perfectly reasonable and coherent and so tends to think of it as being about a fixed and definite world. I believe a quite different account can be given for the reasonableness and coherence of ZFC on the basis of a conception of an ideal world of sets in the cumulative hierarchy, much like the original conception of geometry as being about a world of ideal points (pure positions), ideal lines (perfectly straight and without thickness), etc. This conception of sets can be visualized well enough, just as for the conception of geometrical objects. Such an account would give grounds for the plausibility of the consistency of ZFC without assumption of its truth in some supposed real world of sets. One might even go farther to say that the picture of sets in the cumulative hierarchy is sufficiently clear that the portion C_ω of the cumulative hierarchy consisting of the hereditarily finite sets is well determined. Since the natural numbers are extracted from C_ω in usual developments of set theory, according to this picture every number-theoretical statement provable in ZFC is true. What is cloudy about the picture of the cumulative hierarchy is both the effect of the power set operation in general and the use of "arbitrary" ordinals. But in gross the picture is clear enough to justify confidence in the use of ZFC (and like theories) for deriving number-theoretical results.

[67] I am here excluding V = L from the "remotely plausible assumptions," since, as explained above, it is rejected by set-theoretical platonists as being contrary to the conception of a universe of arbitrary sets existing independently of any means of definition. One should also mention recent work of Freiling (1986), where certain simple "axioms of symmetry" are added to ZFC and used to disprove CH. These new assumptions are supposed to be consequences of a thought experiment involving throwing two (or more) random darts at the real line. They are plausible-looking, but I have not found any support for these assumptions among leading experts in set theory.

Part II

FOUNDATIONAL WAYS

3

The Logic of Mathematical Discovery versus The Logical Structure of Mathematics

1 Introduction

Mathematics offers us a puzzling contrast. On the one hand it is supposed to be the paradigm of certain and final knowledge: not fixed, to be sure, but a steadily accumulating coherent body of truths obtained by successive deduction from the most evident truths. By the intricate combination and recombination of elementary steps one is led incontrovertibly from what is trivial and unremarkable to what can be nontrivial and surprising.

On the other hand, the actual development of mathematics reveals a history full of controversy, confusion, and even error, marked by periodic reassessments and occasional upheavals. The mathematician at work relies on surprisingly vague intuitions and proceeds by fumbling fits and starts with all too frequent reversals. In this picture the actual historical and individual processes of mathematical discovery appear haphazard and illogical.

The first view is of course the currently conventional one which descends from the classic work of Euclid. Following Frege, Russell, and Hilbert it has in this century been given a theoretical formulation in terms of the logical analysis of the structure of mathematics. With *formal systems* as the principal technical object of study, this *metamathematics* has undergone extraordinarily intensive development.

There is also a more isolated tradition which undertakes to discern patterns in the actual dynamic progress of mathematical thought; it dates back to Pappus and can be traced through the writings of Descartes, Leibniz,

"The logic of mathematical discovery versus the logical structure of mathematics" first appeared in P. D. Asquith and I. Hacking (eds.), *PSA 1978: Proceedings of the 1978 Biennial Meeting of the Philosophy of Science Association*, vol. 2 (1981), 309–327 (Feferman 1981). It is reprinted here with the kind permission of the publisher, The Philosophy of Science Association (East Lansing, MI). Some minor corrections and additions have been made.

and Bolzano. Most notable in our time have been the extensive studies by George Pólya of *patterns of plausible reasoning* in mathematical problem solving and demonstration. Imre Lakatos has taken this as one point of departure for his "rational reconstruction" of the growth of mathematical knowledge in what he calls the *logic of mathematical discovery* or *heuristic.* Lakatos' view of mathematics is philosophically much more sweeping and radical than Pólya's. It is situated within a general account of all rationally gained knowledge, which owes its debt to Karl Popper's (so-called) *logic of scientific discovery.* Lakatos rejects the Euclidean deductivist *infallibilist* view and replaces it by one of mathematics as a body of *fallible* knowledge being improved incessantly in response to ongoing critical assaults. To describe this he formulates what is supposed to be a kind of *logic of proofs and refutations.* At first this was directed by him at the detailed examination of particular problem-situations of both historical and mathematical interest. Later, he turned his attack on the search for "certain and final" foundations of mathematics within global formal systems.

Many of those who are interested in the practice, teaching, and/or history of mathematics will respond with eager sympathy to Lakatos' program. (One may add that it fits well with the increasingly critical and anti-authoritarian temper of these times.) Personally, I have found much to agree with both in his general approach and in his detailed analysis. Clearly, logic as it stands fails to give a direct account either of the historical growth of mathematics or the day-to-day experience of its practitioners. It is also clear that the search for ultimate foundations via formal systems has failed to arrive at any convincing conclusion. Nevertheless, the opinion I reach about Lakatos' own program is that it is far too single-minded and much more limited than he tries to make out. Speaking metaphorically, he plays only one tune on a single instrument—admittedly with a number of satisfying variations—where what is wanted is much greater melodic variety and the resources of a symphonic orchestra.

My plan here is to outline Lakatos' general views together with an indication of how they are elaborated in his case studies. The latter half of this chapter is first largely taken up with an extensive critique. This is followed by (i) a brief comparison with Pólya's work and (ii) a defense of logic as a means to analyze the underlying structure of mathematics. In conclusion I try to suggest directions for a possible rapprochement or synthesis of the opposing viewpoints. Fittingly, such would be in accord with Lakatos' own dialectical conception of the progress of human understanding.

2 Lakatos' Writings

There are two principle sources for Lakatos' writings on the nature of mathematics. One is his relatively well-known book *Proofs and Refutations: The Logic of Mathematical Discovery* (Lakatos 1976).[1] The other consists of the

[1]This is not as well known among mathematicians as it ought to be. More recently Hersh (1978) has engaged in bringing Lakatos' ideas, which he largely favors, to the

first five essays in volume 2 of his *Philosophical Papers* (Lakatos, 1978). Both of these appeared after his death in 1974, and include significant portions which had never been published because—true to his philosophical attitude—Lakatos was not completely satisfied with them. He had also planned to improve those parts which had been previously published, including the main body of Lakatos (1976). Nevertheless, there are ideas and themes which are reiterated with sufficient frequency and emphasis that we can be confident that they were well established in his mind.

The body of Lakatos (1976) consists of a case study in the rational reconstruction of mathematical progress. It is a presentation in dialogue form of the amazingly tangled history—spanning most of the nineteenth century—surrounding the (*Descartes-*)*Euler conjecture* for polyhedra. This is a very simple equation (namely, $V-E+F=2$) which expresses an invariable relationship between the number of vertices (V), edges (E), and faces (F) of any polyhedron. The choice of this example as an object of heuristic study (originally suggested to Lakatos by Pólya), has much to recommend it, besides its surprising ins-and-outs. For one thing, the concepts reach back to Greek geometry and, in the form established by Poincaré, bring us forward to the very doorsteps of modern combinatorial topology. In addition, the concepts involved are relatively elementary and the logic of the situation can be followed by anyone having a modicum of appreciation of mathematical proofs. The dialogue is often a delight to read and the entire presentation is a brilliantly sustained tour de force. Eventually, though, the relentless examination and reexamination of concepts, putative results, criticisms, and counterexamples is extremely fatiguing; one must be rather determined to see it through to the end, with little additional insight as reward.

I think one gets a clearer and quicker idea of Lakatos' general views and program by reading the two appendices to Lakatos (1976) and the first two essays of Lakatos (1978). Though these contain illustrations from mathematics of a less elementary conceptual character than the Euler conjecture, they could hardly discourage anyone seriously interested in their subject matter.

3 A Summary of Lakatos' General Views

Modern mathematical philosophy is deeply embedded in general epistemology and is only to be understood with reference to its basic controversy, that is, between the *dogmatists*—who claim that we can know—and the *skeptics*—who claim that we cannot know, or at least cannot know what it is that we know.[2] The skeptical argument that it is hopeless to find foun-

attention of the mathematical community. [Cf. also Hersh (1979).]

[2]In presenting the gist of these views in what follows, I shall move freely between Lakatos (1976) and Lakatos (1978), quoting liberally as well as paraphrasing.

dations for knowledge is based on infinite regress both for meaning and for truth. Three major *rationalist* (dogmatist) enterprises have been developed to try to stop these twin infinite regresses: (1) the *Euclidean program*, (2) the *Empiricist program*, and (3) the *Inductivist program*. The first of these is normally associated with mathematics, the latter two with scientific theories. Each organizes knowledge within (not necessarily formal) deductive systems. In a Euclidean system truth "flows downward through the deductive channels" from the indubitably true axioms, while in an Empiricist system falsity "flows upward" from those basic statements which turn out to be untrue. The Inductivist program also attempts to find conditions for truth to flow upward from the basic statements; this will not concern us further here.

None of the rationalist programs can withstand the criticism of the skeptics. However, there is a fourth program which can answer them, namely, Popper's *critical fallibilism*. This takes infinite regress in proofs and concepts seriously and does not pretend to stop them. In a Popperian theory "we never *know*, we only guess." But guesses can be criticized and then improved. The old problems of reduction and justification of knowledge become pseudoproblems. Instead of asking *How do you know*? one asks *How do you improve your guesses*? There is now no concern if the skeptic complains that you cannot know the answer to that, since in fact your answer itself is only a guess. "There is nothing wrong with an infinite regress of guesses."

The Euclidean program for mathematics is hereby abandoned (indeed rejected), but mathematics can be regarded as a *quasi-empirical* theory under the new stance. By such is meant an empirical theory whose basic statements are not of a singular spatiotemporal character. For example, they may be elementary arithmetical statements or even entire bodies of already accepted informal statements about which one has developed some confidence. Individual mathematical conjectures and whole mathematical theories can be tested for their consequences among such basic statements and be modified or even rejected. A quasi-empirical theory is always conjectural, at best well corroborated. As in empirical theories, the axioms are used to *explain* those basic statements which appear as consequences. Euclidean theories are rigid and antispeculative; by contrast, the quasi-empirical approach is uninhibitedly speculative and advocates a proliferation of "bold, imaginative" hypotheses.

4 The Logic of Proofs and Refutations (Lakatos' Ideas, continued)

Instead of growing through the steady accumulation of indubitably established theorems, mathematics grows through the "incessant improvement of guesses by speculation and criticisms," by the logic of proofs and refutations. While this is a very general pattern of mathematical discovery, it was itself only discovered in the 1840s. Naive conjectures and concepts must pass through the crucible of proofs and refutations. The results are improved conjectures (theorems) and improved (proof-generated or theoretical) concepts. The logic of mathematical discovery "is neither psychology nor logic, it is an independent discipline" also called *heuristic*. While mathematics is a product of human activity, it acquires a certain autonomy with its own laws of growth, its own dialectic. This is the subject of heuristic, which it studies through history and the rational reconstruction of history.

The method of proofs and refutations is also called *proof analysis*. Its skeleton is as follows (cf. Lakatos 1976, pp. 127–128): There is

(1) *a primitive conjecture* and

(2) *an informal proof*.

The latter is a *thought-experiment or argument* which decomposes the primitive conjecture into subconjectures or *lemmas*. Subsequently,

(3) *"global" counterexamples* emerge,

that is, counterexamples to the primitive conjecture. Finally,

(4) *the proof is reexamined for a hidden lemma to which the global counterexample is a local counterexample. This is built into the improved conjecture (theorem); its principal new feature is the proof-generated concept.*

The method frequently has further ramifications. The lemma may be hidden in other proofs and the new concept may be used to improve them; counterexamples open up into new fields of inquiry, and so on.

The method of proof-analysis might not improve a proof. This only happens when the analysis turns up unexpected aspects of the naive conjecture. That might not be the case in mature theories but it is always the case in young, growing theories.

It should also be noted that strategies other than proof-analysis have been used historically to deal with the problems presented by counterexamples. These are the methods of *monster barring* and *exception barring*. The former attempts to restrict concepts involved specifically to exclude pathological cases. In the second, one searches for a "safe" domain of objects for which the conjecture is valid, without seeking the most general domain of validity.

5 Getting Down to Cases

There are only two cases that Lakatos presents in any detail to support his thesis, though features of a number of other cases are taken up, too. These are the Euler conjecture, which, as has already been mentioned, is treated at length in the main body of Lakatos (1976), and Cauchy's theorem on limits of series of continuous functions (Lakatos, 1976, appendix 1). Actually, the latter serves to illustrate the method of proof-analysis in a somewhat clearer way than the former, so we look at it here.

The external history of this theorem runs briefly as follows. Cauchy took the first successful steps to give a rigorous foundation for the calculus without using the troublesome concept of infinitesimal. These became well known through publication of his book *Cours d'Analyse* in 1821. (Actually, much the same achievement had been made by Bolzano several years earlier, but his 1817 publication seems to have received no attention in the mathematical world at that time.[3]) The first concepts which needed reexamination were those of *limit* and *continuous function*, for which Cauchy provided new definitions. These are not as precise as the "ε, δ" definitions we use today, which we owe instead to Weierstrass (around the midnineteenth century). Indeed, there is a kind of ambiguity about them which is disturbing by today's standards of rigor.

Cauchy stated and presented an argument for the following theorem:

> *Suppose* $\Sigma_{n=1}^{\infty} f_n(x)$ *converges to* $f(x)$ *for each* x *and that each* $f_n(x)$ *is continuous; then* $f(x)$ *is continuous.*

The hypothesis is expressed in terms of limits for each x by

$$\lim_{n\to\infty} s_n(x) = f(x) \text{ where } s_n(x) = \Sigma_{m=1}^{n} f_m(x).$$

According to our present-day interpretation of limits and continuity this theorem is false and there are many simple counterexamples. The curious part of this history is that a series which could serve as a counterexample was already known to Cauchy and not recognized as such. This was

$$\sin x - \frac{1}{2}\sin 2x + \frac{1}{3}\sin 3x - \frac{1}{4}\sin 4x + \ldots \left(\text{or } \Sigma_{n=1}^{\infty} \frac{(-1)^{n+1}}{n} \sin nx \right)$$

which appeared in the famous 1807 memoir by Fourier on the propagation of heat. It was shown there to converge to a function with the broken straight line graph:

[3] It is the thesis of Grattan-Guinness (1970) that Cauchy somehow got his ideas from Bolzano without acknowledging them, but this has been disputed. In any case, Bolzano was a bit clearer than Cauchy about basic concepts. His revolutionary work (including anticipations of set theory) was not widely publicized and appreciated until the 1870s.

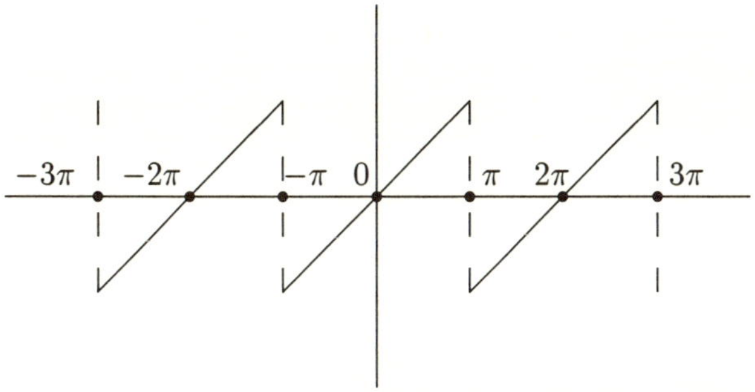

(At the points $n\pi$ the series converges to 0.)

There had been much controversy over Fourier's memoir and his assertion about the representability of "arbitrary" functions by trigonometric series. It seems that Cauchy's theorem, which would hardly have been considered worth stating before, was designed to put Fourier's work definitively into question (this is the view of Grattan-Guinness 1970, p. 78). Be that as it may, no one protested the theorem until 1826, when Abel pointed out that there were "exceptions" such as the series above. But instead of examining Cauchy's putative proof to see where it broke down, Abel took the exception-barring route: he restricted attention to the "safe" domain of power-series functions, for which he obtained definitive convergence results. It was not until 1847 that Seidel (a student of Dirichlet's) reexamined Cauchy's argument and found the "hidden" lemma which makes the conclusion correct. This is analyzed by Lakatos as follows, writing out the concepts involved in modern terms: We assume

(1) (*convergence*) for each x and $\varepsilon > 0$ there exists N such that $|f(x) - s_n(x)| < \varepsilon$ for all $n \geq N$, and

(2) (*continuity of each* f_n, *hence of each* s_n) for each x and $\varepsilon > 0$ there exists $\delta > 0$ such that $|y - x| < \delta$ implies $|s_n(y) - s_n(x)| < \varepsilon$.

The desired conclusion is

(3) (*continuity of* f) for each x and $\varepsilon > 0$ there exists $\delta > 0$ such that $|y - x| < \delta$ implies $|f(y) - f(x)| < \varepsilon$.

The proof idea is to relate $|f(y) - f(x)|$ to $|s_n(y) - s_n(x)|$ for suitably large n. By the triangle inequality we have

$$|f(y) - f(x)| \leq |f(y) - s_n(y)| + |s_n(y) - s_n(x)| + |s_n(x) - f(x)|.$$

Thus $|f(y) - f(x)| < \varepsilon$ if each of $|f(y) - s_n(y)|$, $|s_n(y) - s_n(x)|$, and $|f(x) - s_n(x)|$ is less than $\varepsilon/3$. For any n the second can be arranged using (2). By (1) we can choose N such that $|f(x) - s_n(x)| < \varepsilon/3$ for $n \geq N$ and for any y we can choose N_1 such that $|f(y) - s_n(y)| < \varepsilon/3$ for $n \geq N_1$. But while x in (3) is conceived to be fixed, y must be variable, and the

N_1 associated with y may vary, too. What is needed to carry through the proof is a *uniform* choice of N independent of x (hence applicable to any y) in (1):

(1)′ (*uniform convergence*) for each $\varepsilon > 0$ there exists N such that $|f(x) - s_n(x)| < \varepsilon$ for all x and all $n \geq N$.

This assumption was hidden in Cauchy's argument. The *improved theorem* is that (1)′ *and* (2) *implies* (3). The Fourier series which was a counterexample to Cauchy's formulation of the theorem now becomes a counterexample to the "guilty lemma" that $\Sigma_{n=1}^{\infty} f_n(x)$ converges uniformly to $f(x)$ if it converges: indeed, it converges but not uniformly. The proof-generated concept of uniform convergence is incorporated as the principal feature of the improved theorem.[4]

---

[4]Lakatos continued to be puzzled by Cauchy's failure to acknowledge a difficulty when confronted with the counterexamples. He later wrote a revisionist "history" of Cauchy's misadventure in "Cauchy and the continuum," which appears as the third essay of Lakatos (1978). In this Lakatos seized on A. Robinson's theory of infinitesimals (cf. Robinson 1966) to propose another interpretation; namely, Cauchy was a Leibnizean at heart and still clung to actual infinitesimals. Furthermore, his theorem is *correct* when read as a statement in Robinson's nonstandard analysis.

The following points should be made about this account:

(i) Cauchy defined variables as quantities which "one considers as having to successively assume many values different from one another." For limits he says that "when the successive values attributed to a variable approach indefinitely a fixed value so as to end by differing from it as little as one wishes, this last is called the limit of all the others." Then *infinitesimals* are said to be variables whose "numerical value decreases indefinitely in such a way as to converge to the limit 0" (cf. Kline 1972, pp. 950–951).

(ii) In particular, Cauchy defined *differentials* dx (one principal form in which infinitesimals had previously made their appearance) as any finite variable quantity. Then, given a functional relationship $y = f(x)$, one defines dy to be $f'(x)dx$. This "saves" the equation $dy/dx = f'(x)$, in which $f'(x)$ has been defined independently in terms of limits by $\lim_{h \to 0} [f(x+h) - f(x)]/h$. Infinitesimals are thus treated here as a suggestive notational convenience.

(iii) Nevertheless, Cauchy's position on infinitesimals seems to be equivocal, and it may be said that he continued to "practice infinitesimalism" (Grattan-Guinness 1970, pp. 57ff.).

(iv) Robinson provided the first coherent theory of actual infinitesimals in which a Leibnizean-style calculus could be interpreted. However, that reconstruction involves the use of logical concepts (such as the distinction between internal and external properties in certain formal languages) which are foreign to infinitesimal analysis as it had been practiced.

(v) Robinson himself examined interpretations of earlier statements involving infinitely small and large quantities in analysis and considered possible reinterpretations of them in his system, in particular those given by Cauchy (Robinson 1966, §10.5). Among these is the statement about convergence of series of functions $f_n(x)$ which we considered in the text. In his interpretation, Robinson allows the subscript n to take on infinite values, but considers x to range only over standard real numbers. He found that additional assumptions, such as uniform convergence, are still needed to obtain a correct theorem. By contrast, Lakatos (in the essay mentioned) assumes in his interpretation that x ranges over the full extended real number system (comprising infinitesimals) as well. For this alternative form, he verifies Cauchy's theorem to be correct "as it stands."

Lakatos credits the method of proof-analysis to Seidel. He says that "Seidel discovered the proof-generated concept of uniform convergence and the method of proof-analysis at one blow. He was fully conscious of his methodological discovery which he stated with great clarity."[5]

There is no time here to go into the treatment of the Descartes-Euler conjecture in Lakatos (1976). It should be remarked that a difficulty with the dialogue form of presentation used there is that one is never sure which of the participants' views are shared by Lakatos.[6] Fortunately, Lakatos has provided many (scholarly) footnotes which parallel the text as well as supplementary comments.

6 A Critical Examination of Lakatos' Views and Program

The details of the preceding and other examples spelled out (or indicated) in Lakatos (1976) would seem to show that one must take Lakatos' analysis of mathematical progress rather seriously. Nevertheless, I have a number of questions to raise and criticisms to make.[7] These will concern both what Lakatos tells us and matters about which he says nothing at all.

(vi) My view is that one can hardly credit Cauchy (or his predecessors) with having a coherent use of infinitely small and large quantities which merely awaited a Robinson-style foundation to legitimize it. In his theory of infinitesimals, Cauchy looks forward to the current standard methods for their elimination, while in his practice he slips backward. The type of "rational reconstruction of history" revealed at length by this example seems to me to provide a good illustration of the dangers of Lakatos' free-wheeling "bold, imaginative" approach.

[5]The concept of uniform convergence and its need for rigorous proofs of certain basic theorems of analysis (such as concern also interchange of integration and limits) was independently found by the physicist Stokes (cf. Grattan-Guinness 1970).

[6]According to Hersh (1978) the one criticism Pólya made of Lakatos' treatment of the Euler conjecture was that it is "too witty." But Pólya added the following in a conversation [circa 1978] which I had with him. In his view, Lakatos' method of proofs and refutations simply boils down to the alternating procedure (going back to Pólya and Szegö in 1925) and described in Pólya (1965), vol. 2, pp. 50–51, from which I quote the following portions:

> A problem to prove is concerned with a clearly stated assertion A of which we do not know whether it is true or false: we are in a state of doubt. The aim of the problem is to remove this doubt, to prove A or to disprove it. . . . If we cannot prove the proposed assertion A we try to prove instead a weaker proposition (which we have more chances to prove). And, if we cannot disprove the proposed assertion we try to disprove a stronger proposition (which we have more chances to disprove). . . . In this way, by working alternately on proofs and counterexamples, we may attain a fuller knowledge of the facts.

No doubt Lakatos would have quarreled with this and in particular with the assumption that A is clearly stated; however, much of the Lakatosian dialectic is accounted for in Pólya's alternating procedure.

[7]I must confess to being ignorant of the critical literature on Lakatos' work except for an excellent overall essay review of the two volumes of his *Philosophical Papers* by my colleague Ian Hacking (1979). One specific critical question concerning Lakatos' mathematical philosophy is raised by Hersh (1978), which is otherwise extremely favorable

(i) *What happened before 1847?* According to the quotation given just a moment ago, the *method of proof analysis* appears to be a relative latecomer in the history of mathematics (1847). But this is said to be the same as the *method of proofs and refutations*, which is the only theoretical pattern offered by Lakatos to account for the progress of mathematics. It would seem, then, that Lakatos has nothing to tell us about the growth of mathematics prior to 1847. Actually, he has various things to say; for example, the methods of *monster-barring* and *exception-barring* were practiced before that date as moves to respond to criticism. Shouldn't a logic which is supposed to account for changes in a fallible body of knowledge account for any significant kind of changes? A related question is whether the method of proofs and refutations is supposed to be *descriptive* or *normative*. It seems at best that it could be descriptive of progress since 1847. But much of the tenor of the discussion leads one to view it as normative.

(ii) *Is the method most appropriate to describe mathematics in transitional foundational periods?* The example from Cauchy's rigorization of analysis would seem to suggest that; witness also the statement that the method is more appropriate to young, growing theories. But the example of Euler's conjecture is not of this character, though it turned out that the concepts were less clear than one had imagined. On the other hand, there were a number of foundational moves which took place without response to specific criticism or counterexamples, for example, those establishing the use of imaginary numbers or points at infinity (in projective geometry) or continuous probability measures. Finally, the method tells us nothing about progress by *internal organizational* foundational moves. These proceed by finding suitable abstract concepts around which to wind large parts of the subject in an understandable way. They do *not* arise as responses to critical examination of fallacious proofs. Examples are linear algebra, linear analysis, point-set topology, group theory, etc.

(iii) *How does this "logic of mathematical discovery" relate to working experience?* Most mathematicians throughout the history of theoretical mathematics work at a safe distance from troublesome foundational questions. This is not to say that the concepts used at any given time are all clearly understood (for example, the nature of geometrical objects, infinitesimals, imaginary numbers, sets, etc.). Rather, the mathematician is usually engaged in a project "midstream" which seems hardly affected by foundational considerations. That project usually consists in developing conjectures and seeking proofs of those conjectures. The tests for whether one has succeeded in obtaining such a proof are informal but fairly decisive.[8] It is common experience that proof-attempts proceed by fits and

(the same question is posed here in point (x) below).

[8]Even Euler, that most inductivist of mathematicians, operating at a time of low rigor, knew when he had a proof and when he didn't: "This law, which I shall explain in a moment, is, in my opinion so much more remarkable as it is of such a nature that we can be assured of its truth without giving it a perfect demonstration. Nevertheless, I

starts and involve reversals; (self-)critical examination is an essential element, but this does not necessarily mean that counterexamples form their principal feature.

(iv) *Is there no end to guessing?* Again, what Lakatos suggests here does not square with ordinary experience. The professional mathematician knows rather well what sort of things will work for certain kinds of problems and what won't. So guesswork is minimized from the outset. Moreover, the guesswork finishes with the mathematician's successful struggle to solve a problem or complete a proof. It is true that *results* are viewed in changing perspective over historical periods. Their significance is reassessed; they are generalized and understood in wider settings. (A marvelous example is provided by Pythagoras' Theorem.) But this is quite a different picture from that given by Lakatos of endless guesswork.

(v) *What constitutes improvement in a proof?* Lakatos gives no theoretical criterion for this. He merely produces examples and shows the change which takes place in the situation in response to criticism and/or counterexamples. Evidently—both for him and for us—improvement has taken place. It is my contention, which I shall elaborate below, that in fact *we have informal criteria for what constitutes an adequate proof and that these criteria can be explained in logical terms; improvement is described in the passage from inadequate proofs to adequate ones.* It seems to me that Lakatos must implicitly accept this, or something like it. I believe further that he refuses to say anything explicitly in this direction since doing so would undermine his sweeping rejection of the deductivist account of mathematics. In connection with both this and the preceding point, recall Lakatos' statement that not all proofs can be improved, especially not those in mature theories. In other words, one reaches resting points where there comes *an end to improvement of proofs under criticism.* Of course there is always the possibility of *improvement of results by generalization*, which is quite a different matter (as just described above).

(vi) *What constitutes an initial proof? Where does it come from?* Lakatos tells us that this is a thought-experiment or naive proof-idea. But in the examples he gives us, the idea for the proof is already well advanced. It has significant structure and steps; it pretends to be a rational chain from hypotheses to conclusion. We cannot imagine that such a proof springs full blown, following formulation of a conjecture. Should not a heuristic theory account for the development of such a proof? Indeed, Pólya—but not Lakatos—has significant things to say here (cf. section 7 below.)[9]

(vii) *What is the form of conjectures?* All the examples of conjectures given by Lakatos take the form

---

shall present such evidence for it as might be regarded as almost equivalent to a rigorous demonstration" (Euler, quoted in Pólya 1968, vol. 1, p. 91).

[9]Lakatos tells us that Pólya is concerned with the heuristic leading to conjectures and that *he* takes up where Pólya leaves off. This is false on two counts: Pólya does concern himself as well with the heuristics of finding proofs (cf. section 7 below) while Lakatos says nothing about the big stretch from conjecture to first real proof-ideas.

(1) *for all objects satisfying given hypotheses A, a conclusion B holds,*

which is symbolized logically by

(2) $$\forall x[A(x) \rightarrow B(x)].$$[10]

But there are a number of other forms of statements of mathematical (and historical) interest. For example, there are *singular statements* such as $e^{i\pi} = -1$, or $1 - \frac{1}{3} + \frac{1}{5} - \frac{1}{7} + \ldots = \frac{\pi}{4}$, or that 641 divides $2^{2^5} + 1$. Of course such statements have logical structure when the concepts involved are analyzed, and there is the theoretical possibility of "infinite regress" in such an analysis. But we are interested here in a description of the *naive form* of a statement, that is, as it presents itself to the working mathematician. There are also *existential statements*

(3) $$\exists x A(x)$$

to consider, for example, that the 17-sided regular polygon is constructible by ruler and compass, or that there exists a decision method for the elementary theory of real numbers, or that there exists an equation of degree 5 with rational coefficients which is not solvable by radicals. In refinement of (2), one is very often concerned with statements of the form

(4) $$\forall x[A(x) \rightarrow \exists y B(x,y)],$$

for example, that every complex polynomial of degree > 0 has at least one complex root, or that every Jordan curve in space has a minimal spanning surface. The statement that there exist infinitely many prime numbers and the conjecture that there exist infinitely many twin primes are actually both of this form (for every integer n there exists a larger prime, resp., twin prime), though usually here we think of a representation

(5) $$\forall n \exists m B(n,m)$$

where the variables n, m range over the nonnegative integers. As an example of increasing complexity in the same direction we have the statement of Waring's conjecture:

$$\forall k \exists n \forall m \exists q_1 \ldots \exists q_n [m = q_1^k + \ldots + q_n^k].$$

Finally, there are interesting statements of the form

(6) $$\forall x A(x) \rightarrow \forall x B(x),$$

for example, the statement Con(ZF) $\rightarrow$ Con(ZF + AC + GCH), which says that if the system ZF of Zermelo-Fraenkel Set Theory is consistent,

---

[10] We use here and below the logical symbols $\wedge$, $\rightarrow$, $\forall x$, $\exists x$ for "and," "implies," "for all x," and "there exists x," respectively.

that is, if there is no proof of a contradiction within it, then there is no contradiction to be obtained when we adjoin the Axiom of Choice and the Generalized Continuum Hypothesis. Now one can expect that methods of attack on a conjecture will be sensitive to the (naive or logical) form of the conjecture. We should be suspicious of a supposed logic of mathematical discovery which only concerns itself with statements of the form (2).

(viii) *Can ordinary logical analysis account for the same examples as the method of proofs and refutations?* I believe it can, somewhat as follows. The primitive conjecture considered by Lakatos as we have seen takes the form

(1) $$\forall x[A(x) \rightarrow B(x)].$$

The structure of the informal proof is supposed to decompose (1) into a series of subconjectures or lemmas, that is,

(2) $$\forall x[A(x) \rightarrow A_i(x)] \quad (i = 1, \ldots, n)$$

where

(3) $$\forall x[A_1(x) \wedge \ldots \wedge A_n(x) \rightarrow B(x)]$$

is supposed to hold. There can be various kinds of troubles, including the following two. First, in (2) we may not be clear enough about the concepts involved in $A(x)$ to be sure that the lemmas indeed follow. This is the first issue which is raised concerning Euler's conjecture (cf. Lakatos 1976, p. 8). Second, we may be rather clear about the concepts involved and (2) holds but (3) is not logically valid. This is the case where we look for a "hidden lemma," that is, a property $A_{n+1}(x)$ such that

(3)′ $$\forall x[A_1(x) \wedge \ldots \wedge A_n(x) \wedge A_{n+1}(x) \rightarrow B(x)]$$

is valid. But now the lemmas have to be reexamined, because we do not necessarily have

(2)′ $$\forall x[A(x) \rightarrow A_{n+1}(x)].$$

Indeed, in the situation contemplated by the method there is a global counterexample c, that is, one such that $A(c)$ holds but not $B(c)$. Then (2), (2)′, and (3)′ are of course logically impossible. However, in this case we seek an "improvement" of the conjecture

(1)* $$\forall x[A^{\star}(x) \rightarrow B(x)]$$

for which

(2)* $$\forall x[A^{\star}(x) \rightarrow A_i(x)] \quad (i = 1, \ldots, n+1)$$

now holds, as well as (3)$'$. This may be done by "incorporating" the hidden lemma into the hypothesis (most simply by taking $A^\star(x) = A(x) \wedge A_{n+1}(x)$).

(ix) *Are there no crystal-clear concepts*? Certainly there have been continual historical shifts in what has been regarded as clearly understood. Throughout, though, the structure of positive integers $1, 2, 3, \ldots$ has enjoyed a privileged status. To my mind, this is a crystal-clear mathematical concept. At any rate, if anything is a candidate for being such, this is it. Moreover, there has never been voiced any real concern or confusion on this score in the entire history of number theory (which stretches back at least to Euclid).[11] At no time has the criticism of proofs involved criticism of basic concepts about numbers. A heuristic logic should give some account of progress here. The fact that this subject is ignored by Lakatos is a sign that it threatens his theses, in particular that there are no crystal-clear concepts. (Note that this issue is separate from whether there is such a thing as conceptual finality. For example, the concept of *natural number* is often defined these days in terms of the notion of *set*, thereby reducing a completely clear concept to one that is quite unclear. Of course, in the light of such moves one can always claim there is infinite regress.)

To complete the critique, we ask finally:

(x) *What is distinctive about mathematics*? Lakatos makes no effort to tell us what there is about the *conceptual content* of mathematics or about its *verificational structure* which sets it off from other areas of knowledge. Obviously he has informal criteria, since he chooses to discuss only examples from mathematics. But he offers *no theoretical criteria.* It seems to me that there is nothing he says about the general idea of "proof" which could not apply equally well to "more or less convincing argument"; there is nothing about the "logic of mathematical discovery" which could not be read equally well as a "logic of rational discovery," that is, of the process of reaching convictions rationally. If I am right, then such a logic could hope to account only for a few gross features of the actual growth of mathematics. In any case, all of my preceding comments (i)–(ix) reveal that Lakatos' "logic" hardly begins to be equal to the tasks called for in his grand program.

7 Comparison with Pólya's Work

Pólya has written extensively on heuristic and plausible reasoning in mathematics (cf. Pólya 1957, 1965, 1968). In the context of the present discussion, I would characterize this work of my esteemed [erstwhile] colleague briefly as follows (cf. also footnote 5).

[11] Formalism has bred some skeptics who refuse to be convinced that elementary number theory is consistent. There is even an occasional loner who claims to have established its *inconsistency*; in these cases critical examination by others has only hardened their position (that is, there is no dialectic taking place).

(i) Pólya does not voice philosophical doubts about the certainty of mathematics; he does not raise foundational issues. The concepts and problems with which he deals are supposed to be clearly understood. Moreover, we are supposed to understand what constitutes a demonstration; it is accepted that logic gives a theoretical explanation of that.

(ii) In Pólya (1957) and (1965) he concentrates on tactics and methods for *finding solutions* to problems and, to a lesser extent, on *finding proofs* of theorems. Pólya's motivation here is more toward helping people make their way effectively through mathematics than to establish a theory of heuristic. But in the process he develops well-structured sets of *strategic rules*.

(iii) In (1968) Pólya concentrates on the processes which lead one to formulate general conjectures and to see what counts as *support* for them. In this connection he formulates a logic of plausible reasoning (or *degrees of credibility*); this includes a number of simple rules, of which the following is typical:

$$\frac{A \text{ implies } B \qquad B \text{ is true}}{A \text{ is more credible.}}$$

(iv) Pólya makes use of a wealth of mathematically and/or historically interesting examples to illustrate his points and rules.[12]

Anybody who has read Pólya's works or heard him lecture knows that he is peerless within the framework for which he has set his heuristic. In contrast to Lakatos, he plumbs the relatively safe midstream of mathematics. But this is where most of the day-to-day experience of the subject is going on. Students and teachers could ask for nothing more. What professional mathematicians might want, though, is a continuation along the same lines which concentrated on the ins and outs of finding difficult proofs. *That work is waiting to be carried on.*

There is one aspect of mathematical progress which neither Pólya nor Lakatos have really attempted to deal with, namely, that by *convenient conceptual development*. How does one go about finding the technical but general concepts that help organize masses of material and make difficult proofs understandable? (Cf. 6 (ii) above.) Lakatos' idea of *proof-generated concepts* seems to me a first step in this direction.

[12]In a personal communication, R. Parikh asked me whether Pólya says anything about the rule in (iii) above applied to the case that $A = B \wedge C$ and C is false. To my knowledge he does not. Naturally, this sort of example doesn't arise in practice. It also suggests that there are concepts implicit in the actual situation (knowledge changing over time, relevance of statements) which may need to be made explicit in such a logic in order to give it significance.

8 The Logical Analysis of Mathematics

There is little time here to mount a defense of the logical or metamathematical approach, so I shall simply try to indicate the nature of the position briefly (at least as I see it).

(i) Logic attempts to provide us with a theoretical analysis of the underlying nature of mathematics as physics provides us with a theoretical analysis of the underlying nature of the physical world. Evidently, in both cases, only a part of the experience is accounted for and, in particular, various superficial and/or accidental features cannot be treated at all.

(ii) In the case of logic, this theoretical analysis is supposed to explain what constitutes the underlying *content* of mathematics and what is its *organizational and verificational structure.*

(iii) The study of content has received no final answer. There are a number of conflicting positions about the nature of mathematics: platonist, constructivist, finitist, predicativist among them. However, what logic has succeeded in doing very well is *formulating these positions* in precise terms by a variety of formal systems. It has then gone on to give us significant information about the potentialities and limitations of each of these positions and about their interrelationships. This part of logical achievement has been particularly stressed by Kreisel (cf. 1968 among his other publications).

(iv) The logical analysis of the structure of mathematics has been especially successful. Again, there is not a single analysis, since (for example) ordinary (platonistic) reasoning uses classical two-valued logic while constructive reasoning uses a more restricted ("intuitionistic") logic. There are two parts to this logical analysis. First is the *logical syntax of language*, which gives a description of the structure of mathematical propositions. This accords very well with our informal experience: transforming mathematical statements from informal to logical form and back is a direct matter which is essentially unproblematic. (This contrasts, of course, with the attempts to provide a logical syntax of ordinary language.) Second comes the *logical structure of proofs* as described in certain *deductive systems*. In this case, the relationship with ordinary experience is more or less good: "less" for Hilbert-style systems and "more" for Gentzen-style systems of natural deduction. It is commonly felt that logic gives us a good underlying analysis of the structure of completed proofs (no gaps, no unsure assumptions or steps). Indeed, I believe that the logical analysis of the structure of mathematics comes much closer to explaining our everyday mathematical experience than physics does to explaining our everyday physical experience. (For an elaboration of these views, cf. Feferman 1979b [reproduced as chapter 9 in this volume].)

(v) Though formal systems are not normally conceived to represent "slices" of mathematics in a "frozen" state, so to speak *in vitro* as opposed to *in vivo*, one can use these systems to model growth and change.

First efforts to do so were via *progressions of theories*,[13] but these took on an unreal character when extended into the transfinite. A formalization of predicative mathematics by a *growing system* without use of transfinite progressions I proposed in 1979 (Feferman 1979a) [and, more recently, (1991) and (1996)]. This allows one to expand one's conceptual stock as more and more things are proved which make such extensions admissible.

9 Conclusion

Lakatos' fireworks briefly illuminate limited portions of mathematics conceived as an active growing intellectual endeavor which is subject to confusion, uncertainty, and error. In contrast, logic gives us a coherent picture of mathematics but which at first sight appears ideal and static and which is irrelevant to everyday experience. However, it alone throws light on what is distinctive about mathematics, its concepts and methods. Pólya's heuristic provides one bridge from theory to practice. I believe that Lakatos' successes should inspire us to seek a more realistic theory of mathematics. But his failures and limitations should make us aware that much more from logic will have to be recognized as basic and incorporated into such a theory. It would be best to reserve the name "the logic of mathematical discovery" for that which is yet to come.

[13]By Turing (1939), Kreisel (1960), and myself (1962) [cf. also Feferman (1988b)].

4

Foundational Ways

1 The Foundational Enterprise: Carrying on in Mathematics and Logic

In this century mathematicians have become conscious of the unity and structure of their subject to an unprecedented extent. Early on, logic offered grand claims to explain this and to rescue mathematics from its "crisis," which if it threatened one, seemed to threaten all. Global views about the nature of mathematics were propounded. These then took hold of foundational discussions to such an extent that the logical foundation of mathematics is currently thought of only in terms of such grand positions: logicism, formalism, platonism, and constructivism. They are all rather tired-looking now, if not suffering from senescence and still more basic ills. Long since, most mathematicians have given up worrying about the crises and gone about their daily affairs, attending to logical hygiene only as needed. If asked, they will say they are really formalists, or that Zermelo-Fraenkel meets their needs, or whatever.[1]

Among those mathematicians who still take foundational concerns seriously, there is a growing band who aim to recapture this territory from logic in favor of mathematics. Some offer up a new global scheme, but now one that is purely mathematical such as category theory.[2] Others propose a more phenomenological view: mathematics is as mathematics does. These critics of logical foundations are united in the view that mathematics is more reliable than any of the foundational schemes which have

"Foundational ways" was first published in Willi Jäger, Jurgen Moser, Rheinhold Remmert (eds.), *Perspectives in Mathematics: Anniversary of Oberwolfach 1984*, Birkhäuser Verlag, Basel (1984), 147–158 (Feferman 1984b). It is reprinted here with the kind permission of the publisher. There are some minor corrections and a few bibliographic additions. This article was a "slimmed down" version of "Working foundations," *Synthese* 62 (1985), 229–254, of which chapter 5 is an expanded version.

[1]For a typical expression of such views see Dieudonné (1982).

[2]I have argued particularly against the scheme of categorical foundations in Feferman (1977b).

been propounded by the logicians to "secure" it. They further complain that the logical analysis of mathematics bears little or no relation to actual practice.[3]

Given that the "big" logical positions were extremely overstated and defective in substantial ways, this kind of reaction is natural. However, I believe it has gone too far in the opposite direction. The aim here is to restore the position of the logical or metamathematical approach to the foundations of mathematics but at a more everyday "local" level which is not preoccupied with the grand schemes. On my view, this is a direct continuation of work that mathematicians themselves have carried on from the very beginning of our subject up to the present. The distinctive role of logic lies in its more conscious, systematic approach and its different ways of slicing up the subject.

If one analyzes foundational activity (whether carried on by mathematicians as a matter of course or more consciously by logicians), one finds that it falls into one of five or six characteristic modes. Each of the following sections is devoted to one of these *foundational ways*, and their presentation is the same throughout. Each section begins with a general explanation of that kind of foundational activity. This is then illustrated by some familiar examples from mathematics. It then goes on to give some examples from metamathematics, some of which will be familiar but probably far from all.[4] References will be given only for the (presumably) unfamiliar work.

As usual, a disclaimer: this is not a survey of mathematical logic; it has to do with only that small part of the subject (these days) which has foundational concerns. Even that is hardly surveyed; the illustrative examples are meant to be typical, but evidently depend on my own knowledge and interests. I do believe there is much to interest mathematicians in the rest of logic, but that is another story which is more often told.[5]

2 Conceptual Clarification

The first model of foundational activity we consider (also called *conceptual analysis* or *explanation*) usually takes place at an advanced stage in the organization of a subject (cf. section 6 below on axiomatics). Typically, one has a settled fund of well-understood basic concepts, and there is the question of giving precise definitions of other frequently used informal concepts in terms of these. There are usually some specified requirements to be met, and in some cases it may be shown that only one such definition is

[3]See, for example, Davis and Hersh (1981), Hersh (1979), Lakatos (1976, 1978), and MacLane (1981). Logicians who reached similar conclusions are Goodman (1979), Kreisel (1967, 1976, 1977), and Wang (1974).

[4]Many more such examples are given in the paper (Feferman 1985) from which this chapter is derived [and in the 1991 update of that, reproduced as chapter 5].

[5]A good source of material to begin with is the *Handbook of Mathematical Logic* (1977) edited by J. Barwise.

possible. In other cases, the requirements only partially determine the exact definition, and the criteria for acceptance are subtler and often require the test of practice for complete acceptance.

Some classical examples from mathematical analysis are the definitions of *limit, continuous function, smooth function,* and *area.* More recent examples are the concepts of *dimension* in topology and of *natural mapping* and *universal construction* in algebra and topology.

The best-known examples from logic are the definitions of *satisfaction* and *truth* (by Tarski) and of *mechanically computable function* (by Turing and Post).

Tarski's semantical concepts are all referred to suitably specified *formal languages* L. These were used in the 1950s to explain the informal idea of a *transfer principle* in algebra, the most famous example of which is *Lefschetz' principle* in algebraic geometry. One has a transfer of results in L from one structure $\mathcal{M}$ to another structure $\mathcal{M}'$ if $\mathcal{M} \equiv_{\mathrm{L}} \mathcal{M}'$, which means that $\mathcal{M}$ and $\mathcal{M}'$ satisfy the same statements from L. The strongest present formulation of Lefschetz' principle in these terms has been given by Eklof (1973; following Feferman 1972). A variety of further interesting transfer principles in algebra may be found in Cherlin (1976).

The notions of effective computability mentioned above operate on finitely presented data, for example, the natural numbers or finite strings of symbols. However, they have been the basis for, or suggested as, precise definitions of computability in other situations. For example, the concept of *finite effective construction* as applied in geometry and algebra has been plausibly explained relative to arbitrary mathematical structures as domains in Friedman (1971). Various notions of *infinitary construction* have also been proposed, but with less sureness about the conceptual analysis; even so, the definitions taken have turned out to be exceptionally useful. For one approach and relevant literature see Fenstad (1980).

There are two important concepts in logical work which have not yet been satisfactorily defined. The first is that of *identity of proofs.* Obviously, the significance of this goes far beyond logic; mathematicians constantly compare proofs and judge whether they are (essentially) the same or different. One would think that a *theory of proofs* would establish this as a basic notion (comparable to isomorphism in algebra or homeomorphism in topology). In fact there is a highly developed theory of proofs (inaugurated by Hilbert) but no convincing notion of identity has yet been produced within it. Prawitz (1971) presents interesting relevant notions and results.

A second informal concept which has not yet been defined precisely is that of *natural well-ordering.* The most familiar example of such is the ordering of "figures" $\omega^{\alpha_0} \cdot k_0 + \omega^{\alpha_1} \cdot k_1 + \ldots + \omega^{\alpha_n} \cdot k_n$ in Cantor normal form for all ordinals up to the ordinal ε_0 (the least solution of $\omega^\alpha = \alpha$). This was used by Gentzen (in 1936) to give a semifinitary consistency proof of arithmetic. Since then technical proof theory has been dominated by the use of larger and larger naturally presented effective well-orderings.

Consistency proofs on cooked-up but nonnatural effective well-orderings can always be given, so it is important to know just what makes an ordering natural. This question has been approached through category theory, but without any satisfactory answer thus far; see, for example, Feferman (1968) and Girard (1981).

3 Dealing with Problematic Concepts and Principles (Part 1)

At each stage in the development of mathematics there is a body of ideas and methods generally regarded as well understood, and with which mathematicians in the mainstream confidently carry on their work. It may happen subsequently that the place of such ideas in our understanding undergoes considerable revision (as happened, for example, with the conception of the geometric line). The concern here, though, is with the problems raised *at that stage* by concepts and principles which have a certain plausibility or utility, but about which there is less certainty or security than for the accepted core. In this section, we deal with one foundational way of dealing with such situations, called the method of *interpretation* or *models*; in the next we shall deal with another principal way.

Familiar examples of problematic notions at various stages in the development of mathematics are *zero, negative numbers, imaginary* and *complex numbers*, and *points at infinity.* Examples of troublesome principles are the *parallel postulate* and, more recently, the *Well-Ordering Principle.* A typical example of interpretation is that of the complex numbers as pairs of real numbers. Implicit in the success of this interpretation is that a series of conditions are met about complex numbers, which we now describe as being a field generated from the reals by adjunction of a root of $x^2 + 1 = 0$.

More generally in present-day terms, one has a set of axiomatic requirements containing the problematic notion or principle, and the method of interpretation is to yield a model for these axioms. In the case of a problematic principle, providing a model in addition for its negation shows this to be a nontrivial exercise (otherwise the principle in question is a consequence of the remaining axioms). But there might further be grounds for considering the negated principle to have it own plausibility, for example, as with the parallel postulate. Put in still more logical terms, denote the basic system of axioms (or *axiomatic theory*) S and denote by P the principle in question and ¬P its negation. The adjunction of P (¬P) to S is denoted by S + P (S + ¬P). Providing a model for S + P establishes the *consistency* of P with S, while providing one for S + ¬P establishes *independence* of P from S.

Most mathematicians are familiar with the consistency and independence results in set theory due to Gödel and Cohen. Here the axiomatic theory S taken as basic is the system ZF of Zermelo-Fraenkel Set Theory. The first principle P to consider is AC, the *Axiom of Choice*, which Zermelo had proved to be equivalent to the Well-Ordering Principle. AC is needed to establish a good theory of infinite cardinal numbers and to order them linearly in the transfinite sequence $\aleph_\alpha$. The first question to answer is that of where the cardinality of the continuum $2^{\aleph_0}$ falls in this sequence. Certainly $2^{\aleph_0} \geq \aleph_1$ and the Cantor Continuum Hypothesis CH proposes $2^{\aleph_0} = \aleph_1$. With all efforts to prove or disprove them unsuccessful, CH and its generalization GCH to arbitrary infinite cardinals are problematic principles relative to ZF + AC. As is well known, Gödel established that (i) ZF + AC + GCH has a model, while Cohen established that (ii) ZF + ¬AC has a model and (iii) ZF + AC + ¬CH has a model. In each case, the interpretation is given back in terms of a model M for ZF: Gödel uses M to form a *submodel* M_0 called the *constructible sets*, while Cohen uses it to form an *extension model* $M' = M_0[G]$ by adjunction of certain *generic sets* G (analogous to transcendental extensions in algebra). At any rate, in each case what is achieved can be formulated as a *relative consistency result*; that is, if ZF is consistent then so also is ZF + AC + GCH, etc.

In logic it has been observed that one can usefully sharpen these relative consistency results as follows. Suppose S, T are two axiomatic theories, not necessarily with the same basic language (L_S = language of S, L_T = language of T) but where $L_S \subseteq L_T$. Let Σ be a class of statements in L_S. Suppose further that every axiom of S is a theorem of T (that is, is provable from the axioms of T). Then T is said to be a *conservative extension of* S *for the class* Σ if whenever $A \in \Sigma$ and A is a theorem of T then A is already a theorem of S. Note that as long as Σ contains statements like $0 \neq 0$, each conservation result implies relative consistency. There are usually two kinds of cases of interest, first where $L_S = L_T$ but Σ is properly contained in L_S, and second where $\Sigma = L_S$ and L_S is properly contained in L_T. In the latter case we simply say that T is a conservative extension of S.

Some time ago, Kreisel observed that ZF + AC + GCH is a conservative extension of ZF for the class Σ of purely number-theoretic statements (as expressed in L_{ZF}). The reason is that Gödel's constructible sets model for ZF + AC + GCH, obtained from a model for ZF, is standard for the natural numbers—in other words, though the meaning of "set" changes in the interpretation, the meaning of "natural number" does not. This conservation result proved to be significant in an unexpected way. In their 1965 paper Ax and Kochen proved, using set-theoretical arguments, that a certain effectively given set F_p of first-order axioms for the p-adic numbers $\mathbf{Q}_p$ is complete; that is, every statement or its negation in L_{F_p} is provable from F_p. It follows that there is (in principle) a decision procedure for validity in $\mathbf{Q}_p$; given a statement A in L_{F_p} just run effectively through the theorems of F_p until we hit A or $\neg A$. The statement that (theoremhood in) F_p is decidable is itself equivalent to a statement of elementary number theory, call it D_p. It happens that Ax and Kochen used the unusual hypothesis CH in their argument (of course, with AC). In other words, they showed ZF + AC + CH proves D_p, and observed as a consequence of Kreisel's conservation result that already ZF proves D_p. Later, Cohen (1969) produced an explicit decision procedure for $\mathbf{Q}_p$ by different arguments which could be readily formalized in ZF, thus confirming the prediction in principle by conservation.

Here are some less familiar examples of interpretations in logic. First are the models of the *λ-calculus* produced by Scott (1972) and (following Plotkin) Scott (1976). The *λ-calculus* is a formalism for defining functions, and its basic operation is the binary one of *application*, written fx (or $f(x)$). Formally, this allows any element f of the universe (over which the variables range) to be applied to any other element x; in other words, each member f of the universe acts as a *total function*. It is fairly obvious what axiomatic requirements are to be met under this interpretation—the main one asserts that any expression $\tau[x]$ of the language containing a free variable x determines a function f whose values are given for all x by $fx = \tau[x]$; this function is usually denoted $f = \lambda x.\tau[x]$ (thus explaining the use of "λ-calculus" as designation for this formal system). The problematic feature of the λ-calculus is that it allows *self-application*, that is, formation of ff. This possibility does not exist in familiar universes of functions, where the domain of arguments of a function is always prior in some sense to the function itself. It is by no means obvious how to construct models of the λ-calculus, which is what Scott achieved.

As with the previous example there is further utility to the interpretation, in this case applying to theoretical computer science. Programs involving recursion determine functions as minimal fixed points of functional equations $f = \tau[f]$. It is not hard to derive explicit fixed-points $\varphi = \tau[\varphi]$ in the formalism of the λ-calculus. Scott's models for the λ-calculus give definite meaning to each such expression φ, even when the recursion is not independently analyzed. More generally, Scott has used this to provide what is called a *denotational semantics for programming languages*, in which each element of a program is given meaning as a mathematical object.

Of course the problems raised by *self-reference* (in natural language) and *self-membership* (in the theory of classes) are old, and were at the source of the "foundational crisis" at the beginning of this century. It seems that an essential ingredient of the paradoxes is the combination of negation with self-application. This suggests that use only of *positive* instances of self-application need not lead to inconsistencies. Examples of such are the statement *this statement is true*, and the notion *class of all classes* or *category of all categories*. Axiomatic theories permitting these and other instances of positive self-application have been developed and shown to be conservative extensions of known consistent theories by means of suitable interpretations. For a survey of work in this direction see Feferman (1984). An alternative (and earlier) treatment of the problem of self-membership in category theory is described in the next section.

A number of further examples dealing with problematic notions and principles by the method of interpretation are given in section 3 of chapter 5. One of the most interesting has to do with Brouwer's concept of (*free*) *choice sequence*, which is central to his intuitionistic redevelopment of mathematical analysis. These are supposed to be infinite sequences $\alpha = (a_0, a_1, \ldots, a_n, \ldots)$ generated without any known law, for example, randomly or by a sequence of arbitrary choices, and of which only a finite amount of information is available at any given time. According to Brouwer, this informal understanding leads one to accept certain *continuity principles* concerning statements involving choice sequences. These principles in turn have consequences such as that every function of real numbers is continuous, which prima facie contradict classical statements, but only when one uses the classical Law of Excluded Middle (LEM). Kleene and Vesley (1965) formulated Brouwer's ideas about choice sequences in an axiomatic theory CS with intuitionistic logic (no use of LEM) and showed it to be consistent by means of a complicated *realizability interpretation* (a method of interpretation for intuitionistic systems originally developed by Kleene in 1945).

4 Dealing with Problematic Concepts and Principles (Part 2)

Here we consider a second foundational way of dealing with the problem situation described at the beginning of the previous section. Instead of interpreting the problematic concepts or principles in some sort of direct way, one tries to *replace* or find *substitutes* for them which do the same work, or even to *eliminate* them entirely. In the latter case one wants to preserve the useful consequences, that is, establish a conservation result. In logical work the process of elimination is frequently established by *syntactic transformations.*

The classical example from mathematics of a concept which was eliminated while saving its applications is that of *infinitesimal*; this idea could be dispensed with once the notions of *limit, derivative*, etc., were given "ε, δ" definitions. (As is well known, Abraham Robinson restored infinitesimals as respectable objects of mathematical study; but that is another story, more properly falling under section 5 below.) The replacement of *multivalued function* in complex analysis by *single-valued function on a Riemann surface* is an example of a problematic concept dealt with by substitution.

Some examples from logic of these foundational tacks are the treatments of the notions, the *class of all classes* and the *category of all categories* in the formal framework of the BG (Bernays-Gödel) theory of *sets and classes.* Here the problematic notions are replaced by the *class of all sets* and the *category of all small categories* (respectively), where, following the terminology of MacLane (1961), one uses the distinction between set and class in BG to distinguish between *small* and *large categories.* Another solution to this problem situation had been given in the framework of ZF (a theory of sets only) by the use of *Grothendieck universes.* On the face of it, that required assuming the existence of many inaccessible cardinals.

It was shown in Feferman (1969) how to use the *reflection principle in* ZF (reflecting properties of the universe of all sets down to subuniverses which are themselves sets) in order to avoid such unusual assumptions.

There are a number of frequently used syntactic transformations which serve general procedures of elimination, going back to Hilbert, Gödel, Herbrand, and Gentzen in the late 1920s and early 1930s (see chapter 5, section 4, for more details and references). An interesting recent example of a different kind of syntactic transformation was that introduced by Kreisel and Troelstra (1970) to *eliminate choice sequences.* This yields conservation of the theory CS over a constructive theory without choice sequences, permitting certain strengthenings of the Kleene-Vesley result for CS mentioned in the preceding section.

5 Dealing with Problematic Methods and Results

In the preceding two sections the emphasis was on troublesome concepts or principles which are relatively basic in character and which usually emerge early in the development of a subject. This section concentrates on mathematically more advanced methods or results which, typically, raise questions about their validity at a later stage of development. The problem here is to justify the methods or results, by providing them with a rigorous foundation. This means giving precise *sufficient* conditions under which these methods can be applied or for which the results hold. It is rare that such conditions are *necessary,* and there is then room for improvement. That is generally a process which takes place over a long period of time and involves successive steps of sharpening, extending, and generalizing.

There are many examples from mathematics of this mode of foundational activity, of which the following are just a few: *Descartes-Euler theorem, Fourier series, calculus of variations, Dirichlet's principle, Stokes' theorem, Jordan curve theorem, generalized functions, probability theory,* and *algebraic geometry.* The foundations of the mathematically isolable parts of other subjects follow the same pattern, for example, with *quantum mechanics.*

One finds many fewer examples of this kind from logic. The most well known is for the *theory of transfinite ordinals and cardinals,* initially developed informally by Cantor and currently given precise formulation in axiomatic set theories such as ZF + AC or BG + AC. Of more interest is the work by Robinson (1966) providing foundations for *infinitesimal analysis.* (A good introduction to this is provided by the article Stroyan (1977).) For a few further (less far-reaching) examples, see chapter 5, section 5.

6 Organizational Foundations and Axiomatization

This is one of the most familiar kinds of foundational work, and as a result its significance tends to be overemphasized. It usually takes place at a relatively advanced stage in the development of a subject, when it is ripe for *organization* and *systematic exposition.* What is required is a *choice and ordering of concepts and results.* Basic results which cannot be justified by further appeal within the given framework are taken as *axioms.* Usually, *typical patterns of proof* emerge in mathematical organization, and some attention may be given there to *proof principles*, but the explicit systematization of *methods of proof* is taken up only in logical work. There is frequently some overlap in axiomatic development with the foundational moves described in the preceding sections 2–5.

Euclidean geometry provides, of course, the classical example of axiomatization. But while long considered a model for rational thought, it was relatively isolated of its kind until the nineteenth century. At that time, the growing professionalization of mathematics and the expansion of organized higher mathematical education for engineers and scientists led mathematicians to devote more and more effort to systematic organization and to a renewed involvement in axiomatization. In geometry, this was revived with *non-Euclidean geometry, projective geometry*, and Hilbert's *re-axiomatization of Euclidean geometry.* In algebra, these efforts gave rise in the nineteenth and twentieth centuries to the axioms for *groups, rings, fields, vector spaces, algebras,* etc. Axiomatization took hold in analysis in the twentieth century with *metric spaces, topological spaces, Banach spaces, Hilbert spaces, measure spaces,* etc. Evidently axiomatics is pandemic within present-day mathematics.

The work in the nineteenth century on the axiomatic *characterization of the basic number systems* **N**, **Q**, **R**, and **C**[6] forms a bridge to the logical approach to axiomatic foundations in the twentieth century. This involves a shift away from axiomatics within traditional subject areas (such as geometry, algebra, analysis) to the study of principles concerning ideas, structures, and principles which may be involved in *all* areas. The most extensively developed example of that in the twentieth century is of course *axiomatic set theory.*[7] This is the current expression of the platonistic philosophy of mathematics, and there is considerable agreement about its core from those who accept that position. However, there is also a frontier of active research on the so-called *large cardinal axioms* whose status is far from settled (the article of Kanamori and Magidor (1978) gives a good relatively recent survey [cf. also Kanamori 1994]).

Axiomatization becomes more exciting at the outer fringes of a subject, where one must consider whether propositions that can't be settled by previously accepted axioms should be regarded as new axioms. It is also exciting when there are genuinely competing overall schemes for organization and axiomatization. One of the main sources of this kind of stimulus for logic has been in the *axiomatic foundations of constructive*

[6]By Méray, Weierstrass, Dedekind, Cantor, Peano, Frege, and others.

[7]At the hands of Zermelo, Skolem, Fraenkel, von Neumann, Bernays, Gödel, and others.

mathematics. Initially, constructivity in this century was dominated by Brouwer's ideas, which he propounded in startling and sometimes mystifying ways. However, these have since been captured and made widely accessible by the axiomatic analysis which began with Heyting's work in the 1930s and has stretched to the theories of choice sequences treated more recently by Kleene/Vesley and Kreisel/Troelstra (mentioned in sections 3, 4 above). Other approaches within the constructivist framework were more understandable than Brouwer's to begin with, but raised questions which axiomatization helped to resolve. Two such particularly to be mentioned are those due to Markov and Bishop. Less well known but also distinctive and coherent is the approach due to Martin-Löf. In many respects these schools or approaches account for the same body of mathematics and do so in closely related ways; in other respects they are divergent and in some sense competitors. Not only that, but in each case competitive axiomatizations have been developed for these different approaches (outside of Brouwer's theory of choice sequences). [For detailed descriptions and comparisons, cf. Beeson 1985 and Troelstra and van Dalen 1988.]

Once axiomatic foundations of a given subject are fairly well settled, it is common to undertake a *fine analysis* of the role of different axioms or groups of axioms in different parts of practice. A number of examples and references are given in chapter 5, section 6.

7 Reflective Expansion

One further foundational way is taken up in chapter 5, section 7, and will be indicated only briefly here. This is the idea of *reflective expansion of concepts and principles.* In the historical development of mathematics, new concepts emerge and then become clarified and established as basic ingredients of our thought. Thus, at different stages, one obtained the concepts of *whole number*, of *point, line*, and *plane*, of *ordered pair*, of *function*, and of *set.* Other concepts (and associated principles) are then derived by reflection on these. In a way, that process is a form of generalization, but of the following particular character: at a certain point one reflects on what has led one to accept and work with given concepts, and recognizes that much more is *implicit* in doing so. For example, Euclidean geometry of the plane and space were eventually explained in terms of $\mathbf{R}^2$ ($= \mathbf{R} \times \mathbf{R}$) and $\mathbf{R}^3$ ($= \mathbf{R} \times \mathbf{R} \times \mathbf{R}$) using the real number system $\mathbf{R}$ and pairing and tripling operations. Reflection on that led to the concepts of *n-tuple* and *n-dimensional space* $\mathbf{R}^n$. (The concept of n-tuple is itself obtained by reflection on the notion of ordered pair.) Reflection on the ordinal enumeration of the whole numbers led to the concept of *ordinal number.* Reflection on the formation of derivative and integral as operators on functions led to the idea of *function operators* or *functionals.*

The process of reflective expansion on concepts and principles has been taken up in a theoretical way in logic, starting with Kreisel (1970). A case study on which I have concentrated is that in which one is led from the natural number system $(\mathbf{N}, 0, Sc)$ together with the basic ideas of inductive proof and definition on $\mathbf{N}$, to its reflective expansion into so-called *predicative mathematics*; see Feferman (1979a). Since then I have formulated and studied more generally a concept of the *reflective closure of a theory*, (see Feferman 1991); this is an area of continuing research [cf. Feferman 1996], which should cover ground all the way from finitist mathematics to the generation of extraordinarily large cardinals.

This concludes my roundup of modes of foundational activity, in logic as in mathematics. While each way has a characteristic problem situation and typical examples, the importance overall of foundational work lies in its essential, continuing role in the refinement and improvement of mathematical understanding. I have tried to show that this is carried on almost as a matter of course, independent of any grand philosophical position as to the nature of mathematics. But it is also carried on as a matter of necessity, in a way integral to what makes mathematics such a distinctive body of thought. Any philosophy of mathematics worthy of its name must address itself to the more difficult question of what *that* amounts to.

5

Working Foundations–'91

1 What's the Use of Foundations?

There is currently a general malaise about the logical approach to the foundations of mathematics. One main reason is that foundational thought in this century has been dominated by a few global views about the nature of mathematics—logicism, formalism, platonism, and constructivism—each of which has proved to be defective in substantial ways, while nothing else has come to take their place. In recent years there has been an increasingly steady barrage of criticism directed against these all too familiar positions. Some of that criticism has been very sophisticated, coming from within the field of logic itself. Other criticism has come from mathematicians who would jettison the whole approach via logic and formal systems. Still more specifically on the mathematical side have been movements to replace logical foundations by purely mathematical foundations (for example, using category theory).[1] Though these various critical views diverge from each other in many significant respects, they share two common themes: (i) mathematics is more reliable than any of the foundational schemes which have been proffered by the logicians to "secure" it, and (ii) the logical analysis of mathematics bears little or no relation to actual mathematical practice.[2]

"Working foundations–'91" was first published in Giovanna Corsi, Maria Luisa dalla Chiara, and Gian Carlo Ghiradi (eds.), *Bridging the Gap: Philosophy, Mathematics and Physics*, Kluwer Academic Publishers, Dordrecht (1993), 99–124 (Feferman 1993b). It is reprinted here with the kind permission of the publisher. Some minor corrections have been made and there are a few bibliographic additions.

The article reproduced here is in turn an expanded and updated version of "Working foundations," *Synthese* 62 (1985), 229–254 (Feferman 1985). I am indebted to Dr. Giovanna Corsi for her assistance with the integration of the new material with the text of that article.

[1] My arguments against this have been presented in the first part of Feferman (1977b).

[2] The following is a grab bag of references representing a variety of critical approaches: Davis and Hersh (1981); Goodman (1979); Hersh (1979); Kreisel (1967, 1976, 1977); Lakatos (1976, 1978); MacLane (1981); Tymoczko (1986); Wang (1974).

I believe that the critical evidence must be taken seriously, but disagree with the conclusions. On the contrary, I think a very strong case can be made for the logical approach to the foundations of mathematics, even when we cease to be preoccupied with the grand schemes. As part of the support for this, the present chapter offers a picture of *logical foundations at work*, detailed by a multitude of examples. The main point to be made is that *this work is a direct continuation by more conscious, systematic means of foundational moves which have been carried on within mathematics itself from the very beginning.* Each of the following sections (2–7) is devoted to a particular kind of foundational activity, and begins with familiar examples from mathematics. Each section then goes on to give a number of metamathematical examples, some of which will be familiar to a general audience but most of which will not.[3]

The varieties of foundational work to be exemplified here are categorized as follows: *conceptual clarification; interpretation, reduction, or elimination of problematic concepts and principles; foundations of problematic methods and results; organizational and axiomatic foundations;* and *reflective expansion of concepts and principles.* Except for the last, each of these is a more or less recognized mode of foundational activity for which there are many classical examples.[4] Current research provides new, interesting examples where such moves can be seen in operation. However, the last category—reflective expansion—is a feature of the progress of mathematical thought which has been insufficiently appreciated. Work whose aim is to capture that process in logical terms will be described.

The word "working" in the title is meant to be ambiguous. On one hand it is meant to suggest (as above) the active pursuit of foundational questions without regard to any fixed philosophical standpoint. Another related intention is to suggest the provisional character of what is accomplished, as part of the evolution of *all* mathematical understanding. Finally, it is meant (more personally) to indicate the work itself which has engaged my interests and attention. In presenting the illustrative examples below, I have drawn liberally on my own efforts and on the work of many others that has enlightened me in this sphere of interests.[5]

My original plans for this paper had been somewhat more ambitious. In addition to the material already described, I had wanted to say more in defense of the logical analysis of mathematics and against the argument that it has nothing to do with our everyday experience. My position is that without logic, one will be unable to give a satisfactory answer to the question, *What is it about the concepts and methods of mathematics that makes*

---

[3]Bibliographic references are generally given only for less familiar logical work, but the sources provided are not always the last word on a given subject.

[4]The categorization given here of types of foundational work is my own; at least I have not seen such a side-by-side breakdown elsewhere. The borderlines and application to specific cases are not always clear-cut, and a more refined analysis may be warranted.

[5]I must apologize for the rough-and-ready way in which some of this work is described; to keep the presentation manageable I have had to make some deliberate simplifications.

it such a distinctive body of thought? My ideas about the logical analysis of everyday mathematical experience have previously been advanced, in part, in my articles (1979b) and (1981) [reproduced in this volume as chapters 9 and 3, respectively] but I had hoped to enlarge upon them here. As the present paper [reproduced in this chapter] grew in length and my time to prepare it ran out, I decided to postpone doing that to another occasion. Instead, some indication of the further direction of my ideas is given in a postscript in section 8.

Now, without further ado, let us launch into this survey of types and examples of foundational work.

2 Conceptual Clarification

Conceptual clarification usually takes place at a reasonably advanced stage in the organization of a subject (cf. section 6). After basic concepts have been settled on and are well understood, one looks for precise definitions of other frequently used informal concepts in terms of them. In certain cases it can be shown that only one such definition is possible meeting specified *adequacy conditions.* In other cases, the criteria for accepting an exact definition of an informal concept are subtler and involve a whole range of experience to be accounted for.

Some classical examples from mathematical analysis are the definitions of *limit, derivative, continuous function*, and *area.* Early in this century the concept of *dimension* occupied much attention in topology. More recently, the ideas of *natural mapping* and *universal construction* in algebra and topology have been defined in terms of category theory.

The most outstanding examples from logic are the definitions of *truth* and *satisfaction* (Tarski) and of *mechanically computable function* (Turing). The former also gives precise meaning to the concepts of *definability* and *logical consequence*, each with reference to a suitably specified *formal language.* The notion itself has at its core the idea of *well-formed formula*, that is, a syntactically determinate expression of a property, relation, or statement.

In the 1950s the concept of truth in a language L was used to explain the informal idea of a *transfer principle* from algebra, of which *Lefschetz' principle* in algebraic geometry was the most famous example. Results stated in L can be transferred from one structure $\mathcal{M}$ to another structure $\mathcal{M}'$ if $\mathcal{M} \equiv_{\mathrm{L}} \mathcal{M}'$, that is, if $\mathcal{M}$ and $\mathcal{M}'$ satisfy exactly the same statements from L. For the strongest present formulations of Lefschetz' principle in these terms, see Feferman (1972) and Eklof (1973). A variety of interesting transfer principles in algebra are given in Cherlin (1976).

Nowadays we think of the notions of limit, continuous function, etc., as being given relative to a topological space. Similarly, we have emphasized that the semantical notions just discussed are relative to a given language (and, implicitly, a given class of subject structures). On the other hand, the concept of mechanically computable function seems to have an absolute character. Such a function is supposed to be computed by a *finite algorithmic procedure* (f.a.p.) operating on *finitely presented data.* However, there is an informal relative concept of *finite effective construction* (for example, in geometry or algebra) which does not involve any special assumption about the form in which data is presented. One plausible explanation of f.a.p. in general relative terms, that is, applying to any structure, has been given by Friedman (1971). Less clear are various ideas about *infinitary constructions* which have been explored in a number of generalizations of recursion theory; for one approach and relevant literature see Fenstad (1980).

A ubiquitous mathematical and logical concept is that of *inductively generated set.* The set-theoretic definition of this in terms of *closure operators*, which has now become standard, is reviewed in Aczel (1977); its use also makes clear the ideas of buildup of the set *from below* and definition *from above* ("the least set closed under . . ."). In terms of these notions one can go on to distinguish different kinds of inductive definitions: *finitary, effective, deterministic*, etc. There is a *functional* approach to ordinary recursion theory which uses *inductive schemata*; this can be generalized to much wider contexts, as is advanced in Feferman (1977a [and 1992, 1996]) and Moschovakis (1977 and 1991). An alternative relational approach uses *effectively inductively generated relations*; for generalization of this see (the same references as well as) Fitting (1981). Finally, a notion of *inductively presented formal system* has been advanced in Feferman (1982), improved and extended in Feferman (1989); this is needed particularly for a proper general treatment of Gödel's second incompleteness theorem, where the *canonical* construction of *formal consistency statements* has problematic aspects.

There are several important informal concepts in logical work for which one has not yet arrived at any reasonably convincing precise explanation, though there are technical concepts which serve as partial substitutes. The following are two such examples. First is the idea of *identity of proofs*, which is actually used informally by mathematicians all the time, for example, when they speak of two proofs being essentially the same or of giving a new proof of a known result. Relevant notions from the general theory of proofs are explained in Prawitz (1971).[6] The second is the idea of *natural well-ordering*, of which the most familiar example is the ordering of type ε_0 provided by Cantor's normal form. Ever since Gentzen used this to give a consistency proof of arithmetic, technical proof theory has been dominated by the use of effective transfinite methods and arguments. A precise definition of the concept of natural well-ordering is needed to explain what is accomplished by such consistency proofs (since one can always give trivial demonstrations of consistency by induction on suitable "cooked up"

[6]See my review in *J. Symbolic Logic* 40 (1975), 232–234.

recursive well-orderings). Approaches to this question through category theory can be found in Feferman (1968) and Girard (1981).

3 Interpretation of Problematic Concepts and Principles

At each stage in the development of mathematics there is a body of ideas and methods which are considered to be well understood and accepted (though their place in our understanding may subsequently undergo considerable revision), with which mathematics in the main at that stage is confidently carried on. However, concepts and principles often emerge which have some intrinsic plausibility, or have proved themselves to be particularly useful, and yet which are felt to be less certain or secure than what is then commonly accepted. There are different kinds of foundational moves which are taken to deal with such situations. In this section and the next we separate out two of the main kinds of moves.

Basically this section has to do with cases of troublesome notions or propositions whose use is justified or in which one's confidence is increased by means of interpretations in terms of the body of mathematics accepted at the time. Classical examples of such notions from mathematics are *zero, negative numbers, imaginary and complex numbers, points at infinity;* examples of such statements are the *parallel postulate* (and its negation) and, more recently, the *Well-Ordering Principle.* Nowadays, we do not view such problematic cases in isolation, but rather as part of an axiomatic system of requirements which must be met as a whole. For example, we say that the use of complex numbers is justified by explicit construction of a field **C** extending the field of real numbers by adjunction of a root i of the equation $x^2 = -1 : \mathbf{C} = \mathbf{R}[i]$. But for the original interpretations (by Wessel, Gauss, Argand, and others) which secured the use of complex numbers, the conditions to be met were not explicitly formulated. The exception to this observation, of course, is the case of geometry, with its long-standing Euclidean axiomatic setting. The use of axiomatic organization for other parts of mathematics only began to receive conscious attention in the latter part of the nineteenth century, along with a critical reexamination of axiomatic geometry itself (see section 6).

We turn now to examples from logic; the best-known cases are the *models* given by Gödel (1939, 1940)[7] for the *Axiom of Choice*, AC, as well as the *Continuum Hypothesis*, CH (and its generalization GCH), and by Cohen (1963, 1964, 1966) for their *negations.* Here the usual axiomatic framework taken as basic is ZF, the system of Zermelo-Fraenkel Set Theory, and the results are frequently formulated as *consistency* and *independence*

[7]These papers have been reprinted in Gödel (1990) with an introductory note by R. M. Solovay. All of Gödel's published work has been reprinted, with translations and introductory notes, in Gödel (1986, 1990).

results, for example, that (i) ZF + AC + GCH is consistent, (ii) AC is independent of ZF, and (iii) CH is independent of ZF + AC. Further, analysis of the arguments shows that only the principles of ZF need to be assumed in constructing these models, so one can formulate the conclusions here as *relative consistency results*: if ZF is consistent, then (i)′ so also is ZF + AC + GCH, and (ii)′ so also is ZF + ¬AC, etc. Finally, one can look at these models still more formally as giving a *syntactic interpretation* of ZF + AC + GCH in ZF, etc. The methods of Gödel and Cohen have led to a welter of model constructions and thence relative consistency and independence results for a mass of problematic set-theoretical statements.

Less familiar are the uses to which such models can be put. Here the best way to formulate the matter is in terms of the logical notion of one (axiomatic) theory T being a *conservative extension* of another such theory S for a class of (well-formed) formulas $\mathcal{F}$. This is said to hold if (i) every formula of $\mathcal{F}$ is in the language L_S of S and $L_S \subseteq L_T$; (ii) every theorem of S is a theorem of T; and (iii) every theorem of T which lies in the class $\mathcal{F}$ is already a theorem of S. This notion will be applied frequently in the following.

It was observed by Kreisel some time ago that ZF + AC + GCH is a conservative extension of ZF for the class $\mathcal{F}$ of formulas of elementary number theory (as expressed in ZF). The reason is that Gödel's constructible-sets model for ZF + AC + GCH is *standard* for **N** (but not for sets in general). This conservation result turned out to be applicable to the existence of a decision procedure for first-order statements in p-adic fields. That was established by Ax and Kochen (1965) using model-theoretic methods which can be formalized in ZF + AC + GCH. But the existence of a decision procedure for p-adic fields can be expressed in the given class $\mathcal{F}$ of number-theoretic formulas. Hence by Kreisel's observation, it can already be proved in ZF. Later, Cohen (1969) actually produced an explicit decision procedure for p-adic fields, provably in a weak subtheory of ZF.

We turn now to less familiar examples of interpretations in logic. First are models of the *λ-calculus* produced by Scott (1972) and (following Plotkin) Scott (1976).[8] The λ-calculus is a formalism in which the universe of discourse is conceived to consist of functions defined on the whole universe. The basic operation is that of *application*, written fx (or $f(x)$). There are several basic principles, but the central one is that any well-formed expression (term) $\tau[x]$ in this symbolism containing a free variable x serves to define a function $f = \lambda x.\tau[x]$ whose values are given by $fx = \tau[x]$. The main feature of this formalism which makes it problematic is the possibility of *self-application*, that is, formation of ff. This is not possible for any familiar universe of functions, where the elements of the domain of a function are "prior" to (or of a "lower type" than) the function itself. A form of functional self-application had been met in recursion theory, by using

[8] See also Barendregt (1977) or (1984) on the λ-calculus.

numerical codes f for mechanical algorithms A_f; then fx is defined to be the value of A_f at input x. However, fx need not always be defined in that case; in particular ff might not be defined for a given f. The difference is expressed by saying that the functions in the λ-calculus are supposed to be *total*, while those in recursion theory may be *partial.* Scott's models for the λ-calculus have been particularly useful in providing (so-called *denotational) semantics for programming languages*, which serve to give each element of a program a meaning as a mathematical object, rather than as a procedure; see Stoy (1979) and Mosses (1990).[9]

Different kinds of *self-reference* or *self-membership* appear in ordinary language, in the theory of classes and in algebra. The Liar Paradox and Russell's Paradox show the problems that can arise in the first two cases with *negative* uses of self-application. But is the statement *this statement is true* problematic? And are the notions *class of all classes* and *category of all categories* problematic? It turns out that one can develop formal theories in which these and other such notions are represented and which are conservative extensions of known consistent theories. For a survey of work in this direction see Feferman (1984). An explicit aim there is to develop theories which are mathematically useful as a result of their greater flexibility of expression. See also Martin (1984) for a representative collection of different approaches to the area of self-referential theories of truth and the problem of the Liar Paradox, as well as the expository survey paper by Visser (1989). Feferman (1991) deals further with formal theories for such notions of truth, though its main concern falls under section 7. A workable modification of ordinary set theory allowing self-membership has been developed by Aczel (1988). This has been used by Barwise and Etchemendy (1987) for another approach to the Liar Paradox.

The question of self-membership also arose in connection with an interesting but problematic extension of typed and untyped λ-calculi which emerged in the theory of programming languages in recent years, involving variable types and the notion of *polymorphism.* This borders on having a *type of all types*, and is prima facie *impredicative* (cf. below for the latter idea). In Feferman (1990) and subsequent publications, I have taken an approach to these via axiomatic theories of operations and classes (considered as variable types). It was argued there that a predicatively reducible subtheory accounts for all applications of polymorphism met in practice.

Another interesting example of a problematic mathematical notion is Brouwer's concept of *(free) choice sequence*, in his *intuitionistic redevelopment of mathematics.* The idea is that a choice sequence $\alpha = (\alpha_0, \alpha_1, \dots, \alpha_n, \dots)$ (say, of natural numbers) may be generated without any known law, for example, by some random phenomenon or an arbitrary sequence of choices by an individual. The mathematician working with such a sequence

[9] Mosses (1990) provides an up-to-date survey with a comprehensive list of references. [The texts Gunter (1992) and Winskel (1993) provide still more recent expositions of denotational semantics for programming languages.]

will have only a finite amount of information $\bar{\alpha}(n) = (\alpha_0, \ldots, \alpha_{n-1})$ available about it at any given time. The basic principle for choice sequences used by Brouwer is that if $P(\alpha, m)$ is a property of choice sequences α and numbers m such that $\forall\alpha\exists m P(\alpha, m)$[10] then $\forall\alpha\exists n, m\forall\beta[\bar{\beta}(n) = \bar{\alpha}(n) \rightarrow P(\beta, m)]$. From this and some other (more readily acceptable) principles, Brouwer was able to prove that every function of real numbers is continuous. (Real numbers are represented as choice sequences of rationals satisfying Cauchy's convergence condition.) Since this contradicts the classical statement of existence of discontinuous functions, there must be an escape hatch for Brouwer's principles to be coherent. This is provided by the *elimination* of the *Law of Excluded Middle* as a basic logical principle, so that, for example, $\forall\alpha[\forall n(\alpha_n = 0) \vee \exists n(\alpha_n \neq 0)]$ is not accepted. See Troelstra and van Dalen (1988) for an introduction to intuitionistic theories of constructive mathematics (among others), and in particular volume 2, chapter 12 (*op. cit.*) concerning various approaches to the problematic notion of choice sequences.

In particular, Kleene and Vesley (1965) formulated Brouwer's ideas about choice sequences in an axiomatic theory in intuitionistic logic (denoted here) CS, and proved that CS is consistent by giving what is called a *realizability interpretation* for it. This, then, can be used to justify Brouwer's redevelopment of analysis, which as we have noted is prima facie contradictory with classical analysis. See Troelstra and van Dalen (1988) for an introduction to a variety of forms and applications of the realizability method. (For another way of handling CS, see section 4.)

Brouwer's and other approaches to constructive mathematics suggest consideration of a number of principles which appear plausible in a constructive framework though they contradict classical statements. For example, what is called *Church's thesis* in this context is some form or other of the statement that every constructive number-theoretic function is recursive (that is, is mechanically computable). Since for the constructivist, $\forall n\exists m P(n, m)$ means that m can be found as a (constructive) function of n to satisfy $P(n, m)$, the thesis can be expressed as a scheme CT: $\forall n\exists m P(n, m) \rightarrow \exists e\forall n P(n, \{e\}(n))$, where $\{e\}$ denotes the recursive function with (Gödel-) number e. The method of realizability, originally developed by Kleene, can be used to show consistency of CT with practically all intuitionistically accepted systems; see Troelstra (1977) or Troelstra and van Dalen (1988) for an introduction to this and other such results. Loosely speaking, one significance of the consistency of CT is that theorems proved with its use have in principle an effective mechanizable content.

If one is only willing to accept the constructivist philosophy of mathematics, various principles of *classical* mathematics become problematic in the light of that position. Obviously among such is the *Well-Ordering*

[10]In the following, we use the logical notation $\forall, \exists, \neg, \wedge, \vee, \rightarrow, \leftrightarrow$ for "for all," "there exists," "not," "and," "or," "implies," and "if and only if," respectively.

Principle, but even some more mundane principles like *extensionality* and the *Power-Set Axiom* (existence of the set of all subsets of any set) must be reexamined. For a survey of work on these and other questions that arise from systematic investigations of the foundations of constructive mathematics, see Beeson (1982).

Even within classical mathematics there are different philosophies which accept parts but not all of what is represented in current systems of set theory. For example, the French "semi-intuitionistic" school of Baire, Borel, and Lebesgue accepted transfinitely iterated countable constructions, and (implicitly) the countable Axiom of Choice, but not the full Axiom of Choice; for a good description of their ideas see Moore (1982, section 2.3 and appendix 1). Even more restrictive were Poincaré's views, which influenced Weyl to see how far analysis could be developed using only purely arithmetical definitions (Weyl 1918). (A translation into English of Weyl's 1918 monograph has appeared in Weyl (1987); Feferman (1988a) [reproduced as chapter 13 in this volume] provides a critical exegesis of that work, followed by an explanation of modern developments which constitute a realization of Weyl's program.)

Poincaré had formally rejected *impredicative definition of sets*, which picked out a subset of a given set A by reference to the supposed completed totality of all subsets of A (that is, the power set of A). The simplest example of an impredicative definition of a set is that as the intersection of all sets X satisfying $\Gamma(X) \subseteq X$ where Γ is a set operator; this is used with monotone Γ in the general approach to inductive definitions from above. Feferman (1970, 1982a) shows how to interpret axiomatic theories for the sets successively generated by monotone inductive operators, into corresponding theories for sets explicitly generated from below. For a survey of work on reductions of theories of inductive definitions "from above" to theories of sets inductively generated "from below," see Feferman (1988) [cf. also chapter 10 in this volume]. For work on more strictly predicative principles, see section 7.

We conclude this section with a famous open problem in logic, namely, whether Quine's system NF of *New Foundations* is consistent. This system can be thought of as the image of the Simple Theory of Types obtained by erasing all type distinctions between variables. It permits construction of a universal set V with $V \in V$. Some efforts were made in the past to see whether such features make NF mathematically useful, but it has not been paid much attention in recent years. Jensen (1969) produced an interesting model of NF $-$ Ext, that is, of the system obtained from NF by deleting the axiom of extensionality. However, this has so far given no clue as to how to establish consistency of the full system NF. At present the interest in this problem seems principally to be as a logical challenge.

4 Replacement or Elimination of Problematic Concepts and Principles

The general situation is the same as that brought out at the beginning of the preceding section. Only the foundational moves described here are of a different character. Instead of trying to interpret the problematic concepts or principles (by some kind of model), one tries to find *substitutes* for them which do the same work, or even tries to *eliminate* them entirely. In the latter case the aim is to save the baby from being thrown out with the bath water, that is, to preserve the useful consequences. In modern logic this is accomplished by *conservation results established by syntactic transformations.*

The classical example of a concept which was eliminated while saving its applications is that of *infinitesimal*, when the associated notions of *limit*, *derivative*, etc., which had been explained in terms of it were given substitute "ε, δ" definitions. (For modern theories restoring infinitesimals as mathematical objects, see section 5.) Another example of substitution was for *multiple-valued functions* in complex analysis, which were replaced by *single-valued functions on Riemann surfaces.*

More recent examples tying up with logic are the treatments of the notions: the *class of all classes* and the *category of all categories*, in the context of the formal theory BG (Bernays-Gödel) of *sets* and *classes*; these are replaced, respectively, by the *class of all sets* and the *category of all small categories.* MacLane (1961, 1971) used the distinction between set and class in BG to distinguish between *small* and *large categories*, which led to acceptable substitutes for otherwise problematic general theorems in the subject such as *Yoneda's lemma*, the *adjoint functor theorem*, etc. Another solution to this problem situation had been given in the framework of ZF by the use of *Grothendieck universes.* However, that required the existence of many inaccessible cardinals. The *reflection principle* in ZF was applied in Feferman (1969) to avoid such assumptions; in addition a conservation result was given there which shows how "smallness" can be formally introduced in ZF to facilitate the application to category theory, and then eliminated.

A *syntactic translation* was used by Kreisel and Troelstra (1970) to *eliminate* (Brouwer's notion and principles for) *choice sequences.* This gives conservation of a theory CS over a constructive theory without choice sequences; it also strengthens the results of Kleene and Vesley mentioned in Section 3. The use of syntactic translations in the metatheory of constructive systems and related classical systems goes back some years. In 1933 Gödel and Gentzen independently found the so-called *negative translation* of classical (Peano's) Arithmetic PA into intuitionistic (Heyting's) Arithmetic HA.[11] Since then the same translation has been applied to much

[11] A related translation had first been found for the propositional calculus in 1925,

stronger classical systems; see Troelstra (1977) [or Troelstra and van Dalen (1988)] for references.

Another interesting syntactic transformation was introduced by Gödel (1958) to *eliminate quantified statements* in HA in favor of *quantifier-free* (open) statements about certain *constructive functionals of finite type.* An English version of Gödel (1958) with supplementary notes was worked on by Gödel for publication but never appeared in print. Its final form (as found in his *Nachlass*, and dated 1972) is reproduced in Gödel (1990, pp. 271–280). For a critical examination of this work and survey of further applications of the *Dialectica* functional interpretation, see the introductory note by Troelstra to *1958* and *1972* in Gödel (1990, pp. 271–241).* Gödel's functional interpretation was extended to analysis (second-order arithmetic) in a series of works by Kreisel, Spector, Howard, Girard, and others. In combination with the negative translation, this serves to reduce certain subsystems of classical analysis to constructively accepted quantifier-free systems (see Troelstra 1977, section 11 [and chapter 11 in this volume] for a survey of this work).

Still another relatively early syntactic method of eliminating quantified statements, which had been considered to be problematic in Hilbert's *finitist program*, was that originating with Herbrand in his 1930 thesis. In its simplest form, Herbrand's theorem tells us that if an existential statement $\exists x A(x)$, with A quantifier-free, is provable (=valid) in classical logic, then for some finite sequence of terms $t_1, \ldots, t_n$ we have $A(t_1) \vee \ldots \vee A(t_n)$ provable; moreover, these terms can be found primitive recursively from any given proof of the statement. This has had interesting mathematical applications, for example, to the extraction of primitive recursive bounds from Artin's (prima facie noneffective) solution to Hilbert's Seventeenth Problem; for this and other applications of Herbrand's theorem see Kreisel (1958) [and the articles of Delzell (1996), Feferman (1996a), and Luckhardt (1996)].

A general and very important syntactic method for eliminating statements of complexity higher than that of the conclusion from logical proofs was discovered by Gentzen in 1934 (see his collected works, Gentzen 1969). This is called *cut-elimination* when applied to one kind of formalism introduced by Gentzen (the sequential systems) and *normalization* when applied to another (natural deduction systems). The method of cut-elimination was subsequently extended to classical arithmetic by Gentzen. Modern applications of this method have extended it much further to give constructive elimination proofs for various subsystems of analysis; see Takeuti (1975) and Schütte (1977).[12] Among other things, that work serves to give *precise*

by Kolmogorov.

*[Cf. also chapter 11 in this volume.]

[12]See also my reviews of these books in *Bull. Amer. Math. Soc.* **83** (1977), 351–361, and *Bull. (New Series) Amer. Math. Soc.* **1** (1979), 224–228, respectively. See Takeuti (1987) for the second edition of his 1975 book, which also contains appendixes

ordinal measures to the *effective content* of classical theories in which much of analysis and descriptive set theory can be formalized. The monograph by Buchholz and Schütte (1988) and the survey by Pohlers (1991) provide systematic expositions of (prima facie) uncountably infinitary extensions of Gentzen's cut-elimination procedure, used to establish the ordinal strengths of various systems of analysis and set theory (see also Pohlers (1989) for an introduction).

The results of Jäger and Pohlers (1982) obtained in this direction are noteworthy. Their work serves to reduce the classical theory Σ^1_2-AC + BI to the constructive theory T_0 of functions and classes introduced in Feferman (1979) as well as determine its proof-theoretic ordinal. An easier reduction had previously been established for Σ^1_2-AC to a restricted form of T_0 by Feferman and Sieg in chapter 2 of Buchholz et al. (1981). That volume also presents an arsenal of proof-theoretic methods and results, reducing various classical subsystems of analysis to constructive systems of iterated inductive definitions. (This connects with the work on elimination of impredicative principles described in section 3 above.) The results obtained by Rathjen (1991) are for still stronger systems than those of Jäger and Pohlers; these make use of effective ordinal notation systems incorporating Mahlo cardinals [cf., even further, Rathjen 1995].

5 Foundations of Problematic Methods and Results

The foundational questions in sections 3 and 4 dealt with troublesome but relatively basic ideas or propositions. In this section the focus of concern is on results and methods of some substantial mathematical content but whose application is uncertain. Though more advanced mathematically, such usually appear early in the development of a subject. The problem, often soon recognized, is to provide them with a rigorous foundation. That takes the form of finding precise *sufficient* conditions under which the methods are applicable or for which the results hold. Since the conditions obtained are rarely *necessary*, there is also the constant possibility of improvement. Thus this type of foundation of a subject usually proceeds over a period of time by successive steps of sharpening, extension, and generalization. There is some overlap here with the foundational moves in section 4, since one frequently needs precise concepts to substitute for informal or intuitive notions. For example, in current foundations of *probability theory* the idea of *random variable* is replaced by *measurable real-valued function.*

Mathematics offers us a wealth of examples of this type of foundational situation and treatment. To mention but a few, we have the foundations of *Fourier series, calculus of variations, Dirichlet's principle, Stokes' theorem,*

by Kreisel, Pohlers, Simpson, and myself on further applications and extensions of the method of cut-elimination, as well as other approaches to proof theory. See Sieg (1991) for applications of Gentzen's method combined with Herbrand analyses.

Jordan curve theorem, generalized functions, Descartes–Euler theorem, probability theory, and *algebraic geometry*. When neighboring troublesome subjects have an isolable mathematical part, one sees similar moves as in the *mathematical foundations of quantum mechanics*, etc.

So far, there are many fewer examples of this kind of logic. Two such are quite well known, the first being the *foundations of Cantor's theory of transfinite ordinals and cardinals* as currently given in axiomatic set theory ZFC(= ZF + AC) or in BG + AC. The second is the *foundations of infinitesimal analysis* by Robinson (1966). (See also Stroyan (1977) for an introduction and further references.)

Examples of much lesser scope but still requiring attention were (i) *Gödel's theorem on the unprovability of consistency* (second incompleteness theorem); (ii) *Herbrand's theorem*; and (iii) the *Church-Rosser theorem*. In the case of (i), the first fully worked out proof was given for the case of arithmetic in Hilbert and Bernays (1939). I first gave a precise general form of the theorem in Feferman (1960), and revisited it to give an improved generalization in Feferman (1982) or better (1989). The problem with (ii) is that Herbrand's own proof contained subtle and not easily correctable errors, though nonconstructive proofs (using the completeness theorem for logic) were easy. A constructive reconstruction of Herbrand's work was finally managed properly in Dreben and Denton (1966). As to the third example, the Church-Rosser theorem is fundamental to the λ-calculus and related combinatory calculi; according to it, normal forms of terms are unique when they exist. Early proofs of this were exceptionally difficult and a number of attempts to simplify them and to extend the result to other calculi were bedeviled by errors. A simple, manageable, and generalized proof was finally given by Tait and Martin-Löf; see Barendregt (1984).

6 Organizational Foundations and Axiomatization

This is one of the best-known aspects of foundational work, which is accordingly much overemphasized. It comes at a relatively advanced stage in the development of a subject, when that is ready for *consolidation* and *systematic exposition*. Indeed, it is often undertaken as part of the *pedogogical enterprise*. While this takes considerable experience, skill, and effort, it is hard to get excited about finished work of this kind, certainly in comparison to the top work described in sections 2–5. Excitement comes when one is actively engaged in such an enterprise and especially when there are competing schemes in the offing; current examples are given below.

This type of foundational work involves a *choice and ordering of concepts and results*, with basic results being taken as *axioms*. Usually there is also involved a systematization of *methods of proof*, though that is rarely made explicit outside of logical work. In addition there is overlap with the foundational moves described in the preceding sections 2–5.

The classical example of axiomatic organization is of course provided by *Euclidean geometry*. Though long considered a model for rational thought, the value of axiomatization was not recognized and pursued again until the nineteenth century. No doubt the onset of this had to do with the growing professionalization of mathematics and the institution of organized higher mathematical education for engineers and scientists. Besides Hilbert's reexamination of the axiomatic foundations of geometry, modern mathematics provides a wealth of examples of axiomatic organization. Just consider, in algebra—*groups, rings, fields, vector spaces*, etc.; in analysis—*topological spaces, metric spaces, Banach spaces, Hilbert spaces, measure spaces*, etc. One can go on in field after field, for example, in topology, algebraic geometry, etc.

The work in the nineteenth century on the *axiomatic characterization of the basic number systems* **N**, **Q**, **R**, **C** by Méray, Weierstrass, Dedekind, Cantor, Peano, Frege, and others forms a bridge to twentieth-century work on the logical approach to the foundations of mathematics. Here one has a shift away from axiomatics of traditional subject areas (such as algebra, geometry, analysis) to the study of principles which underlie *all* areas. The best-known example of that is the development of *axiomatic set theory* at the hands of Zermelo, Skolem, Fraenkel, von Neumann, Bernays, Gödel, and others. Relative to the resulting systems (such as ZFC or BGC) there has been much work on principles whose status and justification are less certain, for example, those called *large cardinal axioms*; see Kanamori and Magidor (1978) for a recent survey [and Kanamori 1994].

Axiomatic set theory is the current expression of the platonistic philosophy of mathematics, and there is considerable agreement about its core from those who accept that position. *The axiomatic foundation of constructive mathematics* is not nearly so settled, and is still being actively pursued. One reason is that the body of informal constructive practice in mathematics to be accounted for is much smaller than the body of everyday nonconstructive mathematics accounted for in set theory. Another reason is that the constructive approach in this century was long dominated by Brouwer's ideas, which he propounded in startling and in some cases mystifying ways (see Brouwer's *Collected Works*, 1975). Since then other, more understandable approaches within the constructivist frame have been developed, notably those due to Markov and the Russian school on the one hand (see Aberth 1980) and to Bishop (1967) and his followers on the other hand. However, the various differences in these approaches have not led to a common accepted axiomatization.

As a result of work by Kleene, Vesley, Kreisel, Troelstra, and others already referred to in sections 3 and 4, Brouwer's *constructive* redevelopment of mathematics is much better understood now and has to a good extent been tamed axiomatically. The *axiomatic representation of practice in the Russian school of constructivity* is also well in hand; basically, constructive functions are there identified with recursive functions and so *Church's thesis* is accepted as an axiomatic principle. What makes this approach distinct from so-called "recursive mathematics" is the restriction of argumentation to that which can be carried out in intuitionistic logic plus a special axiom called *Markov's principle.* For some information on the formal systems associated with this school, see Troelstra (1977), Beeson (1981, 1985), and Troelstra and van Dalen (1988).

The development of constructive mathematics along the lines set out by Bishop was continued for analysis in Bishop and Cheng (1972) and Bridges (1979). The basic work of Bishop in analysis was itself updated in Bishop and Bridges (1985). Mines, Richman, and Ruitenberg (1988) provide an exposition of constructive algebra following Bishop's approach. Two, very different, competing ways have been proposed for the *axiomatic representation of practice in Bishop's approach to constructive mathematics.* One, due to Myhill (1975), provides a subsystem CST ("constructive set theory") of IZF (= intuitionistic ZF). The second, due to Feferman (1975), provides a system T_0 of functions and classes in a new specially designed formalism. These systems are compared in Feferman (1979), which also elaborates and extends my earlier work. The kind of competing proposals mentioned and the various considerations which go into axiomatization provide an excellent case study in such foundational work.

Independently of the preceding, Martin-Löf has long been developing his own distinctive viewpoint about constructive mathematics. It has been given axiomatic formulation in his *constructive theories of types*; see Martin-Löf (1982, 1984). Though not motivated by the question of axiomatizing Bishop's approach, Martin-Löf's systems (here denoted ML) offer still another competing candidate. Beeson (1985) contains a detailed comparative study of the systems CST, T_0, ML, and interesting extensions and subsystems (as well as the already mentioned axiomatics of the Russian school).

It is quite common, once axiomatic foundations for a given subject are fairly settled, to undertake a *fine analysis* of the role of different axioms or groups of axioms in different parts of practice. We mention here only a few examples of this type of work. Friedman was engaged for some years in tying down the *exact logical strength* of various results from mathematical analysis, algebra, etc.; see Friedman (1975, 1976). This program has been continued by Friedman et al. (1983) and Simpson (1984). A survey of the work by Friedman, Simpson, and others in the "reverse mathematics" program is given in Simpson (1987). What is surprising is how much everyday mathematics can be done in formal systems which are conservative over arithmetic—PA or HA, according to whether one is looking at nonconstructive or constructive practice. A variety of such systems conservative over PA and explanation of what mathematics can be done in them have been given by Takeuti (1978) (written somewhat earlier), Feferman (1977), and Friedman (1980). See the latter part of Feferman (1988a) [reproduced as chapter 13 in this volume] for further work on this [cf. also chapter 14]. On the constructive side, systems conservative over HA have been given by Friedman (1977) and Feferman (1979) (by the results of Beeson 1980).

One direction of current research is in the role of very weak subsystems of arithmetic; see Simpson (1988). In the opposite direction, Paris and Harrington (1977) opened up the way to show the unexpected *strength of various finite combinatorial partition principles*; see, among others, Friedman et al. (1982). Simpson (1987a) gives a good introductory survey of this work. For a critical discussion of the foundational significance of these results, see Feferman (1987a) [the relevant part is reproduced as chapter 12 in this volume].

7 Reflective Expansion of Concepts and Principles

Mathematical concepts are not fixed once and for all. Every so often a new one appears, and if it establishes itself by its intuitive appeal and great utility, or even necessity, it must be accounted for somehow in mathematics. Sections 2–4 described several ways of doing this. One may try to clarify the concept, defining it precisely in previously understood terms, or may try to interpret it (in those terms) in a way that preserves its expected properties. Finally, one may try to replace it or eliminate it altogether. However, after all this, there are a *few* concepts left over which are *genuinely new at each stage*. Examples of such concepts are those of *natural number*, of *point, line*, and *plane*, of *ordered pair*, of *infinite sequence*, of *function*, and of *set*. The first two arose in the prehistory of mathematics, perhaps with the very emergence of pure mathematics as a separate intellectual discipline. It may appear that a concept is genuinely new at a certain stage and irreducible or uneliminable at that stage, but later reconsideration allows us to shift its place in our conceptual scheme of things. Such was the case with the concepts of *infinitesimal* and of *choice sequence*. Nowadays all the above-mentioned examples are reduced to the concept of set in the set-theoretic scheme of things. But this is not the case in (some) constructive conceptual schemes; for example, the idea of function as given by a *rule* is not reducible to the idea of set as given by a *defining property*, nor is the converse reduction possible. For a discussion of such differences in the two schemes, see Feferman (1979). Be that as it may, if a concept appears genuinely new and forces itself on us, the only way to deal with it is to accept it as a kind of *given* and build around it, all the while trying to sharpen our understanding of it as well as possible. Basically, that falls under axiomatic organization which was just discussed in section 6.

All of the preceding is fairly well recognized in one way or another. The purpose of this section is to concentrate on a different kind of conceptual innovation which frequently takes place, and which I propose to call *reflective expansion*. Basically, this is a form of *generalization*, but of the following particular character. At a certain point one reflects on what has led one to accept work with certain concepts, and sees that a much more general concept is *implicit* in accepting that. For example, Euclidean geometry of the plane and space were eventually explained in terms of $\mathbf{R}^2$ ($= \mathbf{R} \times \mathbf{R}$) and

$\mathbf{R}^3$ ($= \mathbf{R} \times \mathbf{R} \times \mathbf{R}$) by using the real number system $\mathbf{R}$ and the pairing and tripling operations. Reflection on that step led to the concepts of *n-tuple* and *n-dimensional space* $\mathbf{R}^n$. Reflection on the formation of derivatives and integrals as operations on functions led to the idea of *function operators* or *functionals*. Reflection on the idea of the enumeration of the natural numbers in order, where at each stage there is a first element beyond all the previous ones, led to the idea of *ordinal number*; here the novelty lay in passing to the first number ω beyond all the natural numbers. Reflection on the concept of limits of sequences of continuous functions (which may be discontinuous) led one to the idea of iteration of such limits, that is, to the *Baire hierarchy of functions*; a similar process of reflection led about the same time to the *Borel hierarchy of sets*.

All of the foregoing applies *mutatis mutandis* to *principles* in place of *concepts*. An elementary example is provided by the basic principles for pairs, $(a_1, a_2) = (b_1, b_2) \rightarrow a_1 = b_1 \wedge a_2 = b_2$, which is generalized to the corresponding principle for n-tuples. The Cauchy convergence principle for $\mathbf{R}$ is generalized to *completeness* for $\mathbf{R}^n$. Reflecting on the principle of induction for natural numbers leads one to accept the *principle of transfinite induction for ordinals*, and so on.

Note that what is introduced by reflective expansion is rarely genuinely new: the concept of n-tuple may be defined in terms of that of pair, the concept of ordinal number in terms of well-ordered set, the principle of completeness of $\mathbf{R}^n$ can be deduced from completeness for $\mathbf{R}$, etc.

From a logical point of view, our interest here is in whether we can make theoretical sense of *describing all the concepts and principles that one ought to accept if one has accepted given concepts and principles*, or, put more succinctly, *describing all the concepts and principles implicit in given ones*; the general proposal to pursue such characterizations is due to Kreisel (1970). This followed his own work on characterizing *finitist* notions and principles, and the work of Schütte and myself on characterizing *predicative* notions and principles (the leading ideas in these cases had been suggested informally by Hilbert and Poincaré, respectively).

The study of *predicativity* has been carried out rather fully in a series of papers and thus offers a good case study for the general proposal. Here one wants to say just what is implicit in accepting the *structure of natural numbers* $\mathbf{N}$ together with the general *principle of induction* as given, and (in contrast to finitism) with *quantification over* $\mathbf{N}$ as a means of defining properties. The first approach to explaining this used *transfinite autonomous progressions of theories* T_α, where α runs through certain effectively given well-orderings. At each stage $\alpha, \mathrm{T}_\alpha$ incorporates principles obtained by reflecting on what has been accepted in the earlier stages; for example, the induction principle may be applied to properties defined by quantifying over previously determined enumerated collections of properties (technically this is a form of *ramified second-order number theory*). The *autonomy* or *boot-strap* condition allows one to pass to stage α when (and only when) the well-ordering property of the ordering giving α has been established at an earlier stage $\beta < \alpha$. The least nonautonomous ordinal Γ_0 has been characterized by Schütte and myself in terms of hierarchies of normal ordinal functions; see Feferman (1968a) or my appendix to Takeuti (1987) for a survey of this work.

The use of transfinite iteration of reflective processes is a prima facie unrealistic element in the modeling of actual reflective expansion. Thus in my work on predicativity since 1964 I developed alternative single formal systems which serve to characterize the same body of mathematical thought. One which (to my mind) most persuasively embodied the idea of the *reflective closure* of the concept of the structure $(\mathbf{N}, 0, Sc)$ with the principles of induction and definition by quantification over $\mathbf{N}$ is given in Feferman (1979a). Three ideas there are central:

(i) Reflection on what leads one to accept successively the definitions of $+$, $\cdot$, exp, etc., as determining total functions of $\mathbf{N}$ leads one to the definition of the sequence of primitive recursive functions and thence its enumeration; this can be generalized appropriately to recursively defining sequences of operations *and* of properties.

(ii) With any property $X(x, x_1, \ldots, x_n)$ is associated another $Y(x_1, \ldots, x_n) := (\forall x \in \mathbf{N})X(x, x_1, \ldots, x_n)$ obtained by quantifying universally over $\mathbf{N}$.

(iii) The *scheme* of induction rests on the recognition that if one has established a general proposition $A(X)$ of properties $X(x)$, one ought to accept each particular substitution instance B for X by a property $B(x)$ expressed in the system.

It is shown in Feferman (1979a) that a system based on these ideas has the same strength as the autonomous (ramified) progression $\bigcup \mathrm{T}_\alpha (\alpha < \Gamma_0)$ described above.

In 1978–79, I developed a general notion of *reflective closure of a schematic theory* S which for S = PA gives the same result (up to mutual interpretation) for predicativity as that just described.[13] This work finally appeared as Feferman (1991). The notion of reflective closure is formulated for and can be applied to much more general S than PA. In particular, it should be of interest to consider it relative to a system S of set theory, where specifically *set-theoretical reflection principles* have led one to consider stronger and stronger extensions of S; such extensions are currently based on so-called *large cardinal axioms* (see particularly the already mentioned survey article of Kanamori and Magidor 1978 [as well as Kanamori 1994]). The idea of reflective closure should serve to explain how much of this is implicit in the concepts and principles accepted in given S, and what then requires genuinely new considerations in order to

[13]This work was presented in a talk for a symposium on the work of Kurt Gödel at the meeting of the association for Symbolic Logic in San Diego, March 1979.

be accepted. That is a research program which is at present in progress [cf. Feferman 1996].

8 Postscript: Foundational Work and Philosophy of Mathematics

Most presentations of logical work of a foundational character center on one of the grand philosophical schemes, which are taken as their justification (at least implicitly), for example, set-theoretic foundations, constructive foundations, formalist (proof-theoretic) foundations, etc. Instead, current foundational activity has been analyzed here along the lines independent of any particular philosophical position. In each case, I have shown by examples that this is simply a continuation of various types of foundational activity carried on by mathematicians as a matter of course for the progress and improvement of mathematical understanding. Of course, the preoccupations which serve to prod this kind of activity are not, at any given time, universally shared. What one mathematician sees as a need for clarification, for improved organization, or to deal with a problematic concept (resp. principle) is not what may concern another mathematician. Even so, the results of such work in the past have eventually been absorbed into generally accepted mathematics.

It is undeniable that in various of the logical examples given above, the motivating concern *has* been relative to a particular foundational position, but little was needed in each case to appreciate the nature of that concern and the steps taken to meet it. To what extent the results of such work will take their place as part of generally accepted mathematics is a matter for the future to settle. This chapter will have served much of its purpose if the reader has at least been brought to see the *point* of the work arising out of such specific concerns, without trying to judge its eventual significance.

With all this stress on appreciating logical work independently of fixed foundational standpoints, I do not mean to reject interest in reaching a basic philosophical position as to the nature of mathematics. On the contrary, that is required in order to make the full case for the logical approach to foundations which I advanced in the introduction to this chapter. As explained there, elaboration of this will have to wait for another occasion. My only purpose here, in conclusion, is to give an indication of the direction these ideas have taken.

In answer to the question, *What is it about mathematics that makes it such a distinctive body of thought?* logic is clearly most successful in analyzing the underlying characteristics of its *language* and its *verificational methods* (*proofs*). In these respects, it is closer to an analysis of everyday logical experience than its critics realize (cf. Feferman (1979b, 1981) [reprinted here as chapters 9 and 3, respectively] for some arguments in support of this). But to give a full answer to the question just posed, one must also deal with the following more difficult question: *What is the nature of the conceptual content of mathematics*? I agree with the critics of the traditional positions of logicism, formalism, platonism, and constructivism, that each of these has failed to give us a satisfactory, convincing answer to that. In recent years I have been trying to develop and build a coherent case for an alternative and, to my mind, more satisfactory view. Roughly speaking, this comes to the following. I am in agreement with the constructivist position as to the subjective source of basic mathematical conceptions, but for me these are supposed to be conceptions of certain kinds of ideal worlds, including ones which are not countenanced constructively (such as "platonistic" worlds of sets). These worlds (or world-pictures of mathematical structures) are presented more or less directly to the imagination, from which basic principles are derived by examination. All else (in each picture) is obtained by rational reflection on, and from, basic concepts and principles. It may be that at the outset only relatively crude features of a world-picture can be discerned in this way. My main slogan here is that nevertheless, *for mathematics, a little bit goes a long way.*

Part III

GÖDEL

6

Gödel's Life and Work

Kurt Gödel's striking fundamental results in the decade 1929 through 1939 transformed mathematical logic and established him as the most important logician of the twentieth century. His work influenced practically all subsequent developments in the subject as well as all further thought about the foundations of mathematics.

The results that made Gödel famous are the completeness of first-order logic, the incompleteness of axiomatic systems containing number theory, and finally, the consistency of the Axiom of Choice and the Continuum Hypothesis with the other axioms of set theory. During the same decade he made other less dramatic but still significant contributions to logic, including the work on the decision problem, intuitionism, and notions of computability.

In 1940, Gödel emigrated from Austria to the United States, where he became established at the Institute for Advanced Study in Princeton. In the years following, he continued to grapple with difficult problems in set theory and at the same time began to think and write in depth about the philosophy of mathematics. Later in the 1940s he arrived at his unusual but less well-known contribution to relativistic cosmology, in which he produced solutions of Einstein's equations permitting "time travel" into the past. While Gödel's philosophical interests dominated his attention from 1950 to his death in 1978, enormous advances were made in the subject of mathematical logic during the same period. The Institute for Advanced Study became a focal point for much of this activity, largely because of his presence.

"Gödel's life and work" was first published as pp. 1–36 in Kurt Gödel, *Collected Works: Vol. I: Publications 1929–1936*, edited by Solomon Feferman et al., copyright ©1986 by Oxford University Press, New York, and is reprinted here by kind permission of the publisher. I have omitted here the sections pp. 16–28 from the original (Feferman 1986) which were devoted to a more detailed survey of Gödel's work than is found under "Life and career" in the following. However, the sections on his philosophy of mathematics and the significance and impact of his work have been retained.

A full biography of Gödel is to be found in Dawson (1996).

Gödel published comparatively little, but almost always to maximum effect; his papers are models of precision and incisive presentation. In his *Nachlass* [literary estate, presently housed at the Firestone Library, Princeton University] there are masses of detailed notes that he had made on a remarkable variety of topics in logic, mathematics, physics, philosophy, theology, and history. These have begun to reveal further the extraordinary scope and depth of his thought.

This account of Gödel's life and work is divided into three sections. The first is devoted to his life and career, and includes a description of his principle achievements. The second section concentrates on his philosophy of mathematics, and the third provides a summary assessment of the significance and impact of Gödel's work.[1,a]

Life and Career

> Kurtele, if I compare your lecture with the others, there is no comparison.
> —Adele Gödel[b]

In the end we search out the beginnings. Established, beyond comparison, as the most important logician of our times by his remarkable results of the 1930s, Kurt Gödel was also most unusual in the ways of his life and mind. Deeply private and reserved, he had a superb all-embracing rationality, which could descend to a maddening attention to detail in matters of everyday life. Physically, Gödel was slight of build and almost frail-looking. Cautious about food and fearful of illness, he had a constant preoccupation with his health to the point of hypochondria, yet mistrusted the advice of doctors when it was most needed. It was a familiar sight to see Gödel walking home from the Institute for Advanced Study, bundled up in a heavy black overcoat, even on warm days.

Genius will out, but how and why, and what serves to nurture it? What consonance is there with the personality, what determines the particular channels taken by the intellect and the distinctive character of what is achieved? As with any extraordinary thinker, the questions we would really like to see answered in tracing Gödel's life and career are the ones which prove to be the most elusive. What we arrive at instead is a mosaic of particularities from which some patterns clearly emerge, while the deeper ones must be left as matter for speculation, at least for the time being.

Kurt Friedrich Gödel was born 28 April 1906, the second son of Rudolf and Marianne (Handschuh) Gödel. His birthplace was Brünn, in the

---

[1] All documentation of sources is given by lettered notes at the end of the chapter. In particular, note *a* details my main sources of material and the variety of assistance I have received in preparing this chapter. Other footnotes accompany the text and are numbered. [Where no name is clearly attached to a bibliographic reference date, that is to a work of Gödel to be found listed in the References for this volume.]

Austro-Hungarian province of Moravia. This region had a mixed population which was predominately Czech but with a substantial German-speaking minority, to which Gödel's parents belonged. His father Rudolf, an energetic self-made man, had come from Vienna to work in Brünn's thriving textile industry. He was eventually to become managing director and part owner of one of the main textile firms there. The family of Kurt's mother, which came from the Rhineland region, has also been drawn to Brünn for the work in textiles. Marianne Handschuh received a better than ordinary education, partly in a French school in Brünn, in the course of which she developed life-long cultural interests.

Much of our present information concerning the family history comes from Dr. Rudolf Gödel, Kurt's brother and his elder by four years.[c] Rudolf tells us that Kurt's childhood was generally a happy one, though he was timid and could be easily upset. When he was six or seven, Kurt contracted rheumatic fever and, despite eventual full recovery, he came to believe that he had suffered permanent heart damage as well. Here are the early signs of Gödel's later preoccupation with his health. His special intellectual talents emerged early. In the family, Kurt was called *Herr Warum* (Mr. Why), because of his constant inquisitiveness.

Following the religion of his mother rather than that of his father (who was "Old" Catholic), the Gödels had Kurt baptized in the Lutheran church. In 1912, at the age of six, he was enrolled in the Evangelische Volksschule, a Lutheran school in Brünn.[2] From 1916 to 1924, Kurt carried on his school studies at the Deutsches Staats-Realgymnasium, where he showed himself to be an outstanding student, receiving the highest marks in all his subjects; he excelled particularly in mathematics, languages, and religion. (Though the latter was not given much emphasis in the family, Kurt took to it more seriously.[d]) Some of Gödel's notebooks from his young student days are preserved in the *Nachlass*, and among these the precision of the work in geometry is especially striking.

Though World War I took place during Gödel's school years, it had little direct effect on him and his family. The region of Brünn was far from the main fronts and was untouched by the devastation wrought elsewhere by the war in Europe. But the collapse of the Austro-Hungarian empire at war's end and the absorption of Moravia together with Bohemia into the new nation of Czechoslovakia was eventually to affect the German-speaking minority in adverse ways. One of the most immediate signs of the shift in national identity was the displacement of the German name "Brünn" in favor of the Czech name "Brno." For the Gödels, though, life in the years after the war continued much as before, with the family comfortably settled in a villa by that time.

Following his graduation from the Gymnasium in Brno in 1924, Gödel went to Vienna to begin his studies at the University. Vienna was to be

---

[2] A chronology with specific dates of significance in Gödel's life is in Gödel (1986), p. 37; it was prepared by John W. Dawson, Jr. [cf. also Dawson 1996].

his home for the next fifteen years, and in 1929 he was also to become an Austrian citizen. The newly created republic of Austria had entered on a difficult course following the collapse of the Austro-Hungarian empire in 1918. The political, social, and economic upheavals which followed the disappearance of monarchy and empire affected all spheres of activity, what with the economic base enormously shrunk and the *raison d'être* for the swollen bureaucracy gone. The extraordinary cultural and intellectual center that had been Vienna before the war was transformed by the changed conditions, but the Viennese spirit and ambience lived on, now sharing in the general revolutionary ferment and excitement of the 1920s. Before long, Gödel was brought into contact with the Vienna Circle, a hotbed of new thought that proved to be very significant for his work and interests [cf. Menger 1994].

At the university, Gödel was at first undecided between the study of mathematics and physics, though he apparently leaned toward the latter. It is said that Gödel's decision to concentrate on mathematics was due to his taste for precision and to the great impression that one of his professors, the number-theorist Philipp Furtwängler, made on him.[e] A description of the mathematical scene at the University of Vienna in those days is given by Olga Taussky-Todd in her reminiscences (1987) of Gödel, from which the following information is drawn. Besides Furtwängler, the professors were Hans Hahn and Wilhelm Wirtinger. Karl Menger, one of Hahn's favorite students, was an *ausserordentlicher* (associate) professor, and among the *Privatdozenten* (unsalaried lecturers) were Eduard Helly, Walter Mayer, and Leopold Vietoris. Taussky came to know Gödel as a fellow student in 1925, their first real contact coming in a seminar conducted by the philosopher Moritz Schlick on Bertrand Russell's book *Introduction to Mathematical Philosophy* (1919). Gödel hardly ever spoke, but was very quick to see problems and to point the way through to solutions. Though he was very quiet and reserved, it became evident that he was exceptionally talented. Gödel's help was much in demand and he offered such whenever needed. One could talk to him about any problem; he was always very clear about what was at issue and explained matters slowly and calmly.

Hans Hahn became Gödel's principal teacher. He was a mathematician of the new generation, had returned to Austria from a position in Bonn, and was interested in modern analysis and set-theoretic topology, as well as logic, the foundations of mathematics, and the philosophy of science. It was Hahn who introduced Gödel to the group of philosophers around Moritz Schlick, holder of the chair in the Philosophy of the Inductive Sciences (which in earlier years had been held successively by the renowned physicists Ernst Mach and Ludwig Boltzmann). Schlick's group was later baptized the "Vienna Circle" (*Wiener Kreis*) and became identified with the philosophical doctrine called logical positivism or logical empiricism.[3]

---

[3]For general information on Schlick's Circle and its later developments, see the articles on Moritz Schlick and logical positivism in Edwards (1967). Feigl (1969) gives

The aim of this school was to analyze knowledge in logical and empirical terms; it sought to make philosophy itself scientific and rejected metaphysical speculation. Gödel attended meetings of the Circle quite regularly in the period from 1926 to 1928, but in the following years he gradually moved away from it, though he maintained regular contact with some of its members, particularly Rudolf Carnap.

One main reason for Gödel's disengagement from the Circle was that he had developed strong philosophical views of his own which were, in large part, almost diametrically opposed to the views of the logical positivists.[f] Nevertheless, the sphere of concerns that engaged the Circle surely influenced the direction of Gödel's own interests and work. The logical empiricists had combined ideas from several sources, principally Ernst Mach's empiricist-positivist philosophy of science and Bertrand Russell's logicist program for the foundations of mathematics, both filtered through the *Tractatus Logico-philosophicus* (1922) of Ludwig Wittgenstein. The logicist ideas had been developed in great detail by Alfred North Whitehead and Russell in their famous magnum opus, *Principia Mathematica* (1910–1913), over a decade earlier. Hahn, who was at least as important as Schlick in the formation of the Vienna Circle, gave a seminar on this work in 1924 through 1925, but Gödel does not seem to have participated, since he reports first studying the *Principia* several years later.[g]

Hahn's own mathematical interest in the modern theory of functions of real numbers must also have influenced Gödel, as this involved, to a significant extent, set-theoretical considerations deriving from Georg Cantor and passing through the French school of real analysis. However, it seems that the most direct influences on Gödel in his choice of direction for creative work were Carnap's lectures on mathematical logic and the publication in 1928 of *Grundzüge der theoretischen Logik* by David Hilbert and Wilhelm Ackermann. In complete contrast to the massive tomes of Whitehead and Russell, the *Grundzüge* was a slim, unlabored, and mathematically direct volume, no doubt of greater appeal to Gödel, with his taste for succinct exposition. Posed as an open problem therein was the question whether a certain system of axioms for the first-order predicate calculus is complete. In other words, does it suffice for the derivation of every statement that is logically valid (in the sense of being correct under every possible interpretation of its basic terms and predicates)? Gödel arrived at a positive solution to the completeness problem and with that notable achievement commenced his research career. The work, which was to become his doctoral dissertation at the University of Vienna, was finished in the summer of 1929, when he was 23. The degree itself was granted in February 1930, and a revised version of the dissertation was published as Gödel (1930).[4] Al-

---

a lively picture from a more personal point of view and traces the movement of the Circle's members and their ideas. Hahn's role in the Circle is described by Menger in his introduction to the philosophical papers in Hahn (1980). [Cf. also Menger 1994.]

[4]Hahn was nominally Gödel's thesis advisor, but later in life Gödel made it known

though recognition of the fundamental significance of this work would be a gradual matter, at the time the results were already sufficiently distinctive to establish a reputation for Gödel as a rising star.

The ten years from 1929 to 1939 were a period of intense work which resulted in Gödel's major achievements in mathematical logic. In 1930 he began to pursue Hilbert's program for establishing the consistency of formal axiom systems for mathematics by finitary means. The system that had already been singled out for particular attention dealt with the general subjects of "higher" arithmetic, analysis, and set theory. Gödel started by working on the consistency problem for analysis, which he sought to reduce to that for arithmetic, but his plan led him to an obstacle related to the well-known paradoxes of truth and definability in ordinary language.[h] While Gödel saw that these paradoxes did not apply to the precisely specified languages of the formal systems he was considering, he realized that analogous nonparadoxical arguments could be carried out by substituting the notion of provability for that of truth. Pursuing this realization, he was led to the following unexpected conclusions. Any formal system S in which a certain amount of theoretical arithmetic can be developed and which satisfies some minimal consistency conditions is *incomplete*: one can construct an elementary arithmetical statement A such that neither A nor its negation is provable in S. In fact, the statement so constructed is true, since it expresses its own unprovability in S via a representation of the syntax of S in arithmetic.[5] Furthermore, one can construct a statement C which expresses the consistency of S in arithmetic, and C is not provable in S if S is consistent. It follows that, if the body of finitary combinatorial reasoning that Hilbert required for execution of his consistency program could all be formally developed in a single consistent system S, then the program could not be carried out for S or any stronger (consistent) system. The incompleteness results were published in Gödel (1931); the stunning conclusions and the novel features of his argument quickly drew wide attention and brought Gödel recognition as a leading thinker in the field.

One of the first to recognize the potential significance of Gödel's incompleteness result and to encourage their full development was John von Neumann.[i] Only three years older than Gödel, the Hungarian-born von Neumann was already well known in mathematical circles for his brilliant and extremely diverse work in set theory, proof theory, analysis, and mathematical physics. Others interested in mathematical logic were slower to absorb Gödel's new work. For example, Paul Bernays, who was Hilbert's assistant and collaborator, although quickly accepting Gödel's results, had difficulties with his proofs that were cleared up only after repeated correspondence.[j] Gödel's work even drew criticism from various quarters, which was

that he had completed the work before showing it to Hahn and that he made use of Hahn's (essentially editorial) suggestions only when revising the thesis for publication; see Wang (1981), pp. 653–654.

[5] The technical device used for the construction is now called "Gödel numbering."

invariably due to confusions about the necessary distinctions involved, such as that between the notions of truth and proof. In fact, the famous set-theorist Ernst Zermelo interpreted these concepts in such a way as to arrive at a flat contradiction with Gödel's results. In correspondence during 1931 Gödel took pains to explain his work to Zermelo, apparently without success.[k] In general, however, the incompleteness theorems were absorbed before long by those working in the mainstream of mathematical logic; indeed, one can fairly say that Gödel's methods and results came to infuse all aspects of that mainstream.[6]

Gödel's incompleteness work became his *Habilitationsschrift* (a kind of higher dissertation) at the University of Vienna in 1932. In his report on it, Hahn lauded Gödel's work as epochal, constituting an achievement of the first order.[l] The *Habilitation* conferred the title of *Privatdozent* and provided the *venia legendi*, which gave Gödel the right to deliver lectures at the university, but without pay except for fees he might collect from students. As it turned out, he was to lecture only intermittently in Vienna during the following years.

Meanwhile, significant changes had also been taking place in Gödel's personal life. At the age of 21 he met his wife-to-be, Adele Nimbursky (née Porkert), but the difference in their situations led to objections to their developing relationship from his parents, especially his father. Adele was a dancer, had been briefly married before, and was six years older than Kurt. Though his father died not long after, Kurt and Adele were not to be married for another ten years.

The death of Kurt's father in 1929, at the age of 54, was unexpected; fortunately he left his family in comfortable financial circumstances. While retaining the villa in Brno, Gödel's mother took an apartment in Vienna with her two sons. By then Kurt's brother Rudolf had become successfully established as a radiologist. Rudolf never married, and during their period together in Vienna the three of them frequently went out, especially to the theater. According to his brother, at home Kurt went out of his way to "hide his light under a bushel," despite his growing international fame.[m]

In the early 1930s Gödel steadily advanced his knowledge in many areas of logic and mathematics. He took a regular part in Karl Menger's colloquium in Vienna, which had begun meeting in 1929, and he also assisted in the editing of its reports, *Ergebnisse eines mathematischen Kolloquiums.* In the period from 1932 to 1936 he published thirteen short but noteworthy papers in that journal on a variety of topics, including intuitionistic logic, the decision problem for the predicate calculus, geometry, and length of proofs. Some of the results in logic were to be of lasting interest, though not of the same order as his previous work on completeness and incompleteness. During the same period he was an active reviewer for *Zentralblatt für Mathematik und ihre Grenzgebiete* and, less frequently, for *Monatshefte für Mathematik und Physik.*[7]

---

[6]For the influence of Gödel's work on logicians in the 1930s, see Kleene (1981, 1987).
[7]After 1936, Gödel never reviewed again for these or any other journals.

Menger occasionally invited foreign visitors of interest to speak in his colloquium. Among them was the Polish logician Alfred Tarski, who was shortly to become famous for his work on the notion of truth in formal languages and increasingly, later, for his leadership in the development of model theory. In early 1930 Tarski spent a few weeks in Vienna and was introduced to Gödel at that time; Gödel used the occasion to discuss the results of his 1929 dissertation. Tarski returned for a more extensive visit as a guest of Menger's colloquium during the first half of 1935.[n]

Initially, in his unsalaried position as *Privatdozent*, Gödel had to depend on the resources of his family for his livelihood. However, these means were supplemented before long by income from visiting positions in the United States of America. Gödel's first visit was to the Institute for Advanced Study in Princeton during the academic year 1933 through 1934. The Institute had been formally established in 1930, with Albert Einstein and Oswald Veblen appointed its first professors two years later by its original director, Abraham Flexner. Veblen, who was a leader in the development of higher mathematics in America and had played a principal role in building up an outstanding mathematics department at Princeton University, was largely responsible for selecting the further "matchless" mathematics faculty at the Institute: James Alexander, Marston Morse, John von Neumann, and Hermann Weyl.[o] In addition, he helped arrange postdoctoral visits for rising young mathematicians, including Gödel; no doubt Veblen had heard about Gödel from von Neumann, who regarded him as "the greatest logician since Aristotle."[8,p]

Gödel's visit in 1933 through 1934 was the first of three that he was to make to the Institute before taking up permanent residence there in 1940. He lectured on the incompleteness results in Princeton in the spring of 1934. Apparently he had already begun to work with some intensity on problems in set theory; at the same time, he felt rather lonely and depressed during this period in Princeton. Following his return to Europe, he had a nervous breakdown and entered a sanatorium for a time. In the following years there were to be recurrent bouts of mental depression and exhaustion. A scheduled return visit to Princeton had to be postponed to the fall of 1935 and then was unexpectedly cut short after two months, again on account of mental illness. More time was spent in a sanatorium in 1936, and Gödel was unable to carry on at the University of Vienna until the spring of 1937.[q] When he was finally able to resume teaching, he lectured on some of his major new results in axiomatic set theory, the development of which we now trace.

Two problems that had preoccupied workers in the field of set theory since its creation by Cantor beginning in the 1870s concerned the Well-

[8]Apropos of this, Kreisel remarks: "If Gödel's work is to be compared to that of one of the ancients, Archimedes is a better choice than Aristotle (who invented logic, but proved little about it). Archimedes did not invent mechanics, as Gödel did not invent logic. But both of them changed their subjects profoundly" (Kreisel 1980, p. 219).

Ordering Principle and the cardinality of the continuum. Zermelo had examined the first of these, both informally (1904) and then within the framework of his newly introduced system of axioms for set theory (1908, 1908a), and had shown that the Well-Ordering Principle is equivalent to the Axiom of Choice (AC). There was much intense dispute among mathematicians about the evidence for or against this new "axiom." Under its assumption, every infinite set would have a determinate cardinal number in an ordered list of transfinite cardinals. After Cantor proved that the continuum (i.e., the measurement line) is uncountable, he conjectured that its cardinal would be the least among all uncountable cardinal numbers. This conjecture became known as the Continuum Hypothesis (CH).[9]

It was to these problems in set theory that Gödel began to devote himself as his main area of concentration after obtaining the incompleteness results.[r] He considered the statements of AC and CH in the framework of axiomatic set theory (by then enlarged and made more precise through the work of Fraenkel, Skolem, von Neumann, and Bernays), to see whether they could be settled on the basis of the remaining axioms. His major result, finally achieved by the summer of 1937, was that both the Axiom of Choice and the Continuum Hypothesis (even in a natural generalized form, GCH) are consistent with the Zermelo-Fraenkel axioms (ZF) without the Axiom of Choice, and hence cannot be disproved from them if the axioms of ZF are consistent. This result at least provided a minimal guarantee of safety in the use of the seemingly problematic statements AC and GCH.

Underlying Gödel's proof was his definition within ZF of a general notion of constructibility for sets. His plan, which emerged quite early, was to show that the constructible sets form a model for all the axioms of ZF and, in addition, for the Axiom of Choice and the Generalized Continuum Hypothesis. In 1935 he was able to tell von Neumann that he had succeeded in verifying all the ZF axioms together with AC in this model, but, as noted above, it took him two more years to push his work to completion by verifying that GCH holds in the constructible sets as well. With the modest techniques then available, the details that Gödel needed to establish were formidable, and this deep and complicated work caused him much effort, especially in its final part. Perhaps that was one reason for the mental stress he suffered throughout much of the period from 1934 to 1937.[s]

The years 1937 to 1939 brought further significant changes in both Gödel's personal life and career. His mother returned to her home in Brno in 1937, though his brother remained in Vienna to continue his medical practice. That move may have eased the way for Kurt Gödel and Adele Nimbursky to be married finally, in September 1938. The marriage of Kurt and Adele proved to be a warm and enduring one, and for Kurt a source of constant support in difficult times ahead.[t]

---

[9]A full history of the emergence of the Axiom of Choice as a fundamental principle of set theory and of the controversies that surrounded it is provided by Moore (1982).

During the 1930s there were many shifts in the lives of Gödel's friends and colleagues in Vienna. Marcel Natkin and Herbert Feigl, early friends from the Vienna Circle, had already left Vienna at the turn of the decade, the first for Paris and the second for America. Rudolf Carnap left to teach in Prague in 1931; he was eventually to go to America, too. Gödel's teacher Hans Hahn died in 1934, of natural causes. Then in 1936, Moritz Schlick, the central figure of the Vienna Circle, was murdered by a deranged former student on the way to a lecture; naturally, the case created a sensation. Upset by this turn of events and the general situation in Austria, Karl Menger left the following year to take up a position at Notre Dame. Gustav Bergmann and Abraham Wald, two other contemporaries of Gödel's, also left for America in 1938.

These and related changes were taking place in the context of the difficult economic conditions that had been gripping European nations since the severe depression of 1929 and of the political situation created by the advent to power in 1933 of Adolf Hitler and the Nazis in Germany. In 1934 Austria itself fell under the rule of a semifascist regime, led by Engelbert Dollfuss until his assassination by Austrian Nazis later that same year. Dollfuss' murder was a premature attempt by the Nazis to gain power in Austria and to carry out the *Anschluss* (political and economic union) of Austria with Germany, which had been forbidden by the 1919 Treaty of Saint-Germain. There was much sentiment for *Anschluss* among certain groups of Austrians, but the main pressure came from Hitler. That mounted steadily until Hitler's threat of invasion brought down the succeeding Schuschnigg regime in the spring of 1938. Austria thenceforth became a province (*Ostmark*) of a wider Nazi Germany. The year 1938 saw the beginning of a general transformation of Austrian cultural and intellectual life, comparable to that in Germany five years previously. This led to an exodus of intellectuals, particularly those of Jewish background, for whom the move was a matter of survival, while for others emigration was a reaction to the supernationalistic and racist politics characteristic of the Nazi regime. An incidental result of all this was the final disintegration of the Vienna Circle.[u] As for Gödel, his stance was basically apolitical and noncommittal; while he was by no means unaware of what was taking place, he ignored the increasingly evident implications of the transformations around him.

At the urging of Menger, Gödel visited America once more in 1938 through 1939. He spent the fall term at the Institute for Advanced Study, where he lectured on his new results concerning the consistency of the axiom of choice and the generalized continuum hypothesis. For the spring term he joined Menger at Notre Dame, where he lectured again on his set-theoretical work and conducted an elementary course on logic with Menger. Gödel then returned to Vienna to rejoin his wife, whom he had left the previous fall only two weeks after their marriage.

Gödel planned to return to the Institute for Advanced Study in the fall of 1939, but political events intervened; his life was now directly affected by the Nazi regime in two very different ways. He was called up for a military physical examination and much to his surprise (in view of his generally poor health and his conviction that he had a weak heart) found "fit for garrison duty." Then there was the question of his situation at the University of Vienna. The unpaid position of *Privatdozent* had been abolished by the Nazis, who had replaced it by a new paid position called *Dozent neuer Ordnung*. The latter, however, required a fresh application that could be rejected on political or racial grounds. Although Gödel applied for the new position in September 1939, approval was slow in forthcoming. Questions were raised about his associations with Jewish professors (Hahn in particular), and while it was recognized that he was apolitical, his lack of open support for the Nazis counted against him. In this insecure situation and with the likely possibility that he would be drafted (war having begun in September), Gödel wrote Veblen in desperation in November 1939, seeking assistance to leave. Somehow, U.S. nonquota immigrant visas and German exit permits were arranged, and Kurt and Adele managed to leave Vienna in January 1940. As it was too dangerous at that point to cross the Atlantic by boat, they made their way instead by train through Eastern Europe, then via the Trans-Siberian Railway across Russia and Manchuria, and thence to Yokohama. From there they traveled by ship to San Francisco, and in March 1940 they finally proceeded by train to Princeton.[v]

Gödel was never to return to Europe. Ironically, his application for *Dozent neuer Ordnung* was belatedly approved in June 1940.[w] Long afterward he remained bitter about his predicament in Austria in the year 1939 through 1940, apparently blaming it more on Austrian "sloppiness" (*Schlamperei*) than on the outrageous Nazi conditions. In particular, on the occasion of his 60th birthday in 1966, he turned down an honorary membership in the Austrian Academy of Sciences. However, he couched the refusal in pseudolegalistic terms which suggested that his U.S. citizenship might be jeopardized if he were to accept membership in the academy of the country of his former citizenship.[x]

In 1940 Gödel was made an Ordinary Member of the Institute for Advanced Study, and he and his wife settled in Princeton, where they established a quiet social life. Among Gödel's closest friends were Albert Einstein and Oskar Morgenstern. The latter was another ex-Viennese, who had emigrated in 1938 and taken a position at Princeton University. Already established as a mathematical economist, Morgenstern was later to become well known to a wide public through his important and influential work with von Neumann, *The Theory of Games and Economic Behavior* (1944). (Von Neumann himself would have been less accessible to Gödel during the early 1940s, since he was frequently away from the Institute in his capacity as consultant for innumerable government war projects.[y])

Morgenstern had many stories to tell about Gödel. One concerned the occasion when, in April 1948, Gödel became a U.S. citizen, with Einstein and Morgenstern as witnesses.[z] Gödel was to take the routine citizenship examination, and he prepared for it very seriously, studying the U. S. Constitution assiduc [illegible] to appear, Gödel came to Morgenstern in [illegible] have discovered a logical-legal possibility [illegible] ıld be transformed into a dictatorship." [illegible] ever the logical merits of Gödel's argumen[illegible] hypothetical in character, and he urged Gö[illegible]overy at the examination. The next morning, Morgenstern drove Gödel and Einstein from Princeton to Trenton, where the citizenship proceedings were to take place. Along the way Einstein kept telling one amusing anecdote after another in order to distract Gödel, apparently with great success. At the office in Trenton, the official was properly impressed by Einstein and Morgenstern, and invited them to attend the examination, normally held in private. He began by addressing Gödel: "Up to now you have held German citizenship." Gödel corrected him, explaining that he was Austrian. "Anyhow," continued the official, "it was under an evil dictatorship . . . but fortunately, that's not possible in America." "On the contrary," Gödel cried out, "I know how that can happen!" All three had great trouble restraining Gödel from elaborating his discovery, so that the proceedings could be brought to their expected conclusion.

Einstein and Gödel could frequently be seen walking home together from the institute, engaged in rather intense conversations. A number of stories concerning the two have been recounted by the mathematician Ernst Straus, who was Einstein's assistant during the years 1944 through 1948. He summarized appreciatively their unusual relationship in the following passage, take from his reminiscences (Straus 1982, p. 422).

> The one man who was, during the last years, certainly by far Einstein's best friend, and in some ways strangely resembled him most, was Kurt Gödel, the great logician. They were very different in almost every personal way—Einstein gregarious, happy, full of laughter and common sense, and Gödel extremely solemn, very serious, quite solitary, and distrustful of common sense as a means of arriving at the truth. But they shared a fundamental quality: both went directly and wholeheartedly to the questions at the very center of things.

At the institute Gödel had no formal duties and was free to pursue his research and studies as he pleased. During the first years there he continued his work in mathematical logic, along various lines. In particular, he made strenuous efforts to prove the independence of the Axiom of Choice and the Continuum Hypothesis, but only with partial success, and that just on the former problem. His efforts in this direction were never published; they remain to be deciphered (if possible) from notebooks in his *Nachlass*.[10] Another achievement early in this period (though not published

[10] The full independence results were eventually obtained by Paul Cohen (1963).

until 1958) was a new constructive interpretation of arithmetic that proved its consistency, but via methods going beyond evidently finitary means in Hilbert's sense.

From 1943 on, Gödel devoted himself almost entirely to philosophy, first to the philosophy of mathematics and then to general philosophy and metaphysics. The year 1944 marks the publication of his paper on Bertrand Russell's mathematical logic, which was extremely important both for its searching analysis of Russell's work and for its open statement of Gödel's own "platonistic" views of the reality of abstract mathematical objects.[11]

An expository paper on Cantor's continuum problem in 1947 brought out these views quite markedly in the context of set theory. One other writing of a partly philosophical character from this period did not appear until somewhat later, namely, the address in 1946 to the Princeton Bicentennial Conference on Problems of Mathematics. As for general philosophy, Gödel continued his long-pursued reading and study of Kant and Leibniz, turning also to the phenomenology of Edmund Husserl in the late 1950s.[aa] In Gödel's *Nachlass* are many notes on the writings of these philosophers.

An apparent exception to these directions of thought was Gödel's surprising work on the general theory of relativity during the period from 1947 to 1951, in which he produced new and unusual cosmological models that, in theory, permit "time travel" into the past. According to Gödel, this work did not come out of his discussions with Einstein but rather was motivated by his own interests in Kant's philosophy of space and time.[bb] Einstein himself was preoccupied, as he had been for a long time, with constructing a unified field theory, a project about which Gödel was skeptical.[cc] In this work Gödel brought to bear mathematical techniques and physical intuitions that one who was familiar only with his papers in logic would not have expected. The mathematics, however, harks back to his brief contributions to differential geometry in the 1930s, as well as to his studies of analysis with Hahn and in Menger's colloquium.

In addition to reflecting Gödel's primary interests in logic, philosophy and, to a lesser extent, mathematics and physics, the notebooks in his *Nachlass* are unexpectedly wide ranging, revealing, for example, sustained

---

[11] An amusing aside in this respect has its source in a statement by Bertrand Russell to be found in the second volume of his *Autobiography* (1968, pp. 355–356): "I used to go to [Einstein's] house once a week to discuss with him and Gödel and Pauli. These discussions were in some ways disappointing, for, although all three of them were Jews and exiles and, in intention, cosmopolitans, I found that they all had a German bias toward metaphysics [and that] Gödel turned out to be an unadulterated Platonist."

Gödel's attention was drawn to this in 1971 and he drafted a reply that is preserved in the *Nachlass*, though it was never actually sent: "As far as the passage about me [by Russell] is concerned, I have to say *first* (for the sake of truth) that I am not a Jew (even though I don't think this question is of any importance), 2) that the passage gives the wrong impression that I had many discussions with Russell, which was by no means the case (I remember only one), 3) Concerning my 'unadulterated' Platonism, it is no more 'unadulterated' than Russell's own in 1921." Fuller quotations are given in Dawson (1984a), pp. 13 and 15.

interests in history and theology. The latter even included a long-standing fascination with demonology.[dd]

Gödel was made a Permanent Member of the Institute for Advanced Study in 1946. His subsequent promotion to Professor in 1953 required him to take part in some aspects of Institute business.[12] He devoted a good deal of time to the details of these affairs and, in particular, took very seriously the increasingly frequent applications by logicians for visiting memberships.[13] When logic started to flourish in that period, the institute became a Mecca for younger logicians—many of them rising stars—and drew visits as well from older colleagues of the prewar generation, such as Paul Bernays. Gödel limited his contacts with most younger visitors, though he would give serious consideration to their work and interests and would volunteer suggestions. A few of the more advanced logicians were able to establish deeper scientific and personal relations with him and were privy to his thoughts and speculations in extensive conversations; most prominent among these were William Boone, Georg Kreisel, Gaisi Takeuti, Dana Scott, and Hao Wang. Others whose work impressed him and with whom he had some significant (though less extensive) contact were Clifford Spector and Abraham Robinson. But Gödel never had students or disciples in the usual sense of the word.

Beginning in 1951, Gödel received many honors. Particularly noteworthy are his sharing of the first Einstein Award (with Julian Schwinger) in 1951, his choice as Gibbs Lecturer for 1951 by the American Mathematical Society, and his elections to membership in the National Academy of Sciences (1955), to the American Academy of Arts and Sciences (1957), and to the Royal Society of London (1968). In 1975 he was awarded the National Medal of Science by President Ford, but because of ill health he could not attend the ceremony. A complete list of awards and honors is given in Dawson's "A Gödel chronology" (Gödel 1986, pp. 37–43).

In the last fifteen years of his life, Gödel was busy with visitors, institute business, and his own philosophical studies; during this time he returned to logic only rarely. Some papers were revised and a few notes were added to new translations. In particular, he expended a good deal of effort over a period of years on a translation and revision of his 1958 paper, which gave a constructive interpretation of arithmetic. The revised work never reached

---

[12]Questions have been raised about the relative lateness of this promotion in Ulam (1976), p. 80, and Dyson (1983), p. 48. One explanation has it that promotion was held back for Gödel's sake, so as not to burden him with the administrative responsibilities accompanying faculty status. Another has it that there were fears Gödel's exceptional attention to detail and his legalistic turn of mind would hinder the conduct of institute business if he were to assume those responsibilities. (See also Dawson (1984a), p. 15.)

[13]Concerning the latter, Gödel's colleague Hassler Whitney commented as follows: "Gödel was keenly interested in the affairs of the Institute. It was . . . hard to appoint a new member in logic since Gödel could not 'prove to himself that a number of candidates shouldn't be members, with the evidence at hand' " (quotation from *The Mathematical Intelligencer*, 1 (1978), p. 182). For a complementary view, see Kreisel (1980), p. 159.

published form, though it was found in galley proof in his *Nachlass*.[14] In the early 1970s there was a flurry of interest and excitement among logicians about notes by Gödel in which he proposed new axioms for set theory that were supposed to imply the falsity of the Continuum Hypothesis, but essential problems were found in the arguments and the notes were withdrawn. Gödel blamed his having overlooked the difficulties on the drugs he was then taking for his illness. [Cf. Gödel 1995, pp. 405–425.]

In fact, Gödel's health was poor from the late 1960s on. Among other things he had a prostate condition for which surgery had been recommended, but he would never agree to have the operation done. Along with his hypochondriacal tendencies he also had an abiding distrust of doctors' advice. (Back in the 1940s, for example, he delayed treatment of a bleeding ulcer so long that he would have died, had it not been for emergency blood transfusions.) In addition to prostate trouble, he was still convinced that his heart was weak, although there was no medical substantiation. During the last few years of his life, his wife Adele was unable to help him to the same extent as before, since she herself was partially incapacitated by a stroke and was, for a time, moved to a nursing home. Gödel's depressions returned, accompanied and aggravated by paranoia; he developed fears about being poisoned and would not eat. He died in Princeton Hospital on 14 January 1978 of "malnutrition and inanition caused by personality disturbance."[ee] Adele survived him by three years, dying on 4 February 1981. Kurt and Adele had no children, leaving Kurt's brother Rudolf as the sole surviving member of the Gödel family [subsequently deceased] . . . .

Gödel's Philosophy of Mathematics

Gödel is noted for his vigorous and unwavering espousal of a form of mathematical realism (or "platonism"). In this general direction he joins the company of such noted mathematicians and logicians as Cantor, Frege, Zermelo, Church, and (in certain respects) Bernays. These views of mathematics also accord with the implicit working conceptions of most practicing mathematicians (the "silent majority"). However, the preponderance of developed thought on the philosophy of mathematics since the late nineteenth century has been critical of realist positions and has led to a number of alternative (and opposing) standpoints, going under such names as constructivism, formalism, finitism, nominalism, predicativism, definitionism, positivism, and conventionalism. Leading figures identified with one or another of these positions are Kronecker, Brouwer, Poincaré, Borel, Hilbert, Weyl, Skolem, Heyting, Herbrand, Gentzen, and Curry, as well as Carnap. Russell veered from a distinctly realist position in his earlier work, *The Principles of Mathematics* (1903), to a more equivocal predicativist approach in *Principia Mathematica* (1910–1913) coauthored with Whitehead.

[14]It is reproduced in volume 2 of Gödel's *Collected Works* (1990) as Gödel (1972).

Critics of the realist position have raised both ontological and epistemological issues. With respect to the former, the ideas of mathematical objects as independently existing abstract entities and, in particular, of infinite classes as "completed" totalities are considered to be problematic. For the latter, questions have been raised about the admissibility of such principles as that of the excluded middle and the axiom of choice, each in its way leading to nonconstructive existence proofs. Especially in the earlier part of this century, the paradoxes of classes found by Cantor, Burali-Forti, and Russell were felt in addition to require radical reconsideration of the entire set-theoretic, philosophically platonist approach to the foundations of mathematics. This last receded in importance when it was recognized how Zermelo had rescued set theory from the obvious contradictions by means of his axiomatization and its underlying interpretation in the iterative conception of sets.

Hilbert and Brouwer were perhaps the most influential figures proposing alternative foundational schemes during the period in which Gödel was beginning his work in logic. Hilbert had elaborated a program to "secure" mathematics—including, as he hoped, Cantor's set theory—by means of finitary consistency proofs for formal axiom systems. Brouwer rejected nonconstructive existence proofs and Cantorian conceptions of "actual" infinities, seeking to rebuild mathematics according to his own intuitionistic version of constructivism. In Vienna, special attention was naturally also given to the program of the logical empiricists developed by Hahn, Schlick, Carnap, and others. Their aim was to place mathematics in a conventionalist role as the "syntax of language," thus separating it from physical science, which itself was to rest finally on empirical observation. According to his own account much later,[ff] Gödel had arrived at a general platonist viewpoint by 1925, around the time he came to Vienna. When be began to take part in the meetings of the Vienna Circle, he did so primarily as an observer, not openly disputing the approach taken, though disagreeing with it. Gödel did remark critically on the positions of Hilbert and Brouwer in the introduction of his dissertation (1929), but mainly in connection with the completeness problem. In particular, he made some trenchant remarks there concerning the idea of consistency as the criterion for existence. That view could be identified with Hilbert, though it was not a necessary part of Hilbert's program. However, this discussion was omitted from the published version (1930) of the dissertation. Gödel openly criticized the related idea, again deriving from Hilbert, of consistency sufficing for correctness when one extends a system of meaningful statements by a system of "ideal" statements and axioms. These remarks were made at the important symposium on the foundation of mathematics held at Königsberg in 1930 (see 1931a). Gödel made no further published statements on the nature of his position until the appearance of his substantial article (1944) on Russell's mathematical logic. In retrospect, however, one can recognize some brief remarks or footnotes in earlier papers as providing indications of the directions of his thought.

The main published sources for Gödel's views on the philosophy of mathematics are the papers published in 1944, 1946, 1964 (a revised and expanded version of 1947), and 1958, as well as his personal and written communications to Hao Wang reproduced in Wang (1974). Further sources that appear in Gödel (1986, 1990) but were not previously printed are the introduction to the 1929 paper, the 1972 revised and expanded version of the 1958 paper, and finally some brief notes (1972a) in Gödel (1990). All this is amplified but not modified in any significant way by unpublished manuscripts and correspondence found in Gödel's *Nachlass.** In particular, the article "Is mathematics syntax of language?" would have been Gödel's first systematic published attack on the program of the logical positivists, had it appeared in the Carnap volume as intended.[15]

The main features of Gödel's philosophy of mathematics that emerge from these sources are as follows. Mathematical objects have an independent existence and reality analogous to that of physical objects. Mathematical statements refer to such a reality, and the question of their truth is determined by objective facts which are independent of our own thoughts and constructions. We may have no direct perception of underlying mathematical objects, just as with underlying physical objects, but—again by analogy—the existence of such is necessary to deduce immediate sense perceptions. The assumption of mathematical objects and axioms is necessary to obtain a satisfactory system of mathematics, just as the assumption of physical objects and basic physical laws is necessary for a satisfactory account of the world of appearance. An example of mathematical "sense data" requiring this kind of explanation is provided by instances of arithmetical propositions whose universal generalizations demand assumptions transcending arithmetic; this is a consequence of Gödel's incompleteness theorem.[16] While mathematical objects and their properties may not be immediately accessible to us, mathematical intuition can be a source of genuine mathematical knowledge. This intuition can be cultivated through deep study of a subject, and one can thus be led to accept new basic statements as axioms. Another justification for mathematical axioms may be their fruitfulness and the abundance of their consequences; however, that is less certain than what is guaranteed by intuition.

*[See, however, chapter 8 in this volume for a possible reevaluation of this picture.]

[15]In the penultimate section of his introductory note to (1944) in Gödel (1986), Charles Parsons suggests that in one respect, at least, Gödel is more closely engaged with the ideas of the Vienna Circle than is ordinarily viewed. The relationship has to do with the thesis that mathematics is analytic. In (1944) Gödel considers two senses of the notion of analyticity of a statement, respectively (roughly speaking) that of its being true in virtue of the definitions of the concepts involved in it and that of its being true in virtue of the meaning of those concepts. In (1944) Gödel rejects the thesis that mathematics is analytic in its first sense but accepts it in its second sense (at least for the theory of types and axiomatic set theory). [Two versions of the unpublished paper "Is mathematics syntax of language?" have subsequently appeared in Gödel (1995), pp. 334–362; cf. the introductory note to those items by W. Goldfarb, *op. cit.*, pp. 324–334.]

[16]Gödel frequently refers to such propositions as arithmetical problems of Goldbach type (the conjecture that every even positive integer is a sum of two odd primes).

Gödel discussed these ideas most explicitly in connection with set theory and Cantor's continuum problem, particularly his 1947 and 1964 papers (see especially the supplement to 1964). There he argues that Cantor's notion of (infinite) cardinal numbers is definite and unique, and hence that the Continuum Hypothesis CH has a determinate truth value, even though efforts to settle it thus far have failed.[17] One can begin by examining the question of its demonstrability with reference to presently accepted axioms for set theory. These axioms (for example, the Zermelo-Fraenkel system ZF) are evidently true for the iterative structure of sets in the cumulative hierarchy. This is a perfectly self-consistent conception that is untouched by the paradoxes. In Gödel's view, the axiom of choice is just as evident for this notion as are the other axioms, and hence Cantor's cardinal arithmetic is adequately represented in the axiom system.

In his 1940 paper Gödel had shown that AC and CH are consistent with ZF, by use of his model L of constructible sets. But he conjectured in 1947 that CH is false, hence underivable from the true axioms ZF + AC. After Cohen (1963) proved the independence of CH from ZF + AC, Gödel could expand on the anticipated undecidability of CH by presently accepted axioms. This confirmed what he had long expected, namely, that new axioms would be needed to settle CH. In particular, he mentioned the possibility of using strong axioms of infinity (or large cardinal axioms) for these purposes, pointing out once more that in view of the incompleteness theorem, such axioms are productive of arithmetical consequences. But he also thought that axioms based on new ideas may be called for.[18] He argued again that such axioms need not be immediately evident, but may be arrived at only after long study and development of the subject.

There are briefer discussions or indications by Gödel in other of his publications concerning his belief in the objectivity of mathematical notions outside set theory: abstract concepts (in 1944), absolute demonstrability and definability (in 1946), and constructive functions and proofs (in 1958 and 1972). One should also mention the steady interest he showed in intuitionism through several publications in the 1930s and his 1958 paper. Thus his mathematical realism is not necessarily confined to set theory, though that is where it is most thoroughly elaborated.

---

[17] There is one earlier statement by Gödel that apparently presents a different view concerning the questions of definiteness of set-theoretical concepts. Namely, at the end of (1938) he says: "The proposition A [$V = L$] added as a new axiom seems to give a natural completion of the axioms of set theory, insofar as it determines the vague notion of an arbitrary infinite set in a definite way." In a personal communication, Martin Davis has argued: "This is not at all in the spirit of the point of view of (1947), and . . . it suggests that Gödel's 'platonism' regarding sets may have evolved more gradually than his later statements would suggest." There is currently no further evidence available which would help clarify Gödel's intentions in his 1938 remark and its relationship to his later views. [Cf. also chapter 8 in this volume.]

[18] Indeed, he proposed certain new axioms himself in unpublished manuscripts [which have subsequently been published in Gödel (1995) as (*1970a, b, and c)].

In the correspondence reproduced in Wang (1974, pp. 8–11), Gödel credits a large part of his main success, where others had failed, to his realist views; they are said to have freed him from the philosophical prejudices of the times which had shackled others. In this respect he mentions Skolem's failure to arrive at the completeness of predicate logic and Hilbert's failure to "prove" CH in contrast to his own results published in 1930 and 1940; he also mentions his belief in the objectivity of mathematical truth as having led to the incompleteness theorems of 1931. Whatever their final merits, the efficacy of Gödel's views seems in this respect to be indisputable.

A deeper examination of Gödel's ideas on the philosophy of mathematics is given in the introductory notes to a number of Gödel's published and unpublished papers in the three volumes of his *Collected Works* (1986, 1990, and 1995).

Character, Impact, and Influence of the Work

Gödel's main published papers from 1930 to 1940 were among the most outstanding contributions to logic in this century, decisively settling fundamental problems and introducing novel and powerful methods that were exploited extensively in much subsequent work. Each of these papers is marked by a sense of clear and strong purpose, careful organization, great precision—both formal and informal—and by steady and efficient progress from start to finish, with no wasted energy. Each solves a clear problem, simply formulated in terms well understood at the time (though not always previously formulated as such). Their significance, then, was in one sense prima facie evident, though their significance more generally for the foundations of mathematics would prove to be the subject of unending discussion. As he has told us, Gödel was strongly motivated by his realist philosophy of mathematics, and he credited it with much of the reason for his success in being led to the "right" results and methods.[gg] Nevertheless, philosophical questions are given bare notice in these papers. In addition, Gödel made special efforts where possible to extract results of potential mathematical (as opposed to logical or foundational) interest—for example, the compactness theorem for the first-order predicate calculus (1930), the incompleteness of axiomatic arithmetic with respect to (quantified) Diophantine problems (1931), and the existence of nonmeasurable *PCA* sets of reals in the universe of constructible sets (1938).[19]

Concerning Gödel's methods, one may say that many of the constructions and arguments were technically difficult for their time, or at any rate

[19]This must be moderated in two respects. It is certainly the case that Gödel himself never made any use of compactness and that we have the benefit of hindsight in assessing its mathematical value. Moreover, according to Kreisel (1980), p. 197, the existence of nonmeasurable *PCA* sets in *L* was suggested to Gödel by Stanislaw Ulam. [Gödel returned to undecidable Diophantine propositions in an unpublished lecture, which appears as (*193?) in Gödel (1995), pp. 164–175.]

too novel or unexpected to be readily absorbed (though the arguments for completeness were largely anticipated by Skolem, and those for incompleteness and completeness both seem much simpler now). But technical ingenuity is never indulged in or displayed for its own sake in Gödel's papers; rather, it is always there as a means to an end. We have considerable evidence that Gödel worked and reworked his papers many times, partly to arrive at the most efficient means of presentation.

Gödel's contributions bordered on the two fundamental technical concepts of modern logic: *truth for formal languages* and *effective computability*. With respect to the former he stated in his 1934 lectures at Princeton (and elaborated in some correspondence) that he was led to the incompleteness of arithmetic via his recognition of the undefinability of arithmetic truth in its own language, though he took care to credit Tarski for elucidating the exact concept of truth and establishing its undefinability. In the same lectures he offered a notion of general recursiveness in connection with the idea of effective computability; this was based on a modification of a definition proposed by Herbrand. In the meantime, Church was propounding his thesis, which identified the effectively computable function with the λ-definable functions. But Gödel was unconvinced by Church's thesis, since it did not rest on a direct conceptual analysis of the notion of finite algorithmic procedure. For the same reason he resisted identifying the latter with the general recursive functions in the Herbrand-Gödel sense. Indeed, in his Princeton lectures Gödel said that the notion of effectively computable function could serve just as a heuristic guide. It was only when Turing, in 1937, offered the definition in terms of his "machines" that Gödel was ready to accept such an identification, and thereafter he referred to Turing's work as having provided the "precise and unquestionably adequate definition of formal system" by his "analysis of the concept of 'mechanical procedure'" needed to give a general formulation of the incompleteness results.[hh] It is perhaps ironic that the various classes of functions (λ-definable, general recursive, Turing computable) were proved in short order to be identical, but Gödel's initial reservations were justified on philosophical grounds.

In general, Gödel shied away from new concepts as objects of study, as opposed to new concepts as tools for obtaining results. The constructible hierarchy may be offered as a case in point, concerning which Gödel says that he is only using Russell's idea of the ramified hierarchy, but with an essentially impredicative element added, namely, the use of arbitrary ordinals.[ii] Only the concept of effective functional of finite type, which he had arrived at by 1941, comes close to being a new fundamental concept (see Gödel 1958, 1972).

There is a shift in the 1940s that corresponds to Gödel's changed circumstances and interests. Prior to that time, Gödel was understandably cautious about making public his platonist ideas, contrary as they were to the "dominant philosophical prejudices" of the time.[jj] With his reputation solidly established and with the security provided by the Institute for Advanced Study, Gödel felt freer to pursue and publicly elaborate his philosophical vision.

Gödel did the major part of his logical work in isolation, though he had a certain amount of stimulating contact with Menger, von Neumann, and Bernays in the prewar period. As described in the biographical sketch above, in the 1950s the institute increasingly attracted younger logicians, many of them in the forefront of research, as well as older colleagues of the prewar generation. Some of them sought Gödel out and established lengthy scientific relations with him that were also personally comfortable and friendly. Yet Gödel never had any students, never established a school, and never collaborated with others to advance his favorite program, namely, the discovery of essentially new axioms for set theory. Nonetheless, that program was taken up by many others in the wave of work in set theory from the 1960s on. Gödel's main results proved to be absolutely basic—the sine qua non for all that followed in almost all parts of logic—and it is through the work itself that he has had his major impact and influence.

As much as anything, Gödel's achievement lay in arriving at a very clear understanding of which problems in logic could be treated in a definite mathematical way. Along with others of his generation, but always leading the way, he succeeded in establishing the subject of mathematical logic as one that could be pursued with results as decisive and significant as those in the more traditional branches of mathematics. It is for this double heritage of the content and character of his work that we are indebted to him.

Source Notes

[a] In preparing the following I have drawn on a number of sources, of which the main published ones are Christian (1980), Dawson (1983, 1984a), Kleene (1976, 1987a), Kreisel (1980), and Wang (1978, 1981). Some use has also been made of unpublished material from Gödel's *Nachlass*.

For the biographical material I have relied primarily on Kreisel (1980), pp. 151–160, Wang (1981), and Dawson (1984a). Further personal material of value has come from Quine (1979), Zemanek (1978), and Taussky-Todd (1987). I have also made use of personal impressions communicated to me by A. Raubitschek (whose father was one of Hahn's best friends and who himself knew Gödel in Princeton), and of my own impressions (from contacts with Gödel during my visit to the Institute for Advanced Study in 1959 and 1960).

Finally, I am indebted to my co-editors [J.W. Dawson, Jr., S.C. Kleene, G.H. Moore, R.M. Solovay, and J. van Heijenoort] as well as the following of my colleagues for their many useful comments which have helped appreciably to improve the presentation: J. Barwise, S. Bauer-Mengelberg, M. Beeson, G.W. Brown, M. Davis, A.B. Feferman, H. Feigl, J.E. Fenstad, R. Haller, E. Köhler, G. Kreisel, R.B. Marcus, K. Menger, G. Müller, C. Parsons, W.V. Quine, A. Raubitschek, C. Reid, J. Robinson, P.A. Schilpp, W. Sieg, L. Straus, A.S. Troelstra, and H. Wang.

[b] Following Gödel's delivery of the Gibbs lecture in 1951. The story is told by Olga Taussky-Todd in her reminiscences (1987).

[c] (Dr.) Rudolf Gödel wrote up a family history in 1967, with a supplement in 1978. This was made available to Georg Kreisel along with some correspondence between Kurt Gödel and his mother; see Kreisel (1980), p. 151. Gödel also communicated information about his family to Hao Wang for the article Wang (1981).

Another useful source on this and Gödel's intellectual development is a questionnaire which had been put to Gödel in 1975 by Burke D. Grandjean, then an instructor in sociology at the University of Texas. Its purpose was to gather information on Gödel's background in connection with research that Grandjean was doing on the social and intellectual situation in Central Europe during the first third of the twentieth century. The questionnaire was found in Gödel's *Nachlass*, fully filled out, along with a covering letter to Grandjean dated 19 August 1975; however, neither was apparently ever sent. I shall refer to this several times as a source in the following, calling it "the Grandjean interview."

[d]Kreisel (1980), p. 152.

[e]Kreisel (1980), p. 153.

[f]According to Gödel in the Grandjean interview, he had already formed such views before coming to Vienna. See also Wang (1978), p. 183.

[g]The Grandjean interview.

[h]Wang (1981), p. 654. [Cf. also chapter 7 in this volume.]

[i]Wang (1981), pp. 654–655. For information on von Neumann's life and work, see Ulam (1958), Goldstine (1972), and Heims (1980).

[j]See Dawson (1985).

[k]See Grattan-Guinness (1979), Moore (1980), and Dawson (1985, 1985a). An account of Gödel's first (and perhaps only) personal meeting with Zermelo is given in Taussky-Todd (1987).

[l]Hahn's report is quoted in Christian (1980), p. 263.

[m]Kreisel (1980), p. 154.

[n]Information communicated by E. Köhler.

[o]See Montgomery (1963) and Goldstine (1972), pp. 77–79.

[p]Goldstine (1972), p. 174.

[q]The available published information about Gödel's recurrent illness during this period is slim; in this connection see Kreisel (1980), p. 154, Wang (1981), pp. 655–656, and Dawson (1984a), p. 13, as well as Taussky-Todd (1987).

[r]Wang (1981), p. 656.

[s]For more on how Gödel achieved his results on AC and GCH, see Wang (1978), p. 184, Kreisel (1980), pp. 194–198, and Dawson (1984a), p. 13. The dates given in the sources do not always square with each other. It is hoped that study of the correspondence and notes in Gödel's *Nachlass* will be of assistance in clearing this up. Already discovered is a shorthand annotation preceding Gödel's notes on the GCH in his *Arbeitsheft* 1, which has been transcribed (by C. Dawson) as "Kont. Hyp. im wesentlichen gefunden in der Nacht zum 14 und 15 Juni 1937," in other words, that Gödel had "essentially found [the proof for the consistency of] GCH during the night of 14–15 June 1937."

[t]Kreisel (1980), pp. 154–155.

[u]Feigl (1969).

[v]For accounts of these events see Kreisel (1980), pp. 155–156, and Dawson (1984a), pp. 13, 15.

[w]It is a further point of irony that Gödel was listed in the catalogues for the University of Vienna between 1941 and 1945 as "Dozent für Grundlagen der Mathematik und Logik" and under course offerings "wird nicht lesen" (information communicated by E. Köhler).

[x]Gödel's response to the Austrian Academy of Sciences is quoted in Christian (1980), p. 266; Kreisel (1980), p. 155, says that Gödel refused various honors from Austria after the war, "sometimes for mindboggling reasons."

[y]Ulam (1958), pp. 3–4, and Goldstine (1972), pp. 177–182.

[z]This is recounted (in German) in Zemanek (1978), p. 210. Zemanek locates the hearing in Washington but it was more likely Trenton. Since the story is third-hand and translated, quotations are not exact.

[aa]The Grandjean interview, Wang (1978), p. 183, and Wang (1981), pp. 658–659.

[bb]Wang (1981), p. 658. Also, Straus (1982), pp. 420–421, says that "Gödel . . . was really totally solitary and would never talk with anybody while working."

[cc]Unpublished letter from Gödel to Carl Seelig, dated 7 September 1955.

[dd]Kreisel (1980), p. 218.

[ee]See Kreisel (1980), pp. 159–160, Dawson (1984a), p. 16. Information in this paragraph on Gödel's last years also comes from an interview that Dawson had with Gödel's friend and colleague Deane Montgomery. The death of his old friend Oskar Morgenstern in mid 1977 was apparently a shock to Gödel. The quotation giving cause of death is from his death certificate, on file in the Mercer County courthouse, Trenton, New Jersey.

[ff]The Grandjean interview.

[gg]See Gödel's letters to Hao Wang, dated 7 December 1967 and 7 March 1968, quoted in Wang (1974), pp. 8–11. The significance of Gödel's convictions for his work is discussed further in Feferman (1984a) [reproduced in the next chapter].

[hh]See the note, added 28 August 1963, to (1931) and the postscript to (1934).

[ii]See (1944).

[jj]See the letters mentioned in note *gg* for Gödel's characterizations of these "prejudices," and Feferman (1984a) [in the next chapter] for a discussion of Gödel's caution in this respect.

7

Kurt Gödel: Conviction and Caution

In the course of preparing an introductory chapter on Gödel for a comprehensive edition of his works,* I was struck by the great contrast between the deep platonistic convictions which Gödel held concerning the objective basis of mathematics and the special caution that he exercised in revealing those convictions. Gödel said that he had arrived at a position of philosophical realism early in his university years, and he credited his enormous successes in mathematical logic during the 1930s almost entirely to his holding this point of view.[1] Yet there is hardly a word throughout that period giving any indication of his attitude; indeed, the first open expression of it came only in 1944. I was led to seek the reasons for his guarded disposition and to speculate on whether it might have affected his work and choice of problems, especially whether there were things that he *refrained* from doing in consequence. In particular, it seemed to me that he could well have been more centrally involved in the development of the fundamental concepts of modern logic—*truth* and *computability*—than he was; in fact his role turned out to be peripheral in both cases.

What follows, then, is a partly speculative essay on these aspects of Gödel's scientific personality. It is intended to be largely complementary to the more settled and generally accepted picture offered in the piece mentioned above [that is, chapter 6 in this volume]. However, there is some overlap in the data to be interpreted. For the most part, in considering the above questions I drew on Gödel's own publications and published remarks. As it turned out, the views I developed in the process were reinforced by material found in Gödel's *Nachlass* and not previously available.

"Kurt Gödel: Conviction and caution" (Feferman 1984a) was first published in *Philosophia Naturalis* 21 (1984), 546–562, and is reproduced here with the kind permission of the publisher, Vittoria Klostermann GmbH, Frankfurt am Main. Some minor additions and corrections necessary to bring it up to date have been made.

*[The reference is to "Gödel's life and work," which appeared in Gödel (1986) after the article was written on which the present chapter is based. There is thus some overlap between the description here of Gödel's career with that in the preceding chapter.]

[1] Sources for these statements are given below.

It seems to me that the questions raised here are of interest to consider both for the purposes of scientific biography (in this case, of Gödel) as well as for case studies of conceptual analysis (here, of truth and computability).

To provide the needed reference points, it is necessary to begin with a quick survey of Gödel's work [see also the preceding chapter and the parts of Feferman (1986) not reproduced there].[2] Unless otherwise noted, dates of publication are of Gödel's works as listed in the references for this volume.

Gödel's work falls naturally into two parts, with 1940 as the dividing line. This also separates the loci of his activities, namely, Vienna prior to 1940 and Princeton thereafter. Gödel entered the University of Vienna as a student in 1924, finishing with a doctorate in 1930. He continued to be based in Vienna during the period from 1930 to 1939, becoming *Privatdozent* at the university in 1933. But he also made three academic visits to the United States during this time: twice to Princeton and the third to both Princeton and Notre Dame. In 1940, Gödel became a member of the Institute for Advanced Study in Princeton, where he eventually became a professor; he remained there until his death in 1978.

The period to 1940 contained Gödel's most famous contributions to mathematical logic: the completeness theorem (1930); obtained in his dissertation (1929); the incompleteness theorems (1931); and the consistency of the Axiom of Choice and the Generalized Continuum Hypothesis (1938 to 1940). There are further less well known but still important results from this period: on the decision problem, intuitionistic logic and arithmetic, speed-up theorems, and even the propositional calculus. Also of interest are some notes on geometry, most of which had been overlooked until rediscovered by Dawson (1983). [Cf. Gödel (1986) for all these publications.]

In the early 1940s, Gödel continued to work on problems of mathematical logic. He found a new quantifier-free functional interpretation of intuitionistic logic as early as 1941, though the results were not published until 1958 in the journal *Dialectica* (it has thus been dubbed "the *Dialectica* interpretation").[3] He also grappled for some years with the problem of the independence of the Axiom of Choice and the Continuum Hypothesis from the axioms of set theory, attaining only partial success with the first of these problems (never published). Increasingly Gödel turned his attention

[2]See also Kreisel (1980), Wang (1981), and Kleene (1987a) for more analytic surveys as well as biographical information [together with Dawson (1996)].

[3]It is of great interest and perhaps relevant to his work on the *Dialectica* interpretation that Gödel found in the early 1940s some essential errors in Herbrand's arguments for the quantifier-free interpretation which Herbrand had found for the predicate calculus. From material in the *Nachlass* we also know that Gödel sought to correct those errors. This work was never published. The difficulties in Herbrand were eventually rediscovered independently, and the errors corrected a number of years later by Dreben, Andrews, and Aanderaa (1963). For more information on this and further references, see van Heijenoort (1967) p. 525. [For a historical exposition of the *Dialectica* interpretation, see chapter 11 in this volume.]

to the philosophy of mathematics, represented by several publications beginning in 1944. The significance of this philosophy for work in set theory was brought out in the 1947 expository paper on the continuum problem. During the latter part of the 1940s, Gödel made an unusual contribution to relativistic cosmology which had significance for the philosophy of space and time. From Wang (1981) and other sources we know that Gödel was preoccupied with problems of general philosophy and metaphysics in the 1950s and thereafter.

The 1944 article was Gödel's contribution to the volume of the *Living Philosophers* series devoted to Bertrand Russell. He provided there a critique of Russell's mathematical logic both from a technical standpoint and with respect to the underlying interpretation of the theory of types. Russell had held a "no class" interpretation; this led him to adopt a ramified theory, though the prima facie problems of developing analysis in such a theory impelled Russell to adjoin the ad hoc Axiom of Reducibility. The latter move was criticized (in the 1920s) by Ramsey, who pointed out that doing so nullified the effect of ramification and that the whole could be just as well replaced by the simple theory of types. But it was Gödel who emphasized in 1944 that one must look to the conception of classes as entities having an existence independent of human thoughts and constructions in order to provide the proper interpretation of the resulting theory (as well as various theories of sets). Perhaps to ward off criticisms of this strong platonist (philosophically realist) position, Gödel compared the assumption of the existence of underlying mathematical objects with that of underlying physical objects, arguing that such assumptions were needed in both cases to deduce the data of ordinary experience, and were necessary to obtain a satisfactory account of that experience. Incidentally, Gödel mentioned in 1944 that a transfinite extension of the ramified hierarchy had proved useful in his own work, by which he meant the employment of it in 1938 to 1940 to define the constructible hierarchy of sets and to obtain a model of the Axiom of Choice and the Generalized Continuum Hypothesis.

In 1947 Gödel stated his philosophical viewpoint for the framework of Zermelo-Fraenkel Set Theory ZF, in terms of its informal interpretation in the cumulative hierarchy of sets. This is obtained by transfinite iteration of the power-set operation, where at each stage the set of all subsets (or power set) of a given set is supposed to exist independently of human constructions. Gödel says that Cantor's notion of (infinite) cardinal number is definite and unique. In particular, the set of all subsets of ω has cardinal $2^{\aleph_0}$, which is the same as the cardinal of the continuum. Since the Axiom of Choice AC must be granted to be true when there are no limitations on set construction, each set must be in one-to-one correspondence with a definite aleph, $\aleph_\alpha$. Thus the statement $2^{\aleph_0} = \aleph_1$ of the Continuum Hypothesis CH has a determinate truth value. The fact that efforts to prove it or disprove it had previously failed was neither here nor there, according to Gödel. Though he had himself proved CH (and GCH: $2^{\aleph_\alpha} = \aleph_{\alpha+1}$) consistent with ZFC (=ZF + AC), this told us nothing concerning its truth value.

Indeed, Gödel believed that CH was false, since it had (according to him) counterintuitive consequences. Thus, since he took the axioms of ZFC to be true, he anticipated that CH would be independent of ZFC. This was finally established by Cohen in 1963, just in time for Gödel to be able to include that information in a postscript to his 1964 revision of the 1947 paper. Though Gödel may have felt buttressed in his view of CH by this outcome, the question of the truth of CH was still not settled by it. Gödel himself thought that the continuum problem might eventually be settled by adjoining strong axioms of infinity; however, all subsequent work on plausible extensions of ZFC by such axioms has left CH undecided.

Further published expressions of Gödel's platonistic philosophical views are to be found in his Princeton Bicentennial remarks of 1946 (first appearing in Davis (1965), and subsequently in Gödel (1990)) and in the letters quoted in Wang (1974). While set-theoretic notions receive the most attention, a reading of Gödel (1958) and, again, of (1944) would suggest that his view of the objective character of fundamental mathematical entities and the definiteness of questions concerning them seems to go beyond set-theoretical realism to comprehend also such notions as abstract concepts (1944) and constructive functions and proofs (1958).

In the two letters to Wang dated 7 December 1967 and 7 March 1968, quoted in Wang (1974, pp. 8–11), Gödel says in the strongest terms that his philosophical views played an essential role in his work from the very beginning (1929). According to Gödel, though Skolem already had all the machinery needed for the completeness theorem for the first order predicate calculus in 1922, he [Skolem] did not arrive at the theorem itself because he lacked the "required epistemological attitude toward metamathematics and toward non-finitary reasoning."[4]

Gödel goes on in these letters to stress, similarly, the importance of his "objectivist conception of mathematics and metamathematics in general, and of transfinite reasoning in particular" for his further great successes of 1931 and 1938 to 1940. Those passages are elaborated on below.

It is clear from these quotations that Gödel held his "objectivistic" views by the time of his 1929 dissertation [but see the next chapter for anomalous data in this respect]. Unpublished material found in the *Nachlass* allows us to push that date back even further. It happens that Burke D. Grandjean, an Instructor in Sociology at the University of Texas (Austin), wrote to Gödel several times from 1974 to 1975, seeking information on Gödel's background in connection with research he was doing on the social and intellectual situation in central Europe during the first third of the twentieth century. For this purpose he also sent Gödel an individually designed ques-

[4]Gödel asserts that Skolem stated a form of the completeness theorem but gave an "entirely inconclusive argument." However, it may be questioned whether Skolem even appreciated the completeness theorem in the precise sense first formulated by Hilbert and Ackermann in their 1928 book. In particular, the evidence does not show that Skolem had a definite Hilbert-type formal system in hand for which he stated completeness.

tionnaire. In the *Nachlass* was found a typed response from Gödel dated 19 August 1975, together with the questionnaire which was dutifully filled out, though neither was ever sent to Grandjean![5] In his letter, Gödel refers to his (1964 revision of the) 1947 paper and to the quotations from Wang (1974) for a representation of his philosophical views, which he terms a form of conceptual and mathematical realism; he then goes on to say that he held these views since about 1925, in other words, early in his university years. But Gödel relates that his interest in philosophy dates back even further (along with mathematics) to around 1921 to 1922; in particular, he first studied Kant at that time. There is much in the letter and questionnaire concerning his relationship with the Vienna Circle and the standpoint of logical positivism (or empiricism), to which he was directly opposed philosophically. This supports what we know from other sources and is a matter to which we shall return below (cf. ftn. 19).

A general question may be raised about how much to accept of these retrospective reports, in which Gödel speaks of his state of mind and attitudes some forty or fifty years earlier. However, there is no contrary evidence that I know of which would throw doubt on them, and all the information we have available makes a rather coherent picture when assembled [but see chapter 8]. Although it would be a mistake to assume that Gödel's sophisticated views published in 1944 and 1947 were already worked out in his university days, it is safe to say that Gödel already had firm general platonistic views by the time he came in contact with the Vienna Circle.

Gödel was introduced to the Circle, which centered around Moritz Schlick, by his teacher Hans Hahn. This was around 1926; he attended meetings fairly regularly during 1926 to 1928 and after that gradually moved away from the Circle. From Wang (1974, pp. 7–13), Wang (1981, p. 653), and the 1975 letter and questionnaire for Grandjean mentioned above, we know that Gödel disagreed with the fundamental tenet of the Circle according to which mathematics is true by convention as the "syntax of language." [Gödel detailed the reasons for his opposition to this view in a paper entitled "Is mathematics syntax of language?" found in six drafts in his *Nachlass*; two of these drafts have subsequently been published, with an introductory note by W. Goldfarb, in Gödel (1995), pp. 334–362.] Nevertheless, he did not make his opposition to these views openly known at the time. Gödel grants that his interest in foundational problems was influenced by the Vienna Circle, in view of the prominence it gave to logic and the logicist program of Whitehead and Russell in their *Principia Mathematica*.

For the specific problems in mathematical logic and the foundations of mathematics that Gödel came to pursue, the figures who loomed in his

[5]The notation "nicht abgeschickt" is found not infrequently on Gödel's side of his correspondence in the files.

mind would no doubt have been Hilbert and Brouwer. Both had vigorously promoted foundational schemes opposed to the platonism implicit in Cantorian set theory. Hilbert's views had varied over the years but always took a formalist direction, settling in the 1920s into his program to obtain finitary consistency proofs for formal mathematical systems. Earlier he had taken a position equating existence of mathematical concepts with the consistency of axiom systems provided for them. While this was not a necessary part of Hilbert's finitist program, the emphasis on consistency instead of correctness (according to an informal interpretation) carried a residue of that position. Brouwer, on the other hand, rejected nonconstructive existence proofs of the Cantorian concept of "actual" infinities altogether, and attempted to rebuild mathematics on completely constructivist grounds, using his own intuitionistic version of constructivism.

In his 1929 thesis,[6] Gödel solved the problem concerning the completeness of an axiom system for the first-order predicate calculus which had been used by Hilbert and Ackermann in their 1928 book on mathematical logic. He showed that if an axiom system is consistent, then it has a model; Gödel's proof used König's lemma as an essential nonconstructive step. In a sense, Gödel's result could be said to justify the equation between existence and consistency. Gödel takes pains in his introduction to argue against this as an a priori philosophical claim (that the criterion for existence of a mathematical concept is simply the consistency of a system for it). In this respect he implicitly criticized Hilbert, though not identifying the position itself with Hilbert.[7] In other words, he is concerned to explain on this flank why his result is needed at all. Against Brouwer, on the other hand, he first wards off any attempt to interpret the completeness result itself in constructive terms, pointing out that this would lead to the decidability of the predicate calculus.[8] In addition, he says that there is no reason to restrict oneself to constructive methods of proof, since the problem dealt with is a mathematical one like any other.

What is of interest for the present essay is that these incisive (and pregnant) remarks were completely removed when Gödel came to publish his 1930 version of the thesis. Whether this was at Hahn's urging or on his own initiative is not known; in the latter case it would be the first evidence of the caution that is of concern here. But his retrospective comments suggest that from 1929 to 1930 he already saw himself as being opposed to the dominant viewpoint of the times. In his letter to Wang of 1967 (Wang

[6] One copy of this is available at the University of Vienna; a second (practically identical) copy was found in the *Nachlass*. [This appears as (1929) in the original and in English translation in Gödel (1986).]

[7] It is of interest that his criticism is elaborated via an argument, itself inconclusive, which anticipates the possibility of incompleteness results, as eventually obtained in (1931). [Cf. the discussion by B. Dreben and J. van Heijenoort in their introductory note to (1929, 1930), in Gödel (1986).]

[8] Undecidability was not established until 1936 by Church, but Gödel must have viewed decidability as unlikely or, at any rate, of another order of difficulty.

1974, pp. 8–9), Gödel speaks of the "blindness (or prejudice, or whatever you may call it) of logicians" at that time, according to which nonfinitary reasoning was not accepted as a meaningful part of metamathematics. "But now the aforementioned easy inference from Skolem 1922 is definitely nonfinitary, and so is any other completeness proof for the predicate calculus. Therefore these things escaped notice or were disregarded."[9]

Gödel did make one public statement of a philosophical character in the period directly following his thesis work. This was at an important symposium on the foundations of mathematics held in Königsberg in 1930, at which Carnap, Heyting, von Neumann, and Waismann presented the positions of logicism, intuitionism, formalism, and of Wittgenstein, respectively. In the ensuing discussion Gödel openly criticized the adoption of consistency as the criterion of existence. He then went on to announce publicly for the first time the existence of undecidable propositions: "(Assuming the consistency of classical mathematics) one can even give examples of propositions . . . which are really contentually true [*inhaltlich richtig*] but are unprovable in the formal system of classical mathematics." Hence, the negations of such statements could be adjoined to form a consistent system containing false statements; thus consistency would not guarantee existence in the intended sense. These remarks (Gödel 1931a) appeared after the famous 1931 paper. It seems that the significance of Gödel's remarks was appreciated by only a few participants (von Neumann among them) at the meeting itself. For a translation of the full discussion, with background and commentary, see Dawson (1984) [and the introductory note by Dawson to (1931a) in Gödel (1986)].

This brings us to the central topic here, namely, the role of the notion of truth in Gödel's incompleteness results (and, later, in his development of the constructible hierarchy). Let me begin by assembling what Gödel said on the subject in his publications and what he added in other communications.

The 1931 paper commences with a sketch of the proof of the first incompleteness theorem: for a specified formal system PM (adapted from *Principia Mathematica*), the device of (Gödel-)numbering is used to construct a statement $[R(q);q]$ equivalent to $\overline{Bew}[R(q);q]$, which we may interpret as expressing of itself that it is not provable in PM. Then if it is provable it would be true; hence by its interpretation it would not be provable. Thus it is not provable after all, so its negation is not true, hence not provable. Gödel says: "The analogy of this argument with the Richard antinomy leaps to the eye. It is closely related to the 'Liar' too."[10] He goes on to say:

---

[9]Gödel says elsewhere that he did not know Skolem's 1922 paper at the time of his thesis work, referring only to an earlier 1920 paper which achieved less.

[10]We are using the translation of (1931) in van Heijenoort (1967), pp. 596–616. [That is reproduced in Gödel (1986), facing the German original, pp. 144–195.]

> The method of proof just explained can clearly be applied to any formal system that, first, when interpreted as representing a system of notions and propositions, has at its disposal sufficient means of expression to define the notions occurring in the argument above (in particular, the notion of "provable formula") and in which, second, every provable formula is true in the interpretation considered. The purpose of carrying out the above proof with full precision in what follows is, among other things, to replace the second of the assumptions just mentioned by a purely formal and much weaker one.

What this comes down to in the body of the paper is that *consistency* of the system implies nonprovability of $[R(q);q]$, and ω-*consistency* implies the nonprovability of its negation. Evidently both hypotheses are weak consequences of *truth* of the system in a given interpretation.

Gödel lectured on his incompleteness results at the Institute for Advanced Study in the spring of 1934, during his visit for the academic year 1933 to 1934. Notes of the lectures were taken by S. C. Kleene and J. B. Rosser, and after they were reviewed and corrected by Gödel, mimeographed copies were made and circulated fairly widely. But they did not appear in print until they were included in Davis' 1965 collection *The Undecidable.* [They were subsequently reprinted as Gödel (1934) in Gödel (1986), pp. 346–371; see the introductory note by Kleene thereto.] In section 7 of these notes, Gödel discusses the relation of his arguments to the paradoxes, in particular that of Epimenides ("The Liar"). He argues that for person A to say (on a given day) that every statement he makes (on that day) is false can be made precise only if A specifies a language B and says that "every statement that he made in the given time was a false statement in B. But 'false statement in B' cannot be expressed in B, and so his statement was in some other language, and the paradox disappears." In other words, "the paradox can be considered as a proof that 'false statement in B' cannot be expressed in B." To this Gödel appends the following footnote appearing in the 1965, but not in the original 1934 version of the notes: "For a closer examination of this fact see A. Tarski's papers published in *Trav. Soc. Sci. Lettr. de Varsovie*, Cl. III No. 34, 1933 (Polish) [translated in *Logic, Semantics Metamathematics*, *Papers from 1923 to 1938*, by A. Tarski (1956); see in particular pp. 247ff.] and in: *Philosophy and Phenom. Res.* 4, (Tarski 1944), pp. 341–376. In these two papers the concept of truth relating to sentences of a language is discussed systematically." (There is also a reference to Carnap in this footnote.) [Cf. also Gödel (1986), p. 363, ftn. 25].

In the introduction by A.W. Burks to von Neumann (1966) (edited from lectures and manuscripts of von Neumann on automata), Burks refers to correspondence that he had with Gödel on one passage that had puzzled him. Gödel's response (quoted *op. cit.* pp. 55–56) is:

> I think the theorem of mine which von Neumann refers to is not on the existence of undecidable propositions or that on the lengths of proofs but rather the fact that a complete epistemological description of a language A cannot be given in the same language A, because the concept of truth of sentences of A cannot be defined in A. It is this theorem which is the true reason for the existence of undecidable propositions in the formal systems containing arithmetic. I did not, however, formulate it explicitly in my paper of 1931 but only in my Princeton lectures of 1934. The same theorem was proved by Tarski in his paper on the concept of truth published in 1933 in *Act. Soc. Sci. Lit. Vars.*, translated on pp. 152–278 of *Logic, Semantics and Metamathematics* [Tarski 1956].

Let us look next at Gödel's report of how he arrived at the incompleteness results in the first place, as reported by Wang (1981, p. 654). After his thesis, Gödel began studying Hilbert's problem to prove the consistency of analysis by finitary means. Finding this restriction on methods of proof "mysterious" and aiming to "divide the difficulties," "his idea was to prove the consistency of number theory by finitist number theory, and prove the consistency of analysis by number theory, where one can assume the truth of number theory, not only the consistency." In other words, he sought a relative consistency proof of analysis to number theory. Wang (*loc. cit.*) goes on to say:

> [Gödel] represented real numbers by formulas . . . of number theory and found he had to use the concept of truth for sentences in number theory in order to verify the comprehension axiom for analysis. He quickly ran into the paradoxes (in particular, the Liar and Richard's) connected with truth and definability. He realized truth in number theory cannot be defined in number theory and therefore his plan . . . did not work.

But Gödel was then able to go on to realize the existence of undecidable propositions in suitably strong systems.

We can reconstruct and flesh out this account as follows. If one were to try to give a relative consistency proof of analysis (in its form as second-order number theory) in first-order number theory by a formal model or interpretation, the obvious idea would be to interpret the set variables as ranging over the arithmetically definable sets. Equivalently, we could number formulas of arithmetic with one free variable x, say $A_n(x)$ ($n = 0, 1, 2, \dots$) and interpret the set variables as ranging over ω, with $m \in n$ interpreted as $A_n(m)$ is true. But for this model (that is, to interpret $x \in y$ as a formula of arithmetic with two free variables) one needs the *general notion of truth* of sentences of number theory. This would be problematic in view of the classical paradoxes. Gödel's key step was to realize that definite sense could be given to the phrase "this statement" in the formulation of the Liar paradox, by a diagonal substitution construction carried out in arithmetic. While one might question whether "this statement is false" is indeed a statement of natural language, there was no question that by Gödel's argument one could construct a statement of number theory with number q having specified self-referential properties. Thus if truth of number theory *were* definable within itself, one could find a precise version of the Liar statement, giving a contradiction. It follows that truth is not so definable. But provability in the system *is* definable, so the notions of provability and truth must be distinct. In particular, if all provable sentences are true, there must be true nonprovable sentences. The self-referential construction applied to provability (which *is* definable) instead of truth then leads to a specific example of an undecidable sentence. The technical work of 1931 goes into giving precise sense to the notion of definability within a formal system, verifying that provability in the system is so definable, carrying out the diagonal substitution construction, and checking that the hypotheses of consistency respectively ω-consistency, suffice to carry through the argument.[11]

We may conclude from all this that the concept of truth in arithmetic was for Gödel a definite objective notion, and that he had arrived at the undefinability of that notion in arithmetic by 1931. On the other hand, he did not state this as a result (only done so later, and independently, by Tarski in 1933), and he took pains to eliminate the concept of truth from the main results of 1931. This raises a series of questions, the first being: *Why*? Gödel's own answer is contained in part in the correspondence reproduced in Wang (1974, p. 9). Following the passages quoted above on the importance of his philosophical views for arriving at the completeness theorem, Gödel says:

> I may add that my objectivist conception of mathematics and metamathematics in general, and of transfinite reasoning in particular, was fundamental also to my other work in logic. How indeed could one think of *expressing* metamathematics *in* the mathematical systems themselves, if the latter are considered to consist of meaningless symbols which acquire some substitute of meaning only *through* metamathematics . . . it should be noted that the heuristic principle of my construction of undecidable number theoretical propositions in the formal systems of mathematics is the highly transfinite concept of "objective mathematical truth" as *opposed* to that of "demonstrability" (cf. M. Davis, *The Undecidable*, New York 1965, p. 64 [Gödel (1986), pp. 362–363] where I explain the heuristic argument by which I arrive at the incompleteness results), with which it was generally confused before my own and Tarski's work.

[11] As we know, Rosser later showed that consistency alone was sufficient to obtain incompleteness on both sides.

Later he adds (*op. cit.* p. 10), "formalists consider formal demonstrability to be an *analysis* of the concept of mathematical truth, and, therefore were of course not in a position to *distinguish* the two." This explains Gödel's view of why he had succeeded where others had failed (the general theme of the two letters quoted in Wang (1974, pp. 8–11) but only indirectly why he eliminated the concept of truth from his work.

A more direct explanation is found in a letter of reply dated 27 May 1970 to a graduate student named Y. Balas. The reply, which is in Gödel's *Nachlass* in draft form, is unsigned and marked "nicht abgeschickt" (cf. ftn. 6). Gödel explains here how he arrived at the undecidable propositions, including his attempt to give a "relative model-theoretic consistency proof of analysis in arithmetic . . . [by use of] an arithmetical $\in$-relation satisfying the comprehension axiom." He goes on to criticize Finsler's earlier attempt to prove "formal undecidability in an absolute sense," which he terms a "nonsensical aim."[12] But of greatest interest for the question raised here is the following vehement paragraph, which was crossed out in the draft reply:

> However in consequence of the philosophical prejudices of our time 1. nobody was looking for a relative consistency proof because [it] was considered axiomatic that a consistency proof must be finitary in order to make sense, [and] 2. a concept of objective mathematical truth as opposed to demonstrability was viewed with greatest suspicion and widely rejected as meaningless.

Here, in a crossed-out passage in an unsent reply to an unknown graduate student, I think we have reached the heart of the matter. Despite his deep convictions as to the objectivity of the concept of mathematical truth, Gödel feared that work assuming such a concept would be rejected by the foundational establishment, dominated as it was by Hilbert's ideas. Thus he sought to extract results from it which would make perfectly good sense even to those who eschewed all nonfinitary methods in mathematics. In doing so, he got a payoff which apparently even he did not anticipate,[13] namely, the second incompleteness theorem—according to which no sufficiently strong consistent formal system can prove its own consistency. Even more, once Gödel realized the generality of his incompleteness results it was natural that he should seek to attract attention by formulating them for the strong theories that had been very much in the public eye: theories of types such as PM and theories of sets such as ZF (Zermelo-Fraenkel).[14] But if the concept of objective mathematical truth would be rejected in the case of arithmetic, should not one expect an even greater negative reaction to the case of theories of types or sets? All the more reason, then, not to

[12]Cf. van Heijenoort (1967), pp. 438–440.

[13]Cf. Wang (1981), pp. 654–655.

[14]Cf. the first paragraph of Gödel (1931).

have any result depend on it, and no need then to express one's convictions about it.

What is clear so far is that Gödel anticipated Tarski at least in establishing the undefinability of arithmetic truth within arithmetic. No evidence has been presented, and none is available, to show that Gödel considered the problem which Tarski took to be the main one that he (Tarski) had solved: to give a set-theoretic definition of the notion of truth for the first-order language of any structure. Indeed, one may ask whether Gödel thought such a definition would even be necessary; this question should be raised with respect to each of his major results in the 1930s.

To begin with, in his completeness result (1929, 1930), the notion of *validity* of a formula in the first-order predicate calculus is understood in an informal sense. Nowadays, *that* is defined in terms of *satisfaction* in every structure (or interpretation) and that in turn is preceded by Tarski's inductive definition of satisfaction. Of course there was a long tradition of use of the informal notion of satisfiability (through the work of Löwenheim, Skolem, and others). It may be regarded as being well enough understood at the time of the 1928 book by Hilbert and Ackermann so as not to be considered problematic.[15] At any rate, Gödel, with his objectivist conception of truth, would not have thought it so, but would he have considered it a point on which he could be challenged?

Next there is an interesting footnote 48a (evidently an afterthought) in Gödel (1931):

> As will be shown in Part II of this paper, the true reason for the incompleteness inherent in all formal systems of mathematics is that the formation of ever higher types can be continued into the transfinite. . . . For it can be shown that the undecidable propositions constructed here become decidable whenever appropriate higher types are added. . . . An analogous situation prevails for the axiom system of set theory.

Since Tarski's work we analyze this by saying that the inductive definition of truth for a language can be given in an extension of that language to a next higher type, the satisfaction relation being the smallest relation satisfying such and such closure conditions. Formally, within a system of set theory S we can prove the consistency of any fragment S_0 having a natural model in V_α (the sets of rank $< \alpha$) by defining truth in $(V_\alpha, \in \cap V_\alpha)$ set-theoretically and proving that all axioms of S_0 are true in $(V_\alpha, \in \cap V_\alpha)$. Since Gödel did *not* write Part II to his 1931 paper and never commented further on footnote 48a, we do not know whether he saw it necessary to give

[15] There is evidence that Hilbert may have considered it problematic: in his 1928 Bologna speech (ref. (1928a) in van Heijenoort (1967)), Hilbert seeks (in a mistaken way) to replace the semantic notion of satisfiability by a syntactic one. It would be interesting to settle the respective roles of Hilbert and Ackermann in the writing of their book.

a set-theoretic analysis of the concept of truth in order to justify his claim. (The significance of ftn. 48a in this respect was brought to my attention by S. Kripke.)

Finally, there is the implicit use of the set-theoretic definition of truth in Gödel's 1938 and 1939 papers. The constructible hierarchy L_α (α an arbitrary ordinal) is defined inductively, taking $L_{\alpha+1}$ to be the set of all subsets of L_α which are definable in $(L_\alpha, \in \cap L_\alpha)$ (allowing parameters in L_α). But the precise notion of *definability* here requires prior explanation of satisfaction in $(L_\alpha, \in \cap L_\alpha)$. Moreover, to carry through the absoluteness argument outlined in the 1938 and 1939 papers, one would have to formalize the definition of truth in a set-theoretical structure $(a, \in \cap a)$ *within* ZF. Again, Gödel says nothing explicit about the matter. What we do know is that he found in 1940 a way to avoid considering it at all, by working within the Bernays-Gödel system of sets and classes, and replacing all work with formulas by operations on classes. This suggests that he was well aware of what would be needed were he to carry out in detail the proof sketched in 1938 and 1939.

These questions as to whether Gödel saw the need for a truth-definition may reward further pursuit; I hesitate to suggest answers on the basis of the present evidence. Only the slim statement in Davis (1965 ftn. 64) (added in 1965) quoted above—where Gödel credits Tarski for the systematic study of the concept of truth—gives any hint of Gödel's view of the matter.

Limitations of space do not permit me to go into a comparable discussion of Gödel's work bearing on the notion of effective computability. In any case, there is much material on this in the very interesting recent historical papers by Kleene (1981) and Davis (1982). Basically, Gödel was unconvinced by Church's thesis, since the proposed identification of the effectively computable with the λ-definable functions did not rest on a direct conceptual analysis of the notion of finite algorithmic procedure. For the same reason he resisted identifying the latter with the general recursive functions in the sense of Herbrand as modified by Gödel. Indeed, at the time (in 1934) that Church was beginning to propound his thesis, Gödel in his Princeton lectures was saying that the notion of effectively computable function could only serve as a heuristic guide. It was only in 1937 when Turing offered the definition in terms of his "machines" that Gödel was ready to accept such an identification, and thereafter he always referred to Turing's work as having provided the "precise and unquestionably adequate definition of formal system" by his "analysis of the concept of 'mechanical procedure' . . . needed to give a fully general formulation of the incompleteness results." [Cf. the postscript dated 3 June 1964 to the 1965 reprinting of Gödel (1934), also in Gödel (1986) pp. 369–370.] It is perhaps ironic that the various classes of functions (λ-definable, general recursive, Turing computable) were proved in short order to be identical, but Gödel cannot be faulted for his reservations on philosophical grounds at that point. Nevertheless, one may ask why Gödel did not pursue such an analysis *himself.*[16] Surely, given the depth of his understanding and experience, he was in a position to do that in 1934 or even earlier. My guess is that he also feared that no such proposal could be made convincing to the mathematical public of his day, just as the concept of truth would not be taken seriously. If so, the subsequent development showed Gödel to have been mistaken. Though certainly there were controversies about both Tarski's analysis of truth and Turing's thesis, they eventually took their place as accepted cornerstones of mathematical logic.

What I have termed Gödel's caution is in many respects understandable. At least initially, from the standpoint of a relative unknown in Vienna, the reaction of the Hilbertian establishment would be of genuine concern. But von Neumann's quick appreciation of the 1931 paper and Gödel's spreading fame should have reassured him that he would not be laughed offstage if he were to go beyond the purely (logico-)mathematical formulation of his results. What I find striking here is the contrast on the one hand between the depth of Gödel's convictions which underlay his work, combined with his sureness of insight leading him to the core of each problem, and on the other hand the tight rein he placed on the expression of his true thoughts. Only when he became established at the Institute for Advanced Study and the importance of his fundamental contributions was generally recognized did Gödel finally begin to feel free to let out what he really thought all along.[17,18] We know that Gödel was very careful in his personal habits, especially as concerned his health, so the caution discussed here is coherent with other characteristics of his personality. Only an in-depth biography could plumb the common sources and establish the interrelationships of the everyday and the scientific personality [for which, see now Dawson (1996)].

In conclusion, one may wonder how logic might have been different had Gödel been bolder in bringing his philosophical views into play in relation to his logical work, in particular by giving as much importance to concepts as to results. For, throughout the 1930s, he shied away from *new concepts as an object of study* as opposed to *new concepts as a tool for obtaining results* (for example, the constructible hierarchy). From one perspective, his strategy to avoid controversy was exactly right for the times; his work showed that logic could be pursued mathematically with results as decisive, important, and interesting as those from familiar branches of mathematics.

---

[16]This question is raised by Davis (1982) but I think not fully answered there.

[17]Incidentally, Hilbert died in 1943, the year before Gödel (1944) appeared.

[18]The following is curious in this respect. Gödel never did state publicly (until his communications to Wang reported in 1974) what he thought of the views of the Vienna Circle. Actually, he was supposed to contribute an article for the volume of the *Library of Living Philosophers* dedicated to R. Carnap. This volume eventually appeared in 1963 without Gödel's article, despite repeated attempts by the editor of the volume (P.A. Schilpp) to extract it from him. Six versions of this article, entitled "Is mathematics syntax of language?" have been found in the *Nachlass*; Gödel's answer to the question in his title is, of course, No. [As mentioned above, two versions of that article have subsequently been published in Gödel (1995), pp. 334–362.]

But far from leading to endless suspicion and controversy, the fundamental conceptual "philosophical" contributions of Tarski and Turing added to Gödel's work just what was needed to lay the groundwork for the subject of mathematical logic as we know it today.

Addendum

[The following points were added in proof to the article from which this chapter is drawn.]

(i) Hao Wang has reminded me of the exchange of letters in 1931 between Gödel and Zermelo, published in Grattan-Guinness (1979), which is relevant to Gödel's recognition of the undefinability of truth of a formal language within itself. Indeed, this is quite clearly stated by Gödel (*op. cit.*, p. 300), at least with reference to the kinds of languages dealt with in his 1931 paper.

(ii) Martin Davis has advanced the view (in a personal communication to me) that the content of Gödel's platonism underwent an evolution, taking into account the following remark at the conclusion of Gödel (1938): "The proposition A [$V = L$] added as a new axiom seems to give a natural completion of the axioms of set theory, in so far as it determines the vague notion of an arbitrary infinite set in a definite way." Davis argues that "this is not at all in the spirit of the point of view of (1947), and . . . it suggests that Gödel's 'platonism' regarding sets may have evolved more gradually than his later statements would suggest." Unfortunately, Gödel did not enlarge on his 1938 comment, nor did he return to it in any later publication. Certainly the case that Davis makes deserves serious consideration, though it may be difficult to settle conclusively without further evidence. [Cf. also the next chapter, which puts the evolution of Gödel's views in a different light.]

8

Introductory Note to Gödel's 1933 Lecture

This rich and, in certain respects, remarkable article is drawn from the handwritten text for an invited lecture which Gödel delivered (under the same title) to a meeting of the Mathematical Association of America, held jointly with the American Mathematical Society in Cambridge, Massachusetts, 29–30 December 1933. A report of the meeting is in the *American Mathematical Monthly* (vol. 41, 1934, p. 123–131). According to that report, Gödel's paper was supposed to appear in a later issue of the *Monthly*. However, we have no evidence that the paper was ever actually prepared for publication, let alone submitted.[1]

Gödel had arrived on his first trip to the United States in October 1933, for his visit to the Institute for Advanced Study in Princeton, which was to last until May 1934. In the months February to May 1934 he gave a course of lectures (Gödel 1934) at the institute on the incompleteness results.[2] According to these dates the text of Gödel's presentation may well represent the first public lecture which he delivered in English. Only one version of the text was found in the Gödel *Nachlass*; the manuscript is clearly written and there are no ambiguities. The English itself is very readable and the style is close to that of later publications by Gödel in English. There is no evidence of any editorial assistance by native speakers, though it is of course possible that colleagues at the institute provided some help.

The aim of Gödel's lecture is announced in his first paragraph with admirable clarity: "The problem of giving a foundation for mathematics

This chapter was first published as "Introductory note to ⋆*1933o*," pp. 36–44 in Kurt Gödel, *Collected Works: Vol. III: Unpublished Essays and Lectures*, edited by Solomon Feferman et al., copyright ©1995 by Oxford University Press, New York, and is reprinted here by kind permission of the publisher. There are some minor additions and corrections. This piece can be read independently of the lecture entitled "The present situation in the foundations of mathematics," to which it forms an introduction and which is to be found *op. cit.*, pp. 45–53. (Gödel 1933a in the references to this volume.)

[1]Incidentally, the other speakers at the same session were A. N. Whitehead on "Logical definitions of extension, class and number," and Alonzo Church on "The Richard paradox"; only the latter was subsequently published in the *Monthly*. (See Church 1934.)

[2]See the Gödel chronology by John W. Dawson, Jr., in Gödel (1986), p. 39.

. . . can be considered as falling into two different parts. At first [the] methods of proof [actually used by mathematicians] have to be reduced to a minimum number of axioms and primitive rules of inference, . . . and then secondly a justification in some sense or other has to be sought for these axioms."

Gödel says that the first apparent reduction of all of mathematics to a few axioms and rules of inference, as carried out by Frege,[3] was faulty in that it led to contradictions, and some restrictions in dealing with infinite sets (or "aggregates") are necessary. The way such restrictions must be made "seems to be essentially determined by the two requirements of avoiding the paradoxes and retaining all of mathematics," including the theory of sets. Gödel then claims that the first part of the problem of foundations for mathematics has been solved in a completely satisfactory way by means of formalization in the Simple Theory of Types (here abbreviated STT)[4] when all "superfluous restrictions" are removed.

According to Gödel, there are three such restrictions in STT: (1) just pure types 0, 1, 2 . . . are admitted, where the objects of type $(n+1)$ admit only objects of type n as members; (2) thence "$a \in b$" is taken to be meaningless when a, b are not of successive type levels; (3) only finite type levels are admitted. For STT the objects of type level n are interpreted as ranging over sets T_n, where $T_{n+1} = \mathcal{P}(T_n)$ is the set of all subsets of T_n. In place of (1) Gödel considers the inessential modification of replacing the pure type levels T_n by the cumulative type levels, informally interpreted by the collections S_n, where $S_{n+1} = S_n \cup \mathcal{P}(S_n)$. Then in place of (2), he assumes that "$a \in b$" is meaningful for all a, b in the universe of sets under consideration, with "$a \in b$" taken to be false whenever b is of the same or lower cumulative type level than a. Finally, this permits in place of (3) a natural extension to transfinite levels S_α, with $S_\alpha = \bigcup_{\beta<\alpha} S_\beta$ for limit α and $S_{\alpha+1} = S_\alpha \cup \mathcal{P}(S_\alpha)$ for all α. As Gödel states, this kind of generalization of the theory of types is formulated axiomatically in theories of sets (or "aggregates") due to Zermelo, Fraenkel, and von Neumann. He refers specifically to von Neumann (1929) for the motivation to remove the restrictions (1)–(3).[5]

However, Gödel does not fix a specific formal system of set theory as the result of removing the restrictions (1)–(3) from STT. And indeed there is an ambiguity in his description (pp. 8–9) of the transfinite levels α to be admitted, for which he seems to suggest that α must already be "defined" in an earlier system S_β with $\beta < \alpha$. It is not clear what notion of definability

[3]Peano also mentioned in this connection.

[4]See the introductory note to (1930b, 1931, and 1932b) in Gödel (1986), p. 129, as well as Gödel (1931), pp. 176–178, for Gödel's version of STT.

[5]The basic papers are, of course, Zermelo (1908), Fraenkel (1922), and Skolem (1923a). Gödel could have referred to Zermelo (1930) for the first clear statement of the informal interpretation of axiomatic set theory in the cumulative hierarchy.

Gödel has in mind, or whether he is thinking about S_α as a collection of sets or as a formal system true at level α in the cumulative hierarchy.[6]

Though one is freed of the "superfluous restrictions" (1)–(3) in the manner described (Gödel continues), it should not be considered that one has reached an end with this [unspecified] system of axioms for set theory, since the process of "creating" ever higher types can be continued beyond those dealt with in any given system S. While we set out to find a single formal system for mathematics, instead we have found an infinity of such systems. In practice one can confine oneself to a system for the first few levels, where all the mathematical methods and results developed up to now can be formally represented. But, says Gödel, the unending process of construction of formal systems for ever higher types is necessary in principle, as is evidenced by the incompleteness results of his 1931 paper: for each consistent system S based on the theory of types, there are true (and simple) arithmetic propositions which cannot be proved in S but which become provable if one adds axioms to S for the next higher type. A special case of this general theorem is that "there are arithmetic propositions which can be proved only by analytical methods and, further, that there are arithmetic propositions which cannot be proved even by analysis but only by methods involving extremely large infinite cardinals and similar things."[7] Here Gödel harks back to the well-known footnote 48a of his 1931 paper and anticipates his later advocacy, in the 1964 supplement to his 1947 paper and elsewhere, of the need for large cardinal axioms to settle open arithmetical and set-theoretical problems, including the continuum hypothesis.

Gödel turns from this (on p. 15 of his lecture text) to the second part of the foundational problem for mathematics, namely, that of providing a justification for the axioms and rules arrived at in the first part. Here "it must be said that the situation is extremely unsatisfactory." There is no problem about treating the formalism in a purely formal way as a kind of game with symbols, "but as soon as we come to attach a meaning to our symbols serious difficulties arise." These difficulties are essentially of three kinds: (1°) the nonconstructive notion of existence, (2°) the "still more

[6]Under the former reading Gödel may have intended a formulation related to Fraenkel's replacement axiom, namely, that one has a formula $F(x,y)$ which defines a function from members x of a set $a \in S_\beta$ to ordinals y whose range $\alpha = \{y | \exists x(x \in a \wedge F(x,y))\}$ is closed under predecessor and where β is previously obtained. The latter reading in terms of formal systems suggests a notion like that of autonomous progressions of theories, in the sense of Feferman (1962). Charles Parsons has suggested to me that on neither reading is the procedure strong enough to justify the Zermelo-Fraenkel axioms. According to Wang (1974), p. 186, apparently Gödel thought more general considerations are necessary to arrive at the Replacement Axiom.

[7]Note that on p. 13, Gödel calls a proposition "arithmetic" if it states that a decidable property holds for all integers. This kind of proposition is nowadays said to belong to the class Π^0_1 (cf. p. 135 of the introductory note to (1930b, 1931, and 1932b) in Gödel (1986)), while "arithmetical" is reserved for propositions which fall somewhere in the hierarchy of Π^0_m and Σ^0_m classes. Gödel did use "arithmetical" in a sense equivalent to the latter in (1931), §3 (translation in Gödel (1986), p. 181).

serious" general notion of class (of arbitrary type), and (3°) the Axiom of Choice.

As to (1°), Gödel points out that it is possible to infer the existence of an object (even an integer) satisfying a property P, without actually being able to produce (or describe a procedure to produce) a specific instance of P; he rightly traces this back to the assumption of the law of excluded middle in the basic logic of the formalism.

Concerning (2°), Gödel finds problematic the use of impredicative definitions for specifying classes (of any given type), for example, to define a set of integers consisting of all n satisfying a property P, where P is defined by reference to "all" properties of integers (or "all" classes of integers as their extensions), among them P itself. This kind of definition can be considered as unobjectionable only if one assumes that there is a preexistent totality of all properties (or all classes) of integers (and more generally of all properties of a given type).[8] But we have a vicious circle if properties are regarded as "generated by our definitions." The trouble with such a predicative (definitionist) conception of properties is that adhering to it "makes impossible an adequate theory of real numbers, many of whose fundamental theorems appear to depend essentially on the non-predicative definitions."

Gödel says practically nothing about (3°), the Axiom of Choice, as the third "weak spot" in the axioms, on the grounds that "it is of less importance for the development of mathematics." This casual remark is a bit surprising in view of the intense controversy over the legitimacy of the Axiom of Choice AC (and its equivalents such as the Well-Ordering Principle) which figured prominently in foundational discussions in the first third of this century (see Moore 1982). However, it may be that Gödel had already started thinking about the relative consistency of the Axiom of Choice (with the axioms of Zermelo-Fraenkel Set Theory ZF) and had evolved a plan as to how this might be established.[9]

What is most surprising is Gödel's next statement (p. 19): "The result of the preceding discussion is that our axioms, if interpreted as meaningful statements, necessarily presupposes a kind of Platonism, which cannot satisfy any critical mind and which does not even produce the conviction that they are consistent." This does not seem to square with Gödel's unequivocal assertions, quoted in letters of 1967 and 1968 to Hao Wang and reproduced in Wang (1974, p. 8–11), that he had held a platonistic philosophy of mathematics since his student days in Vienna.[10] These are reinforced by

[8]On p. 17, Gödel says that "the notion of 'class of integers' is essentially the same as that of 'property of integers.'" Much later (in 1944) he distinguished between these two notions, but found it unobjectionable to assume a totality of all properties (or all concepts) as well as a totality of all classes of a given type; cf. Gödel (1944), pp. 138ff., and the discussion by Charles Parsons thereto in Gödel (1990), pp. 106ff.

[9]Gödel established the consistency of AC with ZF in 1935 (though not announced until 1938); cf. Feferman (1986), pp. 9 and 21–22.

[10]See also Wang (1978), p. 183.

Gödel's statement, in an unpublished (and unsent!) response of 1975 to a questionnaire from Burke D. Grandjean, that he had held such views since 1925.[11] There is certainly no questioning Gödel's unremitting espousal in print of full-fledged platonism, beginning with his 1944 article on Russell's mathematical logic and continuing (especially in his 1947 article on Cantor's continuum problem) until his death. Of course, one may take Gödel's retrospective assertions about his earlier views *cum grano salis*, colored as they are by his later firm stance. Alternatively, one might seek to make his statement in this article coherent with the rest of the evidence concerning his views by distinguishing varieties of platonism. For example, it is reasonable to say that one holds such a position as regards integers, but not as regards sets; conceivably Gödel could have started as a moderate platonist in this sense and only later shifted to the stronger position.[12] Another possibility is that Gödel had, at the time of this lecture, a temporary period of doubt about set-theoretical platonism. Unless further evidence from the *Nachlass* comes to light that we are not presently aware of, all of this must, unfortunately, remain speculative.

In any case, within the context of this chapter, Gödel's assertion of the unacceptability of the platonist position as justification for the axioms of set theory must be taken at face value. He does go on to say (pp. 19–20) that, at least, it is very likely that these axioms are consistent, since their consequences have been developed extensively without arriving at an inconsistency. Moreover, the question of consistency is a purely combinatorial one about manipulation of symbols according to specified rules, so one might hope to establish the consistency of these axioms by "unobjectionable methods." This leads into a description of Hilbert's foundational program, which was to be carried out by finitary[13] consistency proofs for formal systems for mathematics. However, according to Gödel (p. 25) there is no hope of carrying out Hilbert's program even for classical arithmetic (now referred to as Peano Arithmetic, PA), let alone for set theory, "in view of some recently discovered facts." Namely, by Gödel's second incompleteness theorem (*Satz* XI of (1931) as strengthened in (1932)), no

---

[11]See Feferman (1984a) [reproduced here as chapter 7] for this and further information about Gödel's retrospective claims.

[12]Cf. the suggestion by Martin Davis (quoted in Feferman (1986), p. 31, fn. 21), that Gödel's platonism regarding sets may have evolved more gradually than his later statements would suggest; Davis pointed particularly to the remark at the end of Gödel (1938) as evidence for this.

[13]Gödel equates finitism with intuitionism in its most restricted sense; he denotes by "A" a system in which just such methods are allowed. While he does not specify A in any detail, he does describe its essential features.

According to Bernays (1967), p. 502, the prevailing view in the Hilbert school at the beginning of the 1930s equated finitism with intuitionism (cf. also Sieg (1990), p. 272). This was shaken in part by Heyting's formalization of intuitionistic arithmetic and Gödel's simple reduction (in (1933)) of classical arithmetic to that system. Wilfried Sieg has suggested to me that the kinds of considerations presented by Gödel in the lecture under discussion helped logicians to recognize finitist mathematics as a proper part of intuitionistic mathematics.

formal system S which contains PA can prove its own consistency, unless it is inconsistent. Gödel goes on to say (p. 26) that

> all the intuitionistic [finitary] proofs complying with the requirements of the system A [of finitistically allowable methods] which have ever been constructed can easily be expressed in the system of classical analysis and even in the system of classical arithmetic, and there are reasons for believing that this will hold for any proof which one will ever be able to construct.
>
> So it seems that not even classical arithmetic can be proved to be non-contradictory by the methods of the system A.

This last is especially interesting in view of the cautious statement Gödel had made near the close of his 1931 paper concerning the significance of the second incompleteness theorem for Hilbert's Program, in which he said: "I wish to note expressly that Theorem XI . . . [does] not contradict Hilbert's formalistic viewpoint . . . [which] presupposes only the existence of a consistency proof in which nothing but finitary means of proof is used, and it is conceivable that there exist finitary proofs that *cannot* be expressed in the formalism of P."[14] Here "P" denotes Gödel's system of simple type theory, with the axioms of classical arithmetic taken to hold for the objects of lowest type. Clearly, Gödel had convinced himself in the interim that all consistency proofs using methods that were clearly finitary, as then recognized, could readily be formalized in PA, and that there was nothing in sight that suggested finitary methods would be capable of going beyond arithmetic, and certainly not beyond analysis. Hence, it would be hopeless to carry out Hilbert's program for arithmetic unless some radically new finitary methods not expressible in arithmetic could be found. But concerning this last possibility, as we saw, Gödel added even more strongly that he had "reasons to believe" that no such methods could ever be produced. He gives no indication what such reasons are, and in fact he never provided such in print until 1958, in the introductory discussion of that paper—which, however, makes essential reference to consistency proofs by ordinals as first provided in Gentzen (1936).

It should be noted that Gödel does refer to partial successes of Hilbert's program, "the most far-reaching of which is the following theorem proved by Herbrand," which provides a general method for consistency proofs of certain systems with axioms whose existential quantifiers can be realized constructively.[15] However, this can be used to account only for a fragment of classical arithmetic, namely, that in which induction is restricted to quantifier-free formulas.[16]

---

[14]Translation from the original; cf. Gödel (1986), p. 195.

[15]As pointed out to me by Warren Goldfarb, the theorem stated by Gödel is never stated this way by Herbrand, but it is an easy consequence of a theorem Herbrand formulated in his (1930) (cf. Herbrand 1971, p. 179) and again in (1931) (cf. Herbrand 1971, p. 289).

[16]Cf. Herbrand (1931); that case had previously been treated in a rather complicated

The article concludes with a discussion of the possible use of intuitionistic methods in the wider sense, as developed by the Brouwerian school. This goes beyond the system A of finitary methods by, among other things, permitting negation to be applied to general propositions, as for example, $\neg(x)F(x)$, which can (in certain cases) be established by intuitionistic methods without producing an x such that $\neg F(x)$. For, the meaning given to $\neg p$ is that one has a (constructive) demonstration that any proof of p reduces to absurdity $\perp$, which is the same as the conditions for $p \supset \perp$. Gödel mentions that for the "axioms of the intuitionistic mathematics as stated by Heyting . . . we find that for the notion of absurdity exactly the same propositions hold as do for negation in ordinary mathematics—at least, this is true within the domain of arithmetic" (p. 29). Presumably Gödel is here referring to his results in 1933, which provided a translation of classical arithmetic (PA) into its intuitionistic counterpart (HA), which preserves formulas that do not contain the disjunction or existence operation symbols. But he objects to the intuitionistic meaning of negation (and hence, more generally, of implication) because it refers to the totality of all possible (intuitionistic) proofs of a proposition, and this violates what he considers to be a basic principle of constructivity, namely, that "the word 'any' can be applied only to those totalities for which we have a finite procedure for generating all their elements." Since the supposed totality of intuitionistic proofs is too vague to meet this criterion, Gödel concludes that "this foundation of classical arithmetic by means of the notion of absurdity is of doubtful value." Still he hopes that in the future "one may find other and more satisfactory methods of construction beyond the limits of the system A" to "found" classical arithmetic and analysis. As we see from Troelstra's introductory note to (1958) and (1972) in Gödel (1990, pp. 217–241), this was something Gödel struggled to achieve until late in his life, but he never succeeded to his own satisfaction, even for arithmetic.[17]

No doubt Gödel would have worked on improving this article were he ever to have brought it forward to publication, but it could hardly have been bettered for its time. What is impressive about the piece is how well it holds up after more than half a century's subsequent intensive development of those areas of mathematical logic that most bear on Gödel's foundational concerns, in set theory, proof theory, and constructivity. In only two (related) respects would he perhaps want to modify certain points to take account of technical developments in recent years. The first has to do with the claimed impossibility of an adequate theory of real numbers (and hence of analysis) without the use of impredicative definitions (p. 19). Here Gödel could have referred to Weyl's landmark work *Das Kontinuum* (1918)[18] for

special way by von Neumann (1927).

[17]See also Troelstra's introductory note to (*1941) in Gödel (1995), pp. 186–189 [and chapter 11 in this volume].

[18]Now translated into English, as Weyl (1987).

its pioneering effort to demonstrate the viability of a system of mathematical analysis, based on strictly predicative principles, that could account for the bulk of nineteenth-century analysis. Following World War II, a body of work produced over a period of years by Lorenzen, Kreisel, Takeuti, and others (including the author) has led to the conclusion that much of twentieth-century analysis can also be given a predicative foundation.[19] Indeed, it follows from Feferman (1977), Takeuti (1978), and Feferman (1985a) that this portion of analysis can already be formalized in a system S which is proof-theoretically (finitarily) reducible to classical arithmetic PA (and which is a conservative extension of PA). Moreover, it appears that such an S serves to account for all scientifically applicable portions of analysis.* In a way, this outcome was anticipated by Gödel's remark on page 11 (already noted above) that "all mathematics hitherto developed" can be carried out in much weaker systems than (Zermelo-Fraenkel or von Neumann) axiomatic set theory, since the first few type levels suffice. Indeed, that is correct, but it does not go far enough: what is crucial is Weyl's recognition (in his 1918 monograph) that *sequential (Cauchy) completeness* is adequate for much of analysis where *set-theoretical (Dedekind cut) completeness* is ordinarily applied; the latter is essentially impredicative, whereas the former permits a strictly predicative development.

This ties up with the second respect in which Gödel might have modified his discussion in the light of recent technical work, in this case that carried out by Friedman, Simpson, and their followers under the rubric of "reverse mathematics." According to their results, of which an informative survey is to be found in Simpson (1988), much of nineteenth- and twentieth-century analysis and algebra can already be formalized in a system S_0 that is conservative over the fragment PRA (Primitive Recursive Arithmetic) of PA. It is generally accepted that PRA is a finitistically acceptable system. Moreover, by Sieg (1985) one has a clear finitistic reduction of S_0 to PRA (incidentally, obtained mainly by making use of basic proof-theoretical techniques due to Herbrand and Gentzen). It is in this sense that partial realizations of Hilbert's program are much more successful than could have been recognized at the time of Gödel's lecture. As for the current picture of the more general issues raised by Gödel, one may say that on the one hand, set-theoretical foundations have by and large won the day among working mathematicians, and that on the other hand, the need to provide some sort of constructive foundations for mathematics has not receded among the relatively small group of "critical minds" who reject the platonistic philosophy underlying set theory. Efforts directed at such constructive foundations have taken a great variety of forms, in particular through the massive body of work in proof theory on extensions of Hilbert's program,

[19]Cf. Feferman (1964) for an earlier survey, and Feferman (1988a) [reproduced as chapter 13 in this volume] for a more up-to-date formulation and assessment of Weyl's program.

*[Cf. chapter 14 in this volume in this respect.]

but there is no settled opinion as to what has been achieved thereby. The interested reader should consult such sources as Buchholz et al. (1981), Feferman (1988), Kreisel (1968), Schütte (1977), and Takeuti (1987) (including the appendices by Feferman, Kreisel, Pohlers, and Simpson) [as well as chapter 10 in this volume].[20]

[20]I wish to thank M. Davis, J. W. Dawson, Jr., P. de Rouilhan, W. Goldfarb, C. Parsons, and W. Sieg for their helpful comments on a draft of this note.

Part IV

PROOF THEORY

9

What Does Logic Have to Tell Us about Mathematical Proofs?

Modern logic provides a theoretical analysis of mathematics which cuts across its traditional subdivisions into algebra, geometry, analysis, etc. Actually, there is no single logical theory but a variety of such whose aim is to model the reasoning of, say, an *idealized platonistic* (set-theoretical) *mathematician* or an *idealized constructivist mathematician* at work in one of these areas of mathematics. What is common to the various theories is the use of *formal systems* to describe that activity, namely, as being one of drawing consequences within a specified class of symbolic expressions called *formulas* from certain formulas called *axioms* by use of certain specified *rules of inference.* Some logicians call themselves *formalists* and think that there is nothing more to mathematics than what is pictured as the production of derivations within formal systems. But those who take seriously the platonistic, constructivist, or other such views (for example, finitist, predicativist, etc.) also concern themselves with the *meaning* of what is expressed in formal languages. There is then the question as to which choices of axioms and rules are legitimate and—in case the systems are incomplete—how they might be legitimately extended. This leads one into controversial areas of the *foundations of mathematics.* I stress here the *syntactic* (formal) aspect as opposed to the *semantic* (and/or foundational) aspect of the logical description of mathematical activity. This is more neutral territory, but one within which there are a number of notions and results concerning the logical structure of mathematical proofs that are relevant to the remarks by Y. Manin in "Digression: proof" (Manin 1977, pp. 48–51). It should be added that I found those remarks both refreshing and stimulating.

Logical theories have sometimes been compared with physical theories. In this analogy, formal systems play roughly the same role for the former as

"What does logic have to tell us about mathematical proofs?" was first published in *Mathematical Intelligencer* 2 (1979), 20–24 (Feferman 1979b); it is reprinted here with the kind permission of the publisher, Springer-Verlag GmbH & Co. KG, Heidelberg. There are some minor additions and corrections.

differential equations for the latter. Neither of these theoretical endeavors pretends to be all-encompassing of ordinary experience. Indeed, the explanation of what each kind of theory covers almost requires question-begging answers: logic deals with the logical structure of mathematics as physics deals with the physical structure of the world. An argument can be made (in each case) that it is the essential, underlying structure of that part of experience which is being studied. Be that as it may, the behavior of biological systems cannot be deduced from physical principles and the creative and intuitive aspects of mathematical work evade logical encapsulation.

One part of the analogy with physics is not so apt. The experimental method provides highly sophisticated, refined means to test physical theories and thence lead to their corroboration, modification, or rejection. We have no such tests of logical theories. Rather, it is primarily a matter of individual judgment how well these square with ordinary experience. The accumulation of favorable judgment by many individuals is of course significant. It will be seen, though, that there are some isolated cases where logical theory predicts the existence of a proof or result of a special character and which have been corroborated by further detailed investigation. There seem to be no such tests for negative (impossibility) results. Although the significance of the logical work is thus not conclusive, I hope to convince the reader that there is much of interest which is known or being investigated and to encourage learning more about the proof-theoretical side of logic.

Formal systems were introduced by Frege toward the end of the nineteenth century (1879). Passing through the work of Russell, they reached their first stage of generality at the hands of Hilbert and his collaborators (particularly Ackermann) in the 1920s. A second stage of generality was provided in the 1930s by the work of Gödel, Church, Turing, and Post on effective computational processes. The formulation of a precise notion of such processes in terms of Turing machines is most vivid and most convincing. Here the extent of idealization in comparison with actual computational practice is also readily apparent. Turing machines are conceived to operate on finite sequences of symbols but without limitation of space and time. The resulting class of computable functions—the so-called recursive functions—includes as a small subclass the primitive recursive functions, which in turn are classified into an infinite hierarchy in which the actually computable functions are found at the lower end. In other words, the general theory of effective computation tells us nothing about what is actually feasible, though it provides a theoretical setting in which to study the latter. Even where the theory gives a negative result (and hence is best possible) such as the recursive unsolvability of Diophantine questions in general (cf. Davis, et al. 1976), there is a large gap (measured in terms of number of variables and degrees of polynomials) between the classes of Diophantine equations which can be solved systematically and those which cannot.

Post went so far as to identify formal systems with recursively enumerable (r. e.) sets. This is justified if one simply says that such a system S is prescribed by the choice of a recursive (decidable) set of "axioms" Ax_S and of recursive "rules of inference" $Rule_S$. The resulting sets of "theorems" Thm_S generated from Ax_S by $Rule_S$ are exactly the r. e. sets. But such a conception of formal system has only the most tenuous connection with mathematical practice. It is useful only to state the most general negative results on undecidability of systems which are adequate to computational number theory (cf. Smullyan 1961). (One may add that it has proved useful in recent years to consider concepts of formal systems for infinitely long expressions, where there is not always a natural generalization of recursion-theoretic concepts.)

The more familiar conception of formal system due to Hilbert is closer to practice but, as we shall see, can be improved. Here each S is specified within a given *formal language* L_S, determined by a choice of *basic mathematical vocabulary* and of *logical signs*. The vocabulary will specify variables of different sorts, conceived as having specified ranges (for example, the natural numbers $\mathbf{N}$, or real numbers $\mathbf{R}$, or a vector space V, or sets of natural numbers $\mathcal{P}(\mathbf{N})$, or functions in $\mathbf{N} \to \mathbf{N}$, etc.) and various relations and operations between these (for example, $=, <, \in, +, \cdot$, scalar product, function application, etc.). From these data is generated the set of *terms* (of L_S), which denote objects of various sorts and then the set of *atomic formulas*, which express basic relations between terms.

The logical signs of L_S may be such as $\neg$ (not), $\wedge$ (and), $\vee$ (or), $\to$ (implies), $\forall$ (for all), $\exists$ (there exists). Other common signs are obviously definable in terms of these, for example, $\leftrightarrow$ (is equivalent to) and $\exists!$ (there exists unique). Further reductions depend on the general viewpoint. On the platonistic account of mathematics, propositions assert matters of truth or falsity: $(\neg A)$ is true if and only if A is false. Then both $(A \vee \neg A)$ (excluded middle) and $\neg(A \wedge \neg A)$ are true. Further, $(A \vee B)$ can be identified with $\neg(\neg A \wedge \neg B)$ and $(A \to B)$ with $(\neg A \vee B)$ and finally $(\exists x)A$ with $\neg(\forall x)\neg A$. On the constructivist view, propositions only report assertions of completed (constructive) proofs. We do not assert $(A \vee B)$ unless one or the other of A, B has been established. In particular, we cannot assert $(A \vee \neg A)$ when the status of A has not yet been decided. Similarly, to assert $(\exists x)A(x)$ we must have produced a specific c for which $A(c)$ is proved; to assert $\neg(\forall x)\neg A(x)$, on the other hand, means that the assumption of $(\forall x)\neg A(x)$ leads to a contradiction and in general provides no method for selecting a particular c satisfying A. Thus, no reduction in logical signs is possible here. In any case, once the selection of logical signs is specified, the class of (*well-formed*) *formulas* of L_S is inductively generated from the atomic formulas by repeated application of the associated operations $A \mapsto (\neg A); A, B \mapsto A \wedge B; \ldots ; A \mapsto (\exists x)A$.

It is a matter of common experience that the representation of mathematical predicates and statements by formulas within formal languages and the translation of formulas back again into informal language is straightforward. I believe most would judge that it gives a satisfactory account of the logical structure of mathematical statements. However, the following should be noted: (i) we can construct strange formulas by our rules, for example, $(\exists x)A$ with x not a variable of A, and $(A \wedge B)$ where A, B are unrelated; (ii) we can construct useless combinations, for example, $A \wedge A \wedge A \ldots \wedge A$; and (iii) the process of *explicit definition*, which allows us to expand our basic vocabulary systematically, is necessary to keep formal representations of informal concepts down to manageable size. This last point is most important and is somehow related to psychological features of ordinary understanding. If everything were written out in terms of primitive notions, mathematics would be unlearnable. The process of explicit definition for frequently used concepts reduces complexity. But a trade-off comes when we are required to keep too many definitions in mind!

Having specified the language L_S we can now turn to the description of the proof-structure of S. This is determined by a choice Ax_S of a (decidable) set of formulas as axioms and of a finite collection $Rules_S$ of (decidable) rules of inference

$$\frac{A_1, \ldots, A_n}{A}$$

(in practice $n = 1$ or 2). A *linear proof* (like a sequence of statements in a book) is a finite sequence of formulas, each member of which either is an axiom or is obtained from some previous members by application of one of the rules of inference. A *tree proof* shows the logical structure of the argument more clearly. The topmost nodes of the tree are axioms and at each application of a rule of inference we have a branch:

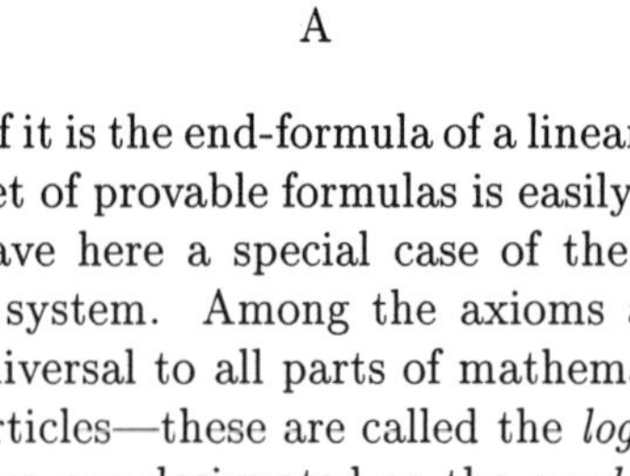

A formula is *provable* if it is the end-formula of a linear proof, or equivalently of a tree proof. The set of provable formulas is easily seen to be recursively enumerable, so we have here a special case of the Turing-Post general conception of formal system. Among the axioms and rules of inference are ones which are universal to all parts of mathematics, and concern the laws of the logical particles—these are called the *logical axioms and rules*, and the remaining ones are designated as the *nonlogical* or *mathematical axioms and rules*. For example, under the former we may have axioms $(A \to A \vee B)$, $(B \to A \vee B)$, $(A \to C) \wedge (B \to C) \to (A \vee B \to C)$, and as rules $\dfrac{A, A \to B}{B}$ and $\dfrac{A(x)}{(\forall x)A(x)}$. Under the latter we may have axioms such as $\forall n \forall m(n+1 = m+1 \rightarrow n = m)$ and $\forall a \forall b[a+b = b+a]$ and rules such as

$$\frac{A(0), [A(n) \rightarrow A(n+1)]}{(\forall n)A(n)}$$

(where "n" ranges over **N**). The choice of axioms and rules is sensitive to the basic point of view; thus, $(A \vee \neg A)$ is an axiom in *classical* (platonistic) *logic* but not in *intuitionistic* (constructive) logic.

The first major variation of this conception of formal system and of proofs was due to Gentzen (1935), who introduced systems of *natural deduction.* (This was done independently by Jaśkowski, but the results obtained by Gentzen have been much more important.) In practice, when establishing an implication $(A \rightarrow B)$ by a direct argument we proceed by assuming A and drawing B as a consequence. (In an indirect argument or proof by contradiction we assume A and $\neg B$ and draw a contradiction $\neg A$.) To explain such procedures we have to analyze how consequences A are drawn from assumptions Γ (finite lists of formulas), where Γ may change as we move from one part of the argument to another. To indicate that this relations holds we write $\Gamma \vdash A$ or Γ above A. Then the direct method of establishing $\Gamma \vdash (A \rightarrow B)$ is to show that $\Gamma, A \vdash B$; this argument is indicated more graphically by

$(\rightarrow)$*introduction*
$$\frac{\begin{matrix}\Gamma, \not{A} \\ B\end{matrix}}{A \rightarrow B}.$$

The part $\begin{matrix}\Gamma, A \\ B\end{matrix}$ of this figure indicates that B has been inferred from Γ and A, while the slash through A (*cancellation*) indicates that we can delete it as hypothesis when now inferring $(A \rightarrow B)$. The whole figure is called the *rule of* $(\rightarrow)$ *introduction.* We have a companion rule:

$(\rightarrow)$*elimination*
$$\frac{A, (A \rightarrow B)}{B}.$$

Similarly for $\vee$ we have the rules

$(\vee)$*introduction*
$$\frac{A}{A \vee B} \text{ and } \frac{B}{A \vee B}$$

and

$(\vee)$*elimination*
$$\frac{(A \vee B) \quad \begin{matrix}\not{A} & \not{B} \\ C & C\end{matrix}}{C}.$$

Natural introduction and elimination rules are further given for each logical operation. A (tree) *derivation* (or *proof*) chains these together; the lowest formula of a proof is a logical consequence of all the uncanceled hypotheses which appear in it.

A little work with the Hilbert- and Gentzen-style systems shows how much closer to practice are the latter as compared to the former. In contrast to the formal representation of sentences, the idea of a complete formal representation in Hilbert style of any nontrivial proof seems forbidding. However, the Gentzen systems would seem to make this more feasible, since reconstruction of the logical part is not forced. Indeed, de Bruijn (1970) and his co-workers have pursued just such a project ("AUTOMATH"), and the results so far indicate that their aim is realistic [cf. also de Bruijn 1980]. (It is to be remarked that, just as with formulas, complexity can be decreased by use of more and "bigger" frequently used logical steps even where they can be derived from more elementary steps.) One of de Bruijn's goals relates to Manin's discussion, namely, possible use of such formal representations as a means for systematic mechanical checking of long and complicated proofs. The obvious question to be raised about this concerns the need to have humans as intermediaries in order to convert informal mathematics into a language such as AUTOMATH in a form to be submitted to a machine. Is such a conversion of really difficult and subtle proofs possible without the human agent understanding in all details what is to be converted? And if he does understand "in all details" isn't the battle over (since complete understanding subsumes checking)?*

To return to the proof theory of Gentzen's system of natural deduction, it is a frequent experience of practice that derivations establish more than just what is needed for the conclusion. We may prove some lemma of the form $(A \to B)$ along the way but only use it later to conclude B, having verified A. This takes the form of an $(\to)$introduction followed by an $(\to)$elimination:

$$\Pi \qquad \dfrac{\begin{array}{c}\Delta\\ A\end{array} \qquad \dfrac{\begin{array}{c}\Gamma, \not{A}\\ B\end{array}}{A \to B}}{B}$$

If we want to extract just what is needed for the conclusion, this part of the proof can be replaced by

$$\Pi^* \qquad \begin{array}{c}\Gamma, \Delta\\ (A)\\ B\end{array}$$

which chains together the proof of A from Δ with that of B from Γ, A. The step from Π to Π^* is called a *reduction* (of Π to Π^*). In general,

*[For recent work on mechanical proof-checking including checked proofs of Gödel's first incompleteness theorem and the Church-Rosser theorem, see Shankar (1994).]

with each logical symbol and proof-figure Π consisting of an introduction followed by an elimination for that symbol, we can associate a reduction Π^*. A proof is said to be in *normal form* if it is *irreducible.* The first main result ("Hauptsatz") of Gentzen's proof theory is that *every proof in the first-order logical calculus (classical or intuitionistic) can be reduced to normal form.*

Actually, Gentzen did his work for a technical variant of the natural deduction system called the *calculus of sequents.* His Hauptsatz for that is frequently referred to as the *Cut-elimination Theorem.* The corresponding work for the natural deduction calculi was first carried out in full by Prawitz (cf. Prawitz 1965). Extensive subsequent work by Prawitz, Martin-Löf, Girard, and others is reported in Prawitz (1971). This includes results of the following form: (i) normal forms are unique, that is, each proof Π reduces to one and only one normal form; (ii) unique normalization also holds for second-order and higher-order calculi (pure theories of sets of finite type); (iii) partial normalization also holds for certain systems with nonlogical axioms and rules (for example, with proofs of numerical statements).

This work is relevant to the informal concept of *identity of proofs.* In practice we are constantly comparing proofs of the same result and judging whether they are the same (though not literally so) or different. (For example, the literature abounds with *new proofs* of previously known results.) It would seem that a relation between formal proofs which reflects this sense of identity should be a basic one for proof theory, just as topological equivalence is the basic relation between the basic objects of topology. Simple examples can be given of formal proofs Π, Π' of the same result which we recognize to be based on different strategies, for which Π and Π' reduce to distinct normal forms. It is plausible that *if* Π, Π' *represent the same informal proof, then they have the same normal form.* Martin-Löf and Prawitz have argued for both this and its converse, hence for *identity of normal form* as the criterion for identity of proofs. Their proposal is by no means generally accepted; however, the discussion of pros and cons cannot be entered into here.* At any rate, there is much work which can be done, perhaps in support of the proposal by more convincing case studies, or as alternatives to it by the investigation of other notions.

A frequent type of result which is obtained using proof theory takes the form S_2 *is conservative over* S_1 *for* $\mathcal{F}$, where S_1 and S_2 are formal systems with languages L_{S_1}, L_{S_2}, respectively, and $\mathcal{F}$ is a class of formulas contained in both languages; this relation holds (by definition) when each A in $\mathcal{F}$ which is provable in S_2 is already provable in S_1. In most cases of interest, S_2 is also an extension of S_1, in which case it is called a *conservative extension.*

Actually, conservation (or conservative extension) results can be obtained by the methods of model theory as well as proof theory. A classic exam-

*[Cf. my review of Prawitz (1971), in *J. Symbolic Logic* 40 (1975), 232–234.]

ple was observed by Kreisel from Gödel's proof that if the system ZF of Zermelo-Fraenkel Set Theory is consistent then it remains consistent when we add AC (the Axiom of Choice) and GCH (the Generalized Continuum Hypothesis). For this Gödel used the model of ZF + AC + GCH in the constructible sets. Now, though the meaning of "set" in general changes in this model to that of "constructible set," there are particular sets which are seen to remain unchanged, for example, the set of finite ordinals (or natural numbers). From this Kreisel concluded that ZF + AC + GCH is conservative over ZF for the formulas of elementary number theory (and, indeed, more). This had interesting consequences in the theory of models. Using AC and GCH, the method of saturated models allows one to prove that certain algebraic theories T are complete and thence decidable, that is, that Thm_{T} is recursive. The conclusion is thus a result which can be formulated in elementary number theory. Hence the use of AC and GCH can be eliminated in principle from the proof of decidability of such T. A specific example was provided by the model-theoretic proof of Ax and Kochen (1965), by which the theory of p-adic fields is decidable. Later Cohen (1969) gave a concrete decision procedure for this theory which was evidently elementary. (More recently Barwise and Schlipf (1976) have shown how to eliminate the use of AC and GCH from various saturated model arguments by use of the refined notion of recursive saturation.)

Hilbert's aim for proof theory had been to use it as a technical tool for his *finitist formalist program* for the foundations of mathematics: one would establish the consistency of formal systems S like ZF by finitist methods, hence prove the conservation of S over finitist systems (elimination of the "ideal" in favor of the "real"). This program was blasted by Gödel's theorem on the impossibility of consistency proofs limited to methods contained in the system to be shown consistent. But Gödel's theorem does not exclude the possibility of conservation results of certain S_2 over S_1 for $\mathcal{F}$ where S_1 and/or $\mathcal{F}$ is prima facie more elementary or more constructive that S_2. I conclude by mentioning some results of this character, as well as some results using proof theory to extract more information from known theorems by showing that they can be proved in certain formal systems.

1. Herbrand's theorem can be applied to axiom systems in special logical form to show that if $(\exists x)A(x)$ is provable, then $A(t_1) \vee \ldots \vee A(t_n)$ is provable for some terms $t_1, \ldots, t_n$ which can be found primitive recursively from a given proof. Kreisel applied this to Artin's solution of Hilbert's Seventeenth Problem, by which if a real polynomial p in one or more variables is positive definite, then it can be represented as a sum of squares of rational functions $p = \Sigma_i r_i^2$. Artin's proof was nonconstructive. Using model theory and Tarski's result that the algebraic theory of real numbers is decidable, A. Robinson had shown there is a (uniform) recursive bound for the number of r_i and their degrees (in the number of variables and degree of p). Using Herbrand's theorem, Kreisel showed how to obtain a primitive recursive bound. (Subsequently he showed how to extract this information from Artin's own arguments.) For this and some other mathematical applications, see Kreisel (1958) [and Delzell (1996), Feferman (1996a)].

2. Bishop (1967) has carried out informally an impressive amount of analysis by constructive means. This can be formalized in various kinds of constructive systems of functions and classes, in particular one (CST) due to Myhill (1975) and a rather different one (T_0) due to the author (Feferman 1975). Friedman (1977) has shown that substantially all of Bishop's work can be formalized in a subsystem of CST which is conservative over intuitionistic elementary number theory $ENT^{(i)}$. There is a corresponding result by Beeson (1987) for a subsystem of T_0 in which the same portion of Bishop (1967) can be formalized.

3. Beeson (1977) has shown that if one proves in T_0 a statement of the form $f : X \to Y$ where X, Y are separable metric spaces and X is complete, then one can prove that f is continuous. More generally, if $(\forall a \in X)\ (\exists b \in Y)A(a, b)$ is proved, then for each $a \in X$ one can find a "stable" solution b, which is not changed much by small perturbations of a. This relates to Brouwer's constructivist doctrines of real functions and to Hadamard's doctrine on well-posed problems in physics.

4. In work under preparation, the author shows that certain *classical* theories of functions and classes are directly adequate to substantial portions of analysis and yet are conservative over classical elementary number theory $ENT^{(c)}$. (This improves work described in Feferman (1977); related theorems have also been reported by Friedman.) Such results connect in principle to the elimination of analytic methods from elementary theorems, as, for example, by Erdös and Selberg. [Cf., further, Feferman (1985), Feferman and Jäger (1993, 1996), and chapters 10, 13, and 14 in this volume.]

Thus, even though Hilbert's program has been discredited, there is much of interest which is obtained from the proof theory arising from his program. The reader who wishes to learn more about the subject to date can profitably consult the chapters on proof theory and constructivity in the *Handbook of Mathematical Logic* (Barwise 1977), which is accessible to a very wide audience. More detailed technical accounts are in Takeuti (1975) and Schütte (1977).* (My reviews of these books in the *Bulletin of the American Mathematical Society* can help as introductions.)** In addition, Kreisel particularly has urged (in a series of publications) cutting loose from the traditional aims and has suggested various new goals and uses, not indicated in the preceding; see for example, Kreisel (1976).

Are any new issues raised for the logical study of mathematical proofs by the Appel-Haken computer-assisted solution of the four-color problem (cf. Tutte 1978)? Not directly, as far as I can see. However, Kreisel's ideas to develop proof theory so as to show the need for abstract methods in order to make proofs *understandable* may hold promise in this direction. Clearly, understandability is related to complexity, which is amenable to a theoretical treatment. Theory aside, for my part I could hardly agree more

---

*[More up-to-date references are given in the next chapter.]

**[Takeuti (1975) was reviewed in *Bull. A.M.S.* 83 (1977), 351–361; Schütte (1977) was reviewed in *Bull.A.M.S.* (*New Series*) 1 (1979), 224–228.]

with Manin's conclusion in "Digression: proof": a good proof is one which makes us wiser.

10

What Rests on What? The Proof-Theoretic Analysis of Mathematics

Introduction

Whenever a subject is organized systematically for expository or foundational purposes (or both), one must deal with the question: *What rests on what?* The way in which this is answered in the case of mathematics depends on whether one is considering it informally or formally, that is, from the point of view of the mathematician or the logician, respectively. The latter usually deals with the question in terms of what specifically follows from what in a given logical/axiomatic setup. Proof theory provides technical notions and results which—when successful—serve to give a more global kind of answer to this question, in terms of reduction of one such system to another; moreover, these results provide a technical bridge from mathematics to philosophy. The purpose of this chapter is to give a picture of what is accomplished in these various respects by reductive proof theory. My own approach to that subject is outlined in section 1, along with a brief comparison with a more standard account. This is then followed in section 2 by a description of some technical results that illustrate the general approach. The chapter concludes in section 3 with a discussion of how reductive proof theory mediates between mathematics and philosophy. A more technical and comprehensive exposition of the material in sections 1 and 2 has previously been given in Feferman (1988).

1 Proof-Theoretical and Foundational Reductions

In the following we use the letters:

$\mathcal{M}$, for an informal part of mathematics (such as number theory, analysis, algebra, etc., or a subdivision of such);

"What rests on what? The proof-theoretic analysis of mathematics" was first published in J. Czermak (ed.), *Philosophy of Mathematics. Proceedings of the Fifteenth International Wittgenstein-Symposium, Part 1* (1993), 147–171 (Feferman 1993). It is reprinted here with the kind permission of the publisher, Verlag Hölder-Pichler-Tempsky, Vienna. There are some minor additions and corrections.

L, for a formal language for a part of mathematics (for example, the language of elementary number theory);
$\phi, \psi, \ldots$ for well-formed formulas or statements of L;
T, for a formal axiomatic system in L (for example, the system of first-order Peano Arithmetic PA in the language of elementary number theory); and
$\mathcal{F}$, for a general foundational framework (for example, finitary, constructive, predicative, countable infinitary, set-theoretical, or uncountable infinitary, etc.).

These categories provide different senses in which we can deal with the question of what rests on what from a logical point of view:

$\mathcal{M}$ *rests on* T, in the sense that $\mathcal{M}$ *can be formalized in* T;
ϕ *rests on* T, in the sense that ϕ *is provable in* T;
T *rests on* $\mathcal{F}$, in the sense that T *is justified by* $\mathcal{F}$; and
T_1 *rests on* T_2, in the sense that T_1 *is reducible to* T_2.

With respect to the last of these, there are different technical notions of reducibility of one axiomatic system to another. We want to contrast, in particular, the notion of T_1 being *interpretable* (or *translatable*) in T_2 with that of T_1 being *proof-theoretically reducible* to T_2, written $T_1 \leq T_2$ (this is defined in section 2). In general, these move in opposite directions from a foundational point of view, since we are mainly concerned with the relation $T_1 \leq T_2$ when T_2 is a *part* of T_1, either directly or by translation. In contrast, T_2 tends to be more comprehensive than T_1 in the case of interpretations; a familiar example is that of Peano Arithmetic PA (as T_1) in Zermelo-Fraenkel Set Theory ZF (as T_2), where the natural numbers are interpreted as the finite ordinals. This is a *conceptual reduction* of number theory to set theory, but not a *foundational reduction*, because the latter system is justified only by an uncountable infinitary framework whereas the former is justified simply by a countable infinitary framework.

The driving aim of the original Hilbert Program (H.P.) was to provide a finitary justification for the use of the "actual infinite" in mathematics. This was to be accomplished by directly formalizing one body or another of infinitary mathematics $\mathcal{M}$ in a formal axiomatic theory T_1 and then demonstrating the consistency of T_1 by purely finitary means; in practice, that would be established by a proof-theoretic reduction of T_1 to a system T_2 justified on finitary grounds. It is generally acknowledged that H.P. as originally conceived could not be carried through even for elementary number theory PA as the system T_1, in consequence of Gödel's (1931) incompleteness theorems. This then gave rise to certain relativized forms of H.P.; the history will be traced briefly below. In our approach, what the results of a relativized H.P. should achieve are best expressed in the following way, from Feferman (1988, p. 364):

(*) A body of mathematics $\mathcal{M}$ is represented directly in a formal system T_1 which is justified by a foundational framework $\mathcal{F}_1$. T_1 is reduced proof-theoretically to a system T_2 which is justified by another, more elementary such framework $\mathcal{F}_2$.

In Hilbert's scheme, $\mathcal{F}_1$ was to be the infinitary framework of modern mathematics featuring (i) the "completed" or "actual" infinite (both countable and uncountable) and (ii) nonconstructive reasoning, while $\mathcal{F}_2$ was to be the framework of finitary mathematics featuring (i)′ only the "potential" infinite of finite combinatorial objects and (ii)′ constructive reasoning applied to quantifier-free statements (typically, equations). According to Hilbert, already the system PA embodies (i) and (ii) by the use of quantified variables which are supposed to range over the set $\mathbf{N}$ of natural numbers and the assumption of the Law of the Excluded Middle, from which follows statements of the form $(\forall x)\phi(x) \vee (\exists x)\neg\phi(x)$. Even for decidable quantifier-free ϕ these require for their justification a survey of the totality of natural numbers, by a kind of infinitary act of omniscience.[1]

The general problem raised by Gödel's incompleteness results (1931) for H.P. is that if finitary mathematics is itself to count as a significant body of informal mathematics, it must be formalizable in a consistent formal axiomatic theory T. Then by Gödel's second incompleteness theorem, the consistency of T would not be provable in T, hence could not be finitarily provable, and so H.P. cannot be carried out for T. Just what T could serve this purpose was not analyzed in the Hilbert school. In fact, the kind of finitary reasoning that had been employed by its workers in the 1920s could evidently be formalized in PA and even in its quantifier-free fragment PRA (Primitive Recursive Arithmetic), and it would be hard to imagine anything at all of the same character which could not be formalized in Zermelo Set Theory or the equivalent finite theory of types P of Gödel (1931).[2] Nevertheless, Gödel was cautious at the time about the significance of his second incompleteness theorem for H.P.: "I wish to note expressly that [this theorem does] not contradict Hilbert's formalistic viewpoint ... it is conceivable that there exist finitary proofs that *cannot* be expressed in the formalism of P."[3] And, in his foreword to Hilbert and Bernays (1934),

[1]This view in the Hilbert school is revealed, for example, by the title of Ackermann (1924), "Begründung des 'tertium non datur' mittels der Hilbertschen Theorie des Widerspruchsfreiheit," and by Hilbert's reference to the axioms for quantification as the "transfinite axioms."

[2]The question of completely formalizing finitary mathematics has been addressed by Kreisel (1960) and Tait (1981); the former arrived at a system of the same strength as PA, while the latter restricted it much more drastically to PRA.

[3]Cf. Gödel (1986), pp. 138–139 and 195. Gödel later sharply revised this opinion in his 1958 paper: "it is necessary to go beyond the framework of what is, in Hilbert's sense, finitary mathematics if one wants to prove the consistency of classical mathematics, or even that of classical number theory" (cf. Gödel 1990, p. 241). There is evidence that he had arrived at that point of view long before 1958.

Hilbert also (implicitly) stressed this possibility, saying that the only consequence of Gödel's incompleteness theorems for his program was that one would have to work harder than anticipated in order to carry it out.

Despite Hilbert's continued optimism, the general feeling after 1931 was that Gödel's second incompleteness theorem doomed H.P. to failure, and that some essentially new idea would be needed to carry out anything like it, even for PA. Yet another result of Gödel in his 1933 paper (independently found by Bernays and Gentzen) forced a further reconsideration of H.P.: this showed that PA could be translated in a simple way into the intuitionistic system HA of Heyting's arithmetic, which differs from PA only in omitting the Law of the Excluded Middle from its basic logical principles. Thus, one specific goal of H.P. was achieved in a single stroke (cf. ftn. 1 above). Moreover, the intuitionists argued that HA is justified on a conception of the natural numbers as a potentially infinite set. From that point of view, the result of Gödel's translation of PA into HA was to eliminate the actual infinite in favor of the potential infinite and thus to meet the main aim of H.P. at least for number theory. However, by its free use of the language of first-order quantification theory, the system HA is not thereby justified on the basis of the more strictly finitary forms of reasoning using only quantifier-free formulas, as demanded by Hilbert. On the one hand, Gödel's 1933 result corrected a common tendency in the Hilbert school prior to 1933 to identify intuitionism with finitism. On the other hand, his reduction of PA to HA fueled the growing search for a wider conception of H.P. As reported by Bernays (1967, p. 502), years later:

> It thus became apparent that the "finite Standpunkt" is not the only alternative to classical ways of reasoning and is not necessarily implied by the idea of proof theory. An enlarging of the methods of proof theory was therefore suggested: instead of a reduction to finitist methods of reasoning, it was required only that the arguments be of a constructive character, allowing us to deal with more general forms of inference.

Here was the germ of a relativized form of H.P.

One strikingly new specific way forward was provided by Gentzen in his 1936 paper in which the consistency of PA was proved by transfinite induction up to Cantor's ordinal ϵ_0 ($\mathrm{TI}(\epsilon_0)$) applied only to decidable quantifier-free predicates and otherwise using only finitary reasoning. Here the ordinals less than ϵ_0 are represented in Cantor normal form $\omega^{\alpha_1} + \ldots + \omega^{\alpha_n}$ ($\alpha_1 \geq \ldots \geq \alpha_n$), and the ordering of these is isomorphic to an effective ordering of the natural numbers by taking the sequence number $p_1^{a_1} \ldots p_n^{a_n}$ to correspond to $\omega^{\alpha_1} + \ldots + \omega^{\alpha_n}$ when a_i corresponds to α_i. The principle $\mathrm{TI}(\epsilon_0)$ may itself be justified on constructive grounds (though no longer clearly finitary grounds, even for decidable predicates). Gentzen showed that his result was best possible by establishing each instance of $\mathrm{TI}(\beta)$ in PA for each $\beta < \epsilon_0$. In modern terms, ϵ_0 was thus identified as the "ordinal of PA," in the sense of being the least nonprovably recursive well-ordering in PA. Gentzen's consistency proof apparently impelled Bernays' acceptance of a further shift away from the original H.P., as is evidenced by its inclusion in Hilbert and Bernays (1939), under the section title "Überschreitung des bisherige methodischen Standpunktes der Beweistheorie."[4]

In the postwar period, Gentzen-style consistency proofs and ordinal analysis of various subsystems of analysis and set theory became the dominant approach in proof theory. These have involved the construction of more and more complicated recursive well-orderings $\prec$ of the natural numbers, obtained from notation systems for larger and larger ordinals α, in each case identified as the least nonprovably recursive well-ordering of the theory T in question; furthermore, the consistency of T is proved by transfinite induction up to $\alpha, \mathrm{TI}(\alpha)$ (that is, on the corresponding recursive $\prec$ relation) applied to decidable predicates, while otherwise using only finitary reasoning. Several modern texts which expound this kind of extension of the Gentzen approach are Schütte (1977), Takeuti (1987), and Girard (1987), as well as Pohlers (1989). These texts are highly technical, and cannot be faulted on mathematical grounds; on the contrary, they contain many deep results. But it is not at all clear what they contribute to an extended H.P. in the sense envisioned by Bernays. The crucial question is: In what sense is the assumption of $\mathrm{TI}(\alpha)$ justified constructively for the very large ordinals α used in these consistency proofs? Indeed, on the face of it, the explanation of which ordinals α are used appeals to the very concepts and results of infinitary set theory that one is trying to account for on constructive grounds. For example, in chapter 9 of Schütte (1977), a notation system is introduced which is defined in terms of a hierarchy of normal functions on the set of ordinals less than the first fixed point of $\aleph_\alpha = \alpha$. But this is only the beginning. In the still more advanced research of Jäger and Pohlers (1982), use is made of a notation system based on the ordinals less than the first (set-theoretically) inaccessible cardinal, to establish the consistency of a moderately strong subsystem of analysis. And Rathjen (1991) has used notation systems based on the ordinals less than the first Mahlo cardinal to prove the consistency of a subsystem of set theory. [Cf., even much further, Rathjen 1995.] The notation systems developed for these purposes are all countable, though they name extremely large uncountable ordinals. Then, by means of an analysis of the ordering relations, one shows in each case that the ordering of the notations is recursive. Moreover, well-ordering proofs of a more or less constructive character can be given which do not appeal to the fact that the notation systems are derived from hierarchies of functions on very large ordinal number classes. However, the conviction that one is indeed dealing with well-ordering relations derives from the latter and not from the well-ordering proofs in which those traces have been obliterated. Finally, there is a prima facie anomaly in the use

[4]The preparation of Hilbert and Bernays (1934 and 1939) had been placed by Hilbert entirely in the hands of Bernays.

of transfinite induction applied to orderings for enormous ordinals while insisting on a restriction to finitary reasoning otherwise. Having given up the original H.P. in favor of a reduction of classical infinitary mathematics to one part or another of constructive mathematics, there seems to be no reason to retain its finitary vestiges.

One more feature of the extended Hilbert-Gentzen program deserves criticism, namely, the focus on consistency proofs: hardly anyone nowadays doubts that Zermelo-Fraenkel Set Theory is consistent or, to be much more moderate, that the system of analysis (full second-order arithmetic) is consistent. And surely the number who doubt that PA is consistent is vanishingly small. It is true that Takeuti (1987, pp. 100–101) and Schütte (1977, p. 3) still pay lip service to the goal of consistency proofs, but their followers have deemphasized that in favor of other goals, such as the ordinal analysis of formal systems (cf. Pohlers 1989, pp. 5–6).

A more thoroughgoing reconsideration of H.P. was initiated by Kreisel in his papers "Mathematical significance of consistency proofs" (1958) and "Hilbert's programme" (1958a). The former emphasized the additional mathematical information that can be gained from successful applications of proof-theoretical methods, by telling what more we know of a statement, beyond its truth, if we know that it has been proved by specific methods. For example, this may take the form of extracting bounds for existential results, or for the complexity of provably recursive functions. The latter paper first suggested the idea of a "hierarchy" of Hilbert programs; this was elaborated in Kreisel (1968), which considered besides reductions to finitary and constructive conceptions also reductions to semiconstructive (for example, predicative) conceptions, and within each a more refined analysis of what principles are needed for various pieces of reductive work (*op. cit.* pp. 323–324).

My own approach in Feferman (1988) to a relativized form of H.P. formulated as (∗) above, is similar in this overall respect to Kreisel (1968) but (to quote myself, *op. cit.* p. 367)

> the details are different both as to the categorization of the conceptions to which the foundational reductions are referred and as to the proof-theoretical work which exemplifies these reductions. Concerning the latter, it is simply that most of the work surveyed has been carried out in the last twenty years. And, with respect to the former, we have tried to seize on *the most obvious features of foundational conceptions* [or *frameworks*] so that, *insofar as possible, what the work achieves will speak for itself.*

Some examples from my 1988 survey which illustrate the scheme (∗) are given in the next section.

2 Hilbert's Program Relativized: Proof-Theoretical and Foundational Reductions

2.1 Proof-Theoretical Reductions

All systems T considered in the following are assumed to contain PRA (described in 2.3). The language L_T of T and the axioms and rules of inference of T are assumed to be specified by primitive recursive presentations via usual Gödel numberings (coding) of expressions; we may identify expressions with their codes. Thus the relation $Proof_T(x, y)$ which holds when y is (the code of) a proof in T of the formula (with code) x, is primitive recursive. The metatheory of primitive recursively presented axiomatic systems can be formalized directly in PA and even in a subsystem of PA which is a conservative extension of PRA (described in 2.4); for details, see Smorynski (1977) or Feferman (1989).

When considering a pair of systems T_1, T_2, we write L_i for $L_{T_i} (i = 1, 2)$. Suppose Φ is a primitive recursive class of formulas contained in both L_1 and L_2, which contains every closed equation $t_1 = t_2$. The basic idea of a proof-theoretic reduction of T_1 to T_2 conserving Φ is that we have an effective method of transforming each proof in T_1 ending in a formula ϕ of Φ into a proof of ϕ in T_2; moreover, we should be able to establish that transformation provably within T_2. More precisely, we say that T_1 is *proof-theoretically reducible to* T_2 *by* f, *conservatively for* Φ, and write

$$f : T_1 \leq T_2 \text{ for } \Phi$$

if f is a partial recursive function such that

(1) whenever $Proof_{T_1}(x, y)$ and x is (the code of) a formula in Φ then $f(y)$ is defined and $Proof_{T_2}(x, f(y))$,

and

(2) the formalization of (1) is provable in T_2.

We write $T_1 \leq T_2$ for Φ, if there is such an f satisfying (1) and (2). In practice, f may be chosen to be primitive recursive and the formalization of (1) can be proved in PRA.

It is immediate that if $T_1 \leq T_2$ for Φ, then T_1 *is conservative over* T_2 *for* Φ in the sense that

(3) $\phi \in \Phi$ and $T_1 \vdash \phi$ implies $T_2 \vdash \phi$.

It then follows by the general assumption on Φ that

(4) if T_2 is consistent then T_1 is consistent,

since if $T_1 \vdash 0 = 1$ then $T_2 \vdash 0 = 1$. Moreover, by (2), the formalization $Con_{T_2} \to Con_{T_1}$ of (4) is provable in T_2 (and, in practice, already in PRA).

Remark. It should be noted that we may have T_1 conservative over T_2 without there being any possible proof-theoretic reduction of T_1 to T_2. For example, if Φ is the class of closed equations $t_1 = t_2$ of the language of PRA and T_1 is any consistent extension of PRA, then T_1 is conservative over PRA for Φ, because if $T_1 \vdash t_1 = t_2$ we must have PRA $\vdash t_1 = t_2$, for otherwise PRA $\vdash t_1 \neq t_2$. Now choose T_1 to be any consistent system which proves Con_{PRA}, so it is not proof-theoretically reducible to PRA (for example, to be extreme, $T_1 = ZF$).

2.2 Foundational Reductions

According to our general scheme (*) of section 1, a proof-theoretical reduction $T_1 \leq T_2$ provides a *partial foundational reduction* of a framework $\mathcal{F}_1$ to $\mathcal{F}_2$, if T_1 is justified directly by $\mathcal{F}_1$ and T_2 by $\mathcal{F}_2$. The reason for the qualification "partial" is that we may well have a system T_1 which is directly justified by $\mathcal{F}_1$ but which is not reducible to *any* T_2' justified by $\mathcal{F}_2$. For example (to be extreme again), the system ZF, which is justified by the uncountable infinitary framework of Cantorian set theory, is not reducible to any finitarily justified system. On the other hand, it may also happen that a system T_1 which prima facie requires for its justification an appeal to an infinitary framework is proof-theoretically reducible to a finitarily justified T_2; in that case we have a partial reduction of the infinitary to the finitary.

In sections 2.4 and 2.6–2.8 I describe some results which exemplify partial foundational reductions for the following pairs of frameworks:

$\mathcal{F}_1$	$\mathcal{F}_2$
Infinitary	Finitary
Uncountable Infinitary	Countable Infinitary
Impredicative	Predicative
Nonconstructive	Constructive

Remarks. (i) The pairs $\mathcal{F}_1$, $\mathcal{F}_2$ are not the only ones which might be considered in this respect (cf. Feferman 1988, p. 367).

(ii) In establishing proof-theoretical reductions which provide partial foundational reductions we may well use results of Gentzen-Schütte-Takeuti style, but these appear behind the scenes. The point is to apply such work to results which speak for themselves (unlike consistency proofs by transfinite induction on very large ordinals).

(iii) The emphasis here on the use of proof theory for a form of relativized H.P. is not meant to diminish other applications of proof theory—on the contrary. But the interest of such is guided by quite different considerations.

2.3 The Language and Basic Axioms of First-Order Arithmetic

In order to describe the reductive results for several systems of arithmetic in the next section, we need some syntactic and logical preliminaries. The language L_0 is a type 0 (or first-order) single-sorted formalism. It contains variables x, y, z, ..., the constant symbol 0, the successor symbol $'$, and symbols $\mathbf{f}_0$, $\mathbf{f}_1$, ... for each primitive recursive function, beginning with $\mathbf{f}_0(x,y)$ for $x + y$ and $\mathbf{f}_1(x,y)$ for $x \cdot y$. Terms t, t_1, t_2, ... are built up from the variables and 0 by closing under $'$ and the $\mathbf{f}_i$. The atomic formulas are equations $t_1 = t_2$ between terms. Formulas are built up from these by closing under the propositional operations ($\neg, \wedge, \vee, \rightarrow$) and quantification ($\forall x$, $\exists x$) with respect to any variable. A formula is said to be quantifier-free, or in the class QF, if it contains no quantifiers. A formula is said to be in the class Σ^0_n if it is given by n alternating quantifiers beginning with "$\exists$," followed by a QF matrix. For example, $(\exists y)\psi \in \Sigma^0_1$ and $(\exists y_1)(\forall y_2)\psi \in \Sigma^0_2$ when $\psi \in$ QF. (The superscript in "Σ^0_n" indicates that these are type 0 variables, and the "Σ" tells us that the quantifier string starts with "$\exists$.") Dually, ϕ is in the class Π^0_n if it is given by n alternating quantifiers beginning with "$\forall$," followed by a QF matrix. For example, $(\forall y)\psi \in \Pi^0_1$ and $(\forall y_1)(\exists y_2)\psi \in \Pi^0_2$ when $\psi \in$ QF.

Unless otherwise noted, the underlying logic is that of the classical first-order predicate calculus with equality. The basic nonlogical axioms Ax_0 of arithmetic are

(1) $x' \neq 0$,
(2) $x' = y' \rightarrow x = y$,
(3) $x + 0 = x \wedge x + y' = (x + y)'$,
(4) $x \cdot 0 = 0 \wedge x \cdot y' = x \cdot y + x$,

and so on, for each further $\mathbf{f}_i$ (using its primitive recursive defining equations as axioms). To Ax_0 may be added certain instances of the *Induction Axiom* scheme:

IA $\quad \phi(0) \wedge \forall x(\phi(x) \rightarrow \phi(x')) \rightarrow \forall x\phi(x)$,

for each formula $\phi(x)$ (with possible additional free variables). Φ-IA is used to denote both this scheme restricted to $\phi \in \Phi$, *and* the system with axioms Ax_0 plus Φ-IA. We use PA (Peano Arithmetic) to denote the system with axioms Ax_0 + IA where the full scheme is used.[5]

The language of PRA (Primitive Recursive Arithmetic) is just the quantifier-free part of L_0. In place of IA it uses an Induction Rule:

---

[5]By a result of Gödel (1931) all primitive recursive functions are explicitly definable in terms of 0, $'$, +, and $\cdot$, and their recursive defining equations are derivable in the subsystem of PA obtained by restriction to formulas in that sublanguage. However, it is more convenient here to take PA in the form described above, so as to include PRA directly.

$$\text{IR} \quad \frac{\phi(0),\ \phi(\mathrm{x}) \rightarrow \phi(\mathrm{x}')}{\phi(\mathrm{x})}, \quad \text{for each } \phi \in \mathrm{QF}.$$

2.4 Reduction of the Countable Infinitary to the Finitary

It is generally acknowledged that PRA is a finitarily justified system, or to be more precise, that each theorem of PRA is finitarily justified.[6] On the other hand, as explained in section 1 above, the use of classical quantificational logic in *any* system containing the base axioms (1) $\mathrm{x}' \neq 0$ and (2) $\mathrm{x}' = \mathrm{y}' \rightarrow \mathrm{x} = \mathrm{y}$ of Ax_0 implicitly requires assumption of the completed countable infinite. The first result which established a partial reduction of the countable infinitary to finitary principles was that the system QF-IA is proof-theoretically reducible to PRA, obtained by Ackermann (1924).[7] Years later this was improved by Parsons (1970) to the following:

Theorem 1. $\Sigma^0_1\text{-IA} \leq \text{PRA}$.

The reduction here is conservative for Π^0_2 formulas in the following sense: If $\Sigma^0_1\text{-IA} \vdash \forall \mathrm{x} \exists \mathrm{y} \psi(\mathrm{x},\mathrm{y})$ with ψ in QF, then $\text{PRA} \vdash \psi(\mathrm{x},\mathrm{t}(\mathrm{x}))$ for some term $\mathrm{t}(\mathrm{x})$.

Remarks. (i) Here, as below, only references are given in lieu of proofs. (ii) As measured by the arithmetical hierarchy, theorem 1 is best possible, since the system Σ^0_2-IA proves the consistency of Σ^0_1-IA (cf. Sieg 1985, pp. 46–47). However, in section 2.6 describe a stronger result obtained by passing to the language of analysis, which is taken up next.

2.5 The Language and Basic Axioms of Analysis

The language L_1 of analysis, or second-order arithmetic, is obtained from L_0 by adjunction of variables for type 1 objects. In some presentations these are taken to be sets, in others they are taken to be functions, while in still others, both kinds of variables are taken to be basic. For simplicity, we shall follow the first choice here, by adjoining to L_0 set variables X, Y, Z, ... and the binary relation symbol $\in$ between individuals and sets. Thus, the atomic formulas are expanded to include $\mathrm{t} \in \mathrm{X}$ for any term t and set variable X. (Equality between sets is considered to be defined extensionally, that is, by $\mathrm{X} = \mathrm{Y} \leftrightarrow \forall x[x \in \mathrm{X} \leftrightarrow x \in \mathrm{Y}]$.) Formulas are now built up using the propositional operations, and the quantifiers applied both to individual variables ($\forall \mathrm{x}, \exists \mathrm{x}$) and set variables ($\forall \mathrm{X}, \exists \mathrm{X}$). The underlying logic is now that of classical two-sorted predicate calculus with equality (in the first sort). The classes QF, Σ^0_n, and Π^0_n are

[6] According to Tait's (1981) analysis, finitary number theory coincides with PRA; if that account is accepted, a finitist would recognize each theorem of PRA as being finitarily justified, but not PRA as a whole.

[7] Actually, Ackermann thought he had accomplished much more, namely, a consistency proof of analysis by finitary means! His error was discovered by von Neumann (1927) and in essence this relatively modest result was extracted. The situation was further clarified by Herbrand (1930), who gave useful sufficient conditions for finitary consistency proofs.

explained in L_1 just as for L_0 in section 2.3 above, but with the understanding that formulas in these classes might contain set variables via the expanded class of atomic formulas. A formula is said to be *arithmetical* if it contains no bound set variables, and we write Arith for the class of all such formulas.[8] In line with the definition of the classifications Σ_n^0 and Π_n^0 in section 2.3, we define classes Σ_n^1 and Π_n^1 as follows: a formula ϕ of L_1 is said to be in Σ_n^1 if it is given by n alternating set quantifiers beginning with "$\exists$," followed by an arithmetical matrix. For example, $(\exists Y)\psi \in \Sigma_1^1$ and $(\exists Y_1)(\forall Y_2)\psi \in \Sigma_2^1$ when $\psi \in$ Arith. (Now the superscript in "Σ_n^1" tells us that we are measuring the type 1 quantifier complexity of a formula ϕ.) Dually, ϕ is in the class Π_n^1 if it is given by n alternating set quantifiers beginning with "$\forall$," followed by an arithmetical matrix. For example, $(\forall Y)\psi \in \Pi_1^1$ and $(\forall Y_1)(\exists Y_2)\psi \in \Pi_2^1$ when $\psi \in$ Arith.

The general set existence axiom is given in L_1 by the *Comprehension Axiom* scheme:

$$\text{CA} \quad \exists X \forall x\, [x \in X \leftrightarrow \phi(x)],$$

where ϕ is a formula of L_1 which does not contain the variable X free but may contain free individual and set variables besides x. The Induction Axiom scheme IA of L_1 is of the same form as in L_0, $\phi(0) \wedge \forall x\, (\phi(x) \rightarrow \phi(x')) \rightarrow \forall x\, \phi(x)$, but where now ϕ may be any formula of L_1. There is another option for the statement of induction in L_1, namely, as the single second-order statement:

$$I_1 \quad \forall X\, [0 \in X \wedge \forall x (x \in X \rightarrow x' \in X) \rightarrow \forall x\, (x \in X)].$$

In the presence of full CA, IA is derivable from I_1. The full system of analysis is given by the axioms Ax_0, CA, and IA (or equivalently I_1). We shall consider subsystems over Ax_0, based on various combinations Φ-CA and Ψ-IA where Φ, Ψ are classes of L_1 formulas. In particular, two extremes are given special attention in combination with a given Φ-CA, namely, adjunction of full IA or adjunction simply of I_1. In the first case the system is denoted Φ-CA, and in the second case it is denoted $\Phi\text{-CA}\!\restriction$ (for which the notation $\Phi\text{-CA}_0$ is also used in some presentations). However, in intermediate cases of adjunction Ψ-IA, the system is designated Φ-CA + Ψ-IA.

The formulas Σ_1^0 of L_0 define the recursively enumerable sets in the standard model $(\mathbf{N}, 0, ', +, \cdot, \ldots)$. The recursive sets S are exactly those which are recursively enumerable and whose complement $\mathbf{N} - S$ is recursively enumerable, that is, which are definable both by a Σ_1^0 formula and a Π_1^0 formula. In L_1, all this relativizes to the set variables of the formulas. For example, if $\phi(x,X)$ is in Σ_1^0 and $\psi(x,X)$ is in Π_1^0 in L_1, and $\forall x(\phi(x,X)$

[8]Every formula ϕ of Arith is logically equivalent to a formula in the class $\bigcup_n \Pi_n^0$, also denoted Π_∞^0.

$\leftrightarrow \psi(x,X))$ holds in the standard model, then $\{x \mid \phi(x,X)\}$ denotes a set recursive in X, and all sets recursive in X are definable in this way. Thus the (relatively) *Recursive Comprehension Axiom* scheme is formulated in L_1 by

$$\text{RCA} \quad \forall x[\phi(x) \leftrightarrow \psi(x)] \rightarrow \exists X \forall x[x \in X \leftrightarrow \phi(x)], \quad \text{for } \phi \in \Sigma^0_1, \psi \in \Pi^0_1$$

(and X not free in ϕ or ψ). RCA is also denoted Δ^0_1-CA. This generalizes in the obvious way to Δ^0_n-CA by taking $\phi \in \Sigma^0_n, \Pi^0_n$. Similarly, we shall consider the scheme

$$\Delta^1_1\text{-CA} \quad \forall x[\phi(x) \leftrightarrow \psi(x)] \rightarrow \exists X \forall x[x \in X \leftrightarrow \phi(x)] \quad \text{for } \phi \in \Sigma^1_1, \psi \in \Pi^1_1$$

(and X not free in ϕ or ψ) and its obvious generalization to Δ^1_n-CA. The classes Δ^1_1 and Δ^1_2 of sets definable in the standard model are very significant in higher recursion theory. In particular, Δ^1_1 coincides with the class of hyperarithmetic sets, obtained by iterating relative arithmetical (or simply relative Π^0_1) definitions through all recursive ordinals (cf. Sacks 1990).

Remark. Second-order arithmetic is called *analysis* because the notion of being a real number can be formally expressed in L_1. First one uses a standard representation of the rational numbers **Q** in terms of the natural numbers, and then real numbers may be defined as Dedekind sections in **Q**, that is, as certain subsets of **Q** and hence of **N**. Alternatively, taking functions as the basic type 1 objects, one may formally represent real numbers as Cauchy sequences of rational numbers. To develop a theory of functions of real numbers in general one must proceed to a third-order language L_2; however, many useful classes of functions, such as the continuous functions of reals, may already be handled in L_1. I discuss the formalization of informal mathematical analysis in subsystems of CA in section 3.

2.6 Reduction of the Uncountable Infinitary to the Finitary

Cantor's diagonal argument for the uncountability of the collection of all subsets of the set of natural numbers is easily carried out formally under very minimal assumptions about sets. First of all, a countable sequence of sets $X_0, X_1, \ldots, X_i, \ldots$ is taken to be represented by a single set X for which $y \in X_i \leftrightarrow \langle i, y \rangle \in X$ (where $\langle , \rangle$ is a primitive recursive pairing function). Then it follows from QF-CA that no countable sequence of sets includes all sets, that is, that $\forall X \exists Y \neg \exists x \forall y[y \in Y \leftrightarrow \langle x, y \rangle \in X]$, by taking $y \in Y \leftrightarrow \langle y, y \rangle \notin X$. Thus if one accepts the Hilbertian point of view that classical quantificational logic over a domain of objects implicitly requires for its justification treating that domain as a completed totality, then assumption of the principle QF-CA (and even much less) requires the uncountable infinitary foundational framework for its justification. Hence if a system T_1 in L_1 includes QF-CA and is proof-theoretically reducible to a system T_2 justified by a finitary framework, then we have a partial reduction of the uncountable infinitary to the finitary. Just such a reduction is given by the following result:

Theorem 2. $\text{RCA} + \Sigma_1^0\text{-IA} \leq \text{PRA}$.

The proof of this is motivated by the fact that RCA has an ω-model in the collection of all recursive subsets of N. Formally, one first obtains a direct translation of $\text{RCA} + \Sigma_1^0\text{-IA}$ into $\Sigma_1^0\text{-IA}$ considered as an L_0 system, using the formalization of elementary recursion theory there, and then applies Theorem 1. Once more, the reduction conserves Π_2^0 formulas.

A very interesting strengthening of Theorem 2 has to do with a set existence principle which cannot be realized recursively in this way, namely, so-called Weak König's Lemma (WKL). König's Lemma (KL) in general says that any finitely branching infinite tree must contain some infinite path; WKL is its restriction to binary branching trees. The formal statement of WKL in L_1 takes the form

WKL $\forall X\, [\text{BinTree}(X) \wedge \text{Inf}(X) \rightarrow (\exists Y)\, \text{Path}(Y,X)]$.

It is well known that there are infinite recursive binary trees which contain no recursive paths, so the recursive sets do not form an ω-model of WKL.

Theorem 3. $\text{RCA} + \text{WKL} + \Sigma_1^0\text{-IA} \leq \text{PRA}$.

The conservation result which follows from this was first established by Friedman in 1977 by a model-theoretic argument (cf. Simpson 1988); the proof-theoretical reduction was established in Sieg (1985).[9,10] This partial reduction of the uncountable infinitary to the finitary is a significant contribution to the original H.P., as has been argued by Simpson (1988).

Remark. Extensions of Theorem 3 which provide partial reductions of the higher uncountable infinitary to the finitary are described in Feferman (1988, p. 375), using systems of analysis with higher-type function(al) variables.

2.7 Reduction of the Uncountable Infinitary to the Countable Infinitary

As was remarked in section 2.4, $\Sigma_2^0\text{-IA}$ is not proof-theoretically reducible to PRA. Without committing oneself to PRA as the limit of finitistically acceptable principles in arithmetic, it may still be said that we have no evident finitary justification (directly or by reduction) for $\Sigma_2^0\text{-IA}$ or beyond.

---

[9]A related earlier result was obtained by Mints (1976).

[10]Model-theoretic arguments for conservation do not on their face provide proof-theoretical reductions. Friedman has shown that certain forms of model-theoretic arguments can be transformed into such reductions, but this is by no means immediate. Cf. the discussion in Feferman (1988), pp. 378–381.

All instances of the first-order induction scheme as embodied in PA are justified in the countable infinitary framework. We concern ourselves in this subsection with results which establish a partial reduction of the uncountable infinitary to the countable infinitary via results of the form T $\leq$ PA where T is a second-order system. The first of these concerns the system ACA, which is taken to abbreviate Arith-CA. Recall our convention that ACA uses full IA in L_1, while ACA$\restriction$ uses only the restricted induction axiom I_1. ACA$\restriction$ contains PA since every arithmetical formula in L_0 defines a set under Arith-CA.

Theorem 4. $\text{ACA}\restriction \leq \text{PA}$

Conservation is for all formulas of L_0. This result is part of the folklore; the earliest reference I'm aware of for a proof (of a more general, but similar) result is Shoenfield (1954); see also Feferman and Sieg (1981, p. 112).

Remarks. (i) It is essential here that ACA$\restriction$ uses only the second-order induction axiom I_1 and not the full induction scheme IA in L_1, since ACA (with full IA) proves Con_{PA}.

(ii) The arithmetically definable subsets of **N** form an ω-model of ACA, but by Remark (i) this cannot be used for a translation argument to establish Theorem 4, as it was for Theorem 2. However, there is an easy model-theoretic proof of conservation, since every model M of PA can be expanded to a model of ACA$\restriction$, by taking the sets to be just those definable by L_0 formulas from elements of M.

While Theorem 4 may not be unexpected (though the matter is delicate as shown by these remarks), the following *is* unexpected:

Theorem 5. $\Delta^1_1\text{-CA}\restriction \leq \text{PA}.$

Here conservation is again for arithmetic formulas. The original proof of the conservation result is due to Barwise and Schlipf (1975) and, independently, Friedman (1976), in both cases by model-theoretic methods. A proof-theoretical reduction was sketched in Feferman (1977, p. 967); see Feferman and Sieg (1981a, pp. 108–112) for a full proof.

In the next section we shall consider what can be said about the full system Δ^1_1-CA, which is much stronger than PA.

2.8 Reducing the Impredicative to the Predicative

A definition of a set S is said to be *predicative* if, roughly speaking, all notions and the ranges of all variables occurring in it are prior to S; otherwise it is called *impredicative* (cf. Feferman 1964). Thus, a definition which singles S out from a totality of sets by reference via quantification to that totality is prima facie impredicative. In particular, this holds for $S = \{x \mid \phi(x)\}$ where ϕ is a formula of L_1 which contains bound set variables; existence of S is given by CA. Thus Π^1_1-CA (or dually Σ^1_1-CA) is justified only on the assumption of the meaningfulness of impredicative definitions, and that in turn implicitly assumes that there is a well-determined totality of subsets of the natural numbers which exists independently of any means of (human) definition or construction.

An axiomatic characterization of what reasoning with sets of natural numbers is predicatively acceptable has been given independently by Feferman (1964) and Schütte (1965). One way of describing the characterization is that it allows iteration of the system based on relative ACA through all ordinals below a certain recursive ordinal Γ_0, giving a system denoted as $\mathrm{ACA}_{<\Gamma_0}$. It is not hard to show that Π^1_1-CA proves the consistency of $\mathrm{ACA}_{<\Gamma_0}$ (via a proof of the well-foundedness of a notation system for Γ_0), hence cannot be reduced to predicative principles under the above characterization. This leaves unsettled the status of Δ^1_1-CA, which asserts the existence of $\{x|\ \phi(x)\}$ when ϕ is Π^1_1 and we have $\forall x(\phi(x) \leftrightarrow \psi(x))$ with ψ in Σ^1_1, because that is still prima facie impredicative. The following theorem settles the matter, since $\epsilon_0 < \Gamma_0$ and so $\mathrm{ACA}_{<\epsilon_0}$ is a predicatively justified system; it thus constitutes a partial reduction of the impredicative to the predicative.

Theorem 6. $$\Delta^1_1\text{-CA} \leq \mathrm{ACA}_{<\epsilon_0}.$$

The conservation part of this result (for Π^1_2 formulas) was first proved by Friedman (1970). A proof-theoretical reduction was outlined in Feferman (1971) (cf. Feferman 1977, p. 965), and a full proof by another method has been given in Feferman and Sieg (1981, pp. 119ff.).

2.9 Reducing the Nonconstructive to the Constructive

We use $T^{(i)}$ to denote the result of substituting intuitionistic logic for classical logic in an axiomatic theory T, otherwise retaining the nonlogical axioms and rules of T. (For a suitable formalization of classical logic, this may simply be obtained by dropping the Law of the Excluded Middle.) While intuitionistic logic is deemed to be justified on constructive grounds, this by no means assures that any such $T^{(i)}$ is constructively acceptable; for example, no constructive justification for $\mathrm{ZF}^{(i)}$ is known. Gödel's (1933) translation of PA into HA provides an instance of a reduction (by translation) of a system T to $T^{(i)}$. Since then, a great number of results of this form have been obtained by an extension of Gödel's translation; see the introductory note by Troelstra to Gödel (1933) in Gödel (1986, pp. 286–287), for a comprehensive survey. Further considerations are necessary in each case to see whether this translation constitutes a reduction of the nonconstructive to the constructive. Even so, not all such foundational reductions can be obtained by this relatively simple method. As an example, we have the following result:

Theorem 7. $$\Delta^1_2\text{-CA} \leq (\Pi^1_1\text{-CA})^{(i)}_{<\epsilon_0}.$$

Here one has conservation for Π^1_2 formulas, as first proved by Friedman (1970). The proof-theoretic reduction, which actually establishes more, namely, a reduction of Δ^1_2-CA to a constructive theory $\mathrm{ID}^{(i)}_{<\epsilon_0}$ of iterated inductive definitions, was established in a series of steps described in my introduction to Buchholz et al. (1981), due to work of Pohlers, Sieg, and myself. For a survey of further such work, see Feferman (1988, pp. 377–378).

Remark. It happens that Theorems 5–7 can be strengthened directly using forms of the *Axiom of Choice* in L_1:

AC $\quad \forall x \exists Y \phi(x,y) \rightarrow \exists X \forall x \phi(x, X_x)$.

We have Σ^1_n-AC $\vdash \Delta^1_n$-CA by a simple argument. Then "Δ^1_1-CA" can be replaced by "Σ^1_1-AC" in Theorems 5 and 6 and "Δ^1_2-CA"' by "Σ^1_2-AC" in Theorem 7.

2.10 What about Full Analysis, Higher Types, and Set Theory?
The original H.P. aimed to conquer arithmetic, analysis, and set theory, more or less in that order. The examples that have been given in the preceding sections seem to fall far short of giving a foundationally informative reduction of analysis, not to speak of a reduction to finitist principles. Indeed, present-day Gentzen-Schütte-Takeuti-style proof-theoretical work on subsystems of analysis has failed so far to provide an informative proof-theoretical reduction of Π^1_2-CA, though not for lack of trying.* Whether this is a temporary block is impossible to tell at this time. On the one hand, it is quite possible that the kind of proof-theoretical reduction used here as illustrations of the scheme $(*)$ simply cannot be extended to Π^1_2-CA, let alone to full CA. On the other hand, some new conceptual breakthrough might succeed in dealing with Π^1_2-CA and then open the way to tackle full analysis.

While the future in this direction is completely cloudy, it is possible to say something useful about formal systems for analysis in higher finite types as well as in systems of set theory. For example, in Feferman (1988), I described a system of finite types which includes WKL and is proof-theoretically reducible to PRA (*op. cit.* p. 375, section 4.3); this constitutes a partial reduction of the higher uncountable infinite to the finitary. I also described a system of finite types which includes the system $\Delta^1_1\text{-CA}{\restriction}$ and is proof-theoretically reducible to PA (*op. cit.* p. 375, section 5.3); this constitutes a partial reduction of the higher uncountable infinite to the countable infinite. And in Feferman (1985) I showed how to formulate a flexible theory $\mathrm{VT}_\mu{\restriction}$ of *variable* finite types[11] which also includes $\Delta^1_1\text{-CA}{\restriction}$ and is proof-theoretically reducible to PA. This is a theory of functions and classes which is directly interpretable in Zermelo Set Theory, so the reduction $\mathrm{VT}_\mu{\restriction} \leq$ PA constitutes a partial reduction of the set-theoretical

*[This situation has now changed, due to the work described in Rathjen (1995).]

[11]Denoted Res-VT $+(\mu)$, *op. cit.*; cf. alternatively the closely related system W of Feferman (1988a), section 8. [Cf. chapters 13 and 14 in this volume.]

infinitary to the countable infinitary. [These results have now been improved through the work of Feferman and Jäger (1993 and 1996).]

One also obtains a partial reduction of the higher set-theoretical infinitary framework, including transfinite number classes, to the constructive (and in some cases to the predicative framework) through the results of Jäger (1986). These provide reductions of various theories of iterated admissible sets, which are contained in ZF, to subsystems of Δ^1_2-CA + BI, where BI is the principle of bar induction.[12] Though these theories lack the Power-Set Axiom, which is used in ZF with Replacement to establish the existence of the transfinite (accessible) number classes, they build in the existence of such classes by axioms of iterated admissibility.[13] (Speaking in set-theoretical terms, one has the existence of the "alephs," that is, the cardinals $\aleph_\alpha$ for α accessible in these theories, but not the "beths," that is, the cardinals $\beth_\alpha$ for which $\beth_{\alpha+1} = 2^{\beth_\alpha}$).

It would take us too far afield to try to describe the above systems of higher type, variable type, and set theory in any detail, for which the interested reader should refer to the articles mentioned in this section and in the footnotes.

3 From Mathematics to Philosophy via Reductive Proof Theory

Two matters remain to be dealt with in this final section[14] in order to fill out the scheme (*) of section 1. The first is to say something about the passage by formalization from a body of mathematics $\mathcal{M}$ to a formal theory T, especially with reference to the systems presented in section 2. The second is to indicate the philosophical significance of the kind of reduction of T_1 to T_2 illustrated by the results in section 2. We take these up in that order.

The scheme (*) calls for $\mathcal{M}$ to have a direct formalization in T_1; at the same time we should expect that T_1 does not go beyond $\mathcal{M}$ in any essential respects. It is worth elaborating what is required, following the criteria for formalization set forth in Feferman (1979, pp. 171–172), for any $\mathcal{M}$ and T, as follows:

(i) T is an *adequate formalization* of $\mathcal{M}$ if every concept, argument, and result of $\mathcal{M}$ may be represented by a (basic or defined) concept, proof, and theorem, respectively, of T.

(ii) T is in *accordance with* (or *faithful to*) $\mathcal{M}$ if every basic concept of T corresponds to a basic concept of $\mathcal{M}$ and every axiom and rule of T corresponds to, or is implicit in, the assumptions and reasoning followed

[12] BI expresses that one can carry out transfinite induction on any well-founded ordering; Δ^1_2-CA + BI is weaker than Π^1_2-CA.

[13] This direction of work has been pushed up to Mahlo cardinals by Rathjen (1991); what is not evident is whether that gives a reduction to constructive principles. [Cf. also Rathjen 1995.]

[14] With apologies to Hao Wang for borrowing the title of Wang (1974), but with a different meaning.

in $\mathcal{M}$ (in other words, if T does not go beyond $\mathcal{M}$ conceptually or in principle).

The idea of T being *directly adequate to*, respectively *directly in accordance* with $\mathcal{M}$ is clear. We would say that T is *indirectly adequate to* $\mathcal{M}$ if a theory directly adequate to $\mathcal{M}$ is reducible to T in an elementary way (for example, by a translation or proof-theoretic reduction), while it is *indirectly in accordance with* $\mathcal{M}$ if T is reducible to a theory directly in accordance with $\mathcal{M}$. There is a second way in which a theory T may be indirectly adequate to $\mathcal{M}$, that is, to reformulate the concepts, proofs, and theorems of $\mathcal{M}$ informally in such a way that the resulting $\mathcal{M}'$ can be directly formalized in T.

Obviously these criteria are not precise and there may be reasonable differences of opinion as to their application in specific cases. The idea, again, is to say what strikes us as a just ascription on the basis of general experience. Detailed work of formalization may then lead us to modify such an attribution. In particular, it is a common result of such work that a system T which appears to us to provide an adequate and faithful formalization of a body of mathematics $\mathcal{M}$ goes far beyond what is actually needed to represent $\mathcal{M}$ in practice.

Consider, first, elementary (nonanalytic, nonalgebraic) number theory $\mathcal{M}$. At first sight, this has PA as an adequate and faithful formalization in the way we have presented it in the language L_0 with symbols for all primitive recursive functions and (thence) relations. But evidently in practice one makes use of only a small stock of these, for example, $+$, $\cdot$, exp, Σ, Π, p_i, $<$, $|$, $\equiv$, etc. The fact from Gödel (1931) that we can make do with only $+$ and $\cdot$ shows the indirect adequacy of *that* restricted version of PA to informal elementary number theory. While the general principle of induction is in accordance with this body of mathematics, detailed work of formalization shows that the complexity of inductive arguments used in practice is very low and rarely goes beyond Σ^0_2-IA (or, equivalently, Π^0_2-IA); that is, low parts of PA are still directly adequate to $\mathcal{M}$. Then for more specific results or groups of results one can verify that the system Σ^0_1-IA is adequate to a significant portion of $\mathcal{M}$ and hence, by Theorem 1, that PRA is indirectly adequate to that part of elementary number theory.[15] For recent case studies and extensive references to earlier work, see Hájek and Pudlák (1993).

Turning next to analysis, we find a much more complicated picture, both as to the categorization of mathematical practice and choice of related formal languages and systems. One immediate way to break up practice so as to begin to make this more manageable is by means of the separation between concrete analysis, based on finite-dimensional real and complex

[15] Dedicated finitists, beginning with Skolem (1923), had made an effort to see what part of elementary number theory could be directly formalized in PRA; cf. also Goodstein (1957). However, it is much easier to work directly in Σ^0_1-IA to see what can be justified by PRA.

spaces, and abstract analysis, based on various kinds of general spaces such as metric, Hilbert, Banach, etc.

In considering the direct formalization of concrete analysis, one must be able to deal among other things with

(i) the basic number systems (integers, rationals, reals, complexes),

(ii) finite and infinite sequences of numbers,

(iii) specific operations such as sum and product of finite and infinite sequences,

(iv) sets and functions of numbers (n-ary, for various n),

(v) specific operations on functions and infinite sequences of functions such as differentiation and integration, and

(vi) specific operations on sets and infinite sequences of sets, such as complement, union and intersection.

Taking first-order (type 0) arithmetic as basic, and considering real numbers to be represented as sets or sequences of rationals and thence (by reduction to $\mathbf{N}$) as sets or sequences of natural numbers, we need to consider reals as second-order (type 1) objects. Then functions and sets of reals are located at the next type level (third-order or type 2 objects), and specific operations on such are located at still one type higher. Thus for the direct representation of concrete analysis, the language of functions and sets of finite type is adequate, but already the part with first-, second-, and third-order variables serves most purposes comfortably. The question then is: What principles (axioms, rules) are appropriate for this formalization?

On the face of it, concrete analysis makes use of the ideas of "arbitrary" set and "arbitrary" function (of n-tuples of numbers) and, for these, the general forms of the Comprehension Axiom CA and the Axiom of Choice AC, would seem to be the appropriate set and function existence principles. But what special forms of these principles are actually required for and/or in accord with practice is only revealed by closer attention to the details of formalization. In fact, concrete analysis deals with relatively narrow classes of functions and sets, such as piecewise continuous functions (or, more generally, Lebesgue measurable functions) and open and closed sets (or, more generally, Borel sets of finite level), whose properties are individually determined by a countable amount of information. Such functions and sets may then be represented or "coded" by second-order objects, and thus the second-order system CA (or AC) is indirectly adequate to concrete analysis. The recognition of this is often credited to Hilbert and Bernays (1939, supplement 4). However, it was already essentially realized by Weyl in his 1918 monograph that far weaker assumptions about second-order objects, given by arithmetical closure conditions, that is, on the order of ACA, account at least for the familiar (nineteenth-century) concrete analysis of piecewise continuous functions.[16]

Modern continuation of Weyl's work has given much more precise information about what can be done in ACA and related systems. Here we

[16]Cf. Feferman (1988a) [chapter 13 in this volume] for an exegesis of the principles which underlie Weyl's work.

take note of the result that as formulated with the second-order axiom of induction I_1, ACA$\restriction$ is reducible to PA. Now Feferman (1977) and Takeuti (1978) described theories of finite type extending ACA$\restriction$ which are still reducible to PA, and in which the concrete analysis of piecewise continuous functions and of more general classes of functions (for example, Lebesgue measurable) can be carried out directly. These systems have not been described explicitly in section 2 above, so we can't plug them in as examples for the scheme (*); for that, see Feferman (1988). On the other hand, considerable detailed work on the formalization of analysis in second-order systems has been carried out by Friedman and Simpson and their students within Friedman's "Reverse Mathematics" program; see Simpson (1987) for a survey.[17] This achieves its end by reformulating and/or coding various higher-type notions in second-order terms and is thus an example of the second way (described above) in which a theory T may be indirectly adequate to a body of mathematics.

The work of the Friedman-Simpson school gives quite precise information about what principles are needed to prove various statements of concrete (and even abstract) analysis, when these can be reformulated in second-order terms. Combined with the earlier work mentioned above, the general conclusions are as follows:

(i) That part of concrete analysis which does not require set-theoretical (transfinite) notions and constructions in any essential way can be comfortably formalized in systems like ACA$\restriction$ which are reducible to PA.

(ii) Those parts of analysis that require essential use of transfinite ordinals such as "descriptive" set theory of the reals (Borel and projective hierarchies, etc.) can be mostly carried out in systems reducible to Π^1_1-CA and at worst in Δ^1_2-CA or Δ^1_2-CA + BI.

On the other hand, a striking result of the work on the Reverse Mathematics program is that

(iii) a great deal of what can be done in ACA$\restriction$ can already be formalized in the system RCA + WKL + Σ^0_1-IA, which is reducible to PRA by Theorem 3 of section 2.6.

Abstract analysis goes beyond concrete analysis by its use of arbitrary spaces of various kinds, for example, metric spaces, Banach spaces, Hilbert spaces, etc. Moreover, for functional analysis, applications often involve spaces of functions such as the L^p spaces, and, more abstractly, spaces of functions between give spaces. Thus the mathematics prima facie requires *variable types* $X, Y, \ldots$ to represent talk about arbitrary spaces (of the kinds indicated) and higher-type constructions such as of the type $X \to Y$ of all functions from X to Y, from which function spaces are constructed. It happens that for most of the applications of abstract analysis one deals with separable spaces, that is, those X for which there is a countable dense

[17] Full details will become available in a book on subsystems of second-order arithmetic in preparation by Simpson. [Added in press: This has now appeared as Simpson (1998).]

subset X_0 whose completion (under a suitable metric or norm) is X. Talk about such X may informally be recast by talk about X_0, and the latter, being countable, may be coded as a subset of the natural numbers. It is in this way that parts of abstract analysis are indirectly accounted for in the Friedman-Simpson approach. However, a direct formalization requires something like the theories of variable finite type $VT_\mu\restriction$ or W described in section 2.10 (cf. ftn. 11), again reducible to PA by Feferman (1985) [see also Feferman and Jäger 1993, 1996]. More information about the part of concrete and abstract analysis that is directly accounted for in this way is given in that paper and in Feferman (1988a). There I conjectured that *all of scientifically applicable mathematics can be directly formalized in* W; further discussion of this conjecture is in Feferman (1993a) [chapter 14 in this volume].

What, finally, is to be said about the philosophical significance of this work? It seems to me that the information provided by the kind of case studies described here involving the formalization of considerable tracts of everyday mathematics in appropriate systems, in combination with the results of their proof-theoretical reductions such as those presented in section 2, must be taken into account in the continuing efforts to develop a relevant and sustainable philosophy of mathematics. To take just one example, the Quine-Putnam (scientific) indispensability arguments for a form of mathematical realism (cf. Maddy 1992) are considerably weakened when confronted with the kind of information described in the preceding paragraphs.[18] The question in consequence of that is whether the indispensability arguments retain sufficient force to be maintained in the arsenal that has been mounted in defense of realism in mathematics, as, for example, by Maddy (1990).

In general, the kinds of results presented here serve to sharpen what is to be said in favor of, or in opposition to, the various philosophies of mathematics such as finitism, predicativism, constructivism, and set-theoretical realism. Whether or not one takes one or another of these philosophies seriously for ontological and/or epistemological reasons, it is important to know which parts of mathematics are in the end justifiable on the basis of the respective philosophies and which are not. The uninformed common view—that adopting one of the nonplatonistic positions means pretty much giving up mathematics as we know it—needs to be drastically corrected, and that should also no longer serve as the last-ditch stand of set-theoretical realism. On the other hand, would-be nonplatonists must recognize the now clearly marked sacrifices required by such a commitment and should have well thought-out reasons for making them. Though I personally believe that the kind of results described here on the whole strengthen the case for a nonplatonistic philosophy of mathematics and further undermine the case for set-theoretical realism, they do not speak for themselves to that extent,

[18] I argue this at greater length in Feferman (1993a) [reproduced as chapter 14 in this volume].

and it is at that point that well-informed critical philosophical discussion must take over.

11

Gödel's *Dialectica* Interpretation and Its Two-way Stretch

1 The *Dialectica* Paper

In 1958, Gödel published in the journal *Dialectica* an interpretation of intuitionistic number theory in a quantifier-free theory of functionals of finite type; this subsequently came to be known as Gödel's functional or *Dialectica* interpretation. The article itself was written in German for an issue of that journal in honor of Paul Bernays' seventieth birthday. In 1965, Bernays told Gödel of a plan to publish an English translation by Leo F. Boron of his 1958 paper, again in *Dialectica.* However, Gödel was dissatisfied with certain aspects of the original, and set out to revise the translation. A year after doing so to his apparent satisfaction, Gödel changed his mind and decided instead to add a new series of extensive footnotes by way of improvement and amplification. The result was sent to the printer in 1970 after much help and encouragement by Bernays and Dana Scott, but when the proof sheets were returned, Gödel was again dissatisfied, especially with two of the added notes. Though he apparently worked on rewriting these until 1972, the paper was never returned in final form for publication. The corrected proof sheets found in his *Nachlass* were reproduced for the first time in volume 2 of Gödel's *Collected Works* (1990), where they appear as (1972). The full story of the vicissitudes of this paper is told by A. S. Troelstra in his introductory note to (1958) and (1972) in that volume.

There is also a long and interesting prior history to the development of Gödel's functional interpretation, much of which has only emerged in recent years through the study of previously unpublished lecture texts going back to 1933 and found in Gödel's *Nachlass*; these texts have now appeared in

"Gödel's *Dialectica* interpretation and its two-way stretch" was first published in G. Gottlob et al., (eds.), *Computational Logic and Proof Theory*, Lecture Notes in Computer Science 713 (1993), 23–40 (Feferman 1993c). It is reprinted here with the kind permission of the publisher, Springer-Verlag GMbH & Co. KG, Heidelberg. There are some minor additions and corrections, and references have been brought up-to-date.

volume 3 of his *Collected Works* (1995). It seemed to me fitting to use the present occasion to trace the development of these ideas to which Gödel devoted repeated attention over such a long stretch of time.

2 Toward the Functional Interpretation: 1933–1938

The previously unpublished lecture texts referred to above are three in number; they are

A. "The present situation in the foundations of mathematics"—an invited lecture delivered in December 1933 to a meeting of the Mathematical Association of America held jointly with the American Mathematical Society in Cambridge, Massachusetts. (Gödel (1933a), = *1933o in Gödel (1995), pp. 45–53.)

B. "Vortrag bei Zilsel"—a lecture in January 1938 for an informal seminar organized by Edgar Zilsel in Vienna. (*1938a in Gödel (1995), pp. 86–113.)

C. "In what sense is intuitionistic logic constructive?"—a lecture at Yale University in April 1941. (Gödel (1941), = *1941 in Gödel (1995), pp. 189–200.)

The texts for the Cambridge and Yale lectures were found fully written out in Gödel's *Nachlass* and required very little editorial work to be included in volume 3 of the Gödel *Works.* That for the Zilsel seminar was a different matter altogether and required a great deal of arduous effort; the notes in this case were found in the *Gabelsberger* shorthand employed by Gödel. They were initially transcribed by Cheryl Dawson and worked up in collaboration with her to a relatively coherent text by Wilfried Sieg and Charles Parsons, who then translated it into English. Introductory notes to all of these items appear in Gödel (1995): the first of these is mine,* the second is by Sieg and Parsons, and the third is by Troelstra. I have drawn on all three of these notes in the following, and am indebted to Parsons, Sieg, and Troelstra for what I have learned from their exegetical work. I will give a synopsis of the relevant portions of the Cambridge and Vienna lectures in this section; the Yale lecture is taken up in the next section.

In the 1933 Cambridge lecture (A), Gödel says that the problem of providing a foundation for mathematics falls into two parts: the first is to represent the methods of proof actually used by mathematicians in a deductive system reduced to a minimum number of axioms and rules of inference, and the second is to give a justification in some sense or other for these axioms. After arguing that the first part has been successfully accomplished via systems of axiomatic set theory (such as that of Zermelo-

*[The introductory note to item A also appears in this volume as chapter 8.]

Fraenkel), Gödel turns to the second part of the foundational project with the surprising statement:

> The result . . . is that our axioms, if interpreted as meaningful statements, necessarily presuppose a kind of Platonism, which cannot satisfy any critical mind and which does not even produce the conviction that they are consistent. (Gödel 1995, p. 50)[1]

While Gödel says that it is very likely these axioms *are* consistent, he says a proof of freedom from contradiction must use utterly unobjectionable methods—and thus must exclude such problematic features of set theory as the use of nonconstructive reasoning in existence proofs and of impredicative definitions. Thus one must seek a consistency proof by constructive methods; however, here there is a choice, as there are different layers of constructivity. The lowest of these is that of finitism, which is distinguished by three features:

> 1. The application of the notion of "all" or "any" is to be restricted to those infinite totalities for which we can give a finite procedure for generating all their elements [such as the integers]. . . .
> 2. Negation must not be applied to propositions stating that something holds for all elements, because this would give existence propositions. . . . [These] are to have a meaning in our system only in the sense that we have found an example but, for the sake of brevity, do not state it explicitly. . . .
> 3. And finally we require that we should introduce only such notions as are decidable for any particular element and only such functions as can be calculated for any particular element. (Gödel 1995, p. 51)

But, contrary to Hilbert's expectation, it appears to Gödel hopeless to demonstrate even the consistency of classical arithmetic [Peano Arithmetic, PA] by finitist methods, in view of his second incompleteness theorem and the fact that all known or prospective such methods can easily be carried out in that system. This leads one to consider the possible use of intuitionistic methods in the wider sense of Brouwer and Heyting. And, although one has a reduction of classical to intuitionistic arithmetic [Heyting Arithmetic, HA] by Gödel (1933), this does not meet the desired goals, since intuitionistic principles violate the above criteria in two essential respects.

---

[1] This doesn't seem to square with Gödel's unequivocal assertions in various sources from the 1960s and 1970s that he had held a platonistic philosophy of mathematics since his student days in Vienna. The discrepancy is discussed further in chapter 8.

Namely, in intuitionism one allows the formation of negation of arbitrary propositions by giving to $\neg p$ the meaning that one has a (constructive) demonstration that any proof of p leads to an absurd conclusion. This violates criterion 1, since "any" here ranges over the inherently vague totality of arbitrary constructive proofs (which cannot be limited to formal proofs in any one formal system), and it violates criterion 2 by allowing the formation of (and reasoning with) negations of universal propositions, $\neg\forall x\phi(x)$. At the end of his 1933 Cambridge lecture, Gödel concludes that the "foundation of classical arithmetic by means of the notion of absurdity is of doubtful value," but he is hopeful that "one may find other and more satisfactory methods of construction beyond the limits of [finitism]" to found at least classical arithmetic and then analysis.

Gödel returned to modified forms of Hilbert's program in his lecture for the Zilsel seminar (B above), to meet certain criteria such as 1–3 above, but now with more definite candidates for constructive consistency proofs. The possible routes considered are (1°) the use of higher-type functionals, (2°) a "modal-logical" approach, and (3°) use of transfinite induction and transfinite recursion. In the first of these, Gödel follows Hilbert (1926) by giving examples of schemata for primitive recursive functionals of finite type; however, he gives no indication that he is in possession of an interpretation of HA in such a system. Nevertheless, this should be considered a first step toward the *Dialectica* interpretation. Under (2°), Gödel sketched an abstract theory of constructive proofs as a foundation of intuitionistic reasoning. In this he anticipates Kreisel (1962) (which, however, met various difficulties; cf. Goodman 1970). Finally, under (3°), Gödel considers Gentzen's consistency proof for classical arithmetic (Gentzen 1936). Here he explains the essentials of Gentzen's reduction procedures in terms of a functional interpretation using functionals just of type level ≤ 2. The idea is illustrated by means of an example; consider

$$\forall x \exists y \forall z \exists u R(x, y, z, u) \tag{1}$$

where R is decidable. Constructively one would have a counterexample to (1) if for some a, f we have

$$\forall y, u \neg R(a, y, f(y), u). \tag{2}$$

Then an interpretation of (1) is that any proposed counterexample can be blocked by computable functionals $Y(f, a), U(f, a)$, such that

$$\forall f, a[R(a, Y(f, a), f(Y(f, a)), U(f, a))]. \tag{3}$$

Moreover, Gödel suggests how Gentzen's assignments of ordinals in his proof can be used to define such realizing functionals for provable statements of classical arithmetic by transfinite recursion on ordinals up to ϵ_0. Here Gödel anticipated the "no-counterexample interpretation" (n.c.i.) introduced by Kreisel (1951, 1952) for arithmetic (itself obtained by an analysis of Ackermann's version of Gentzen's consistency proof). It should also be noted that the n.c.i. was later shown (by Kreisel) to be a consequence of Gödel's translation of PA into HA followed by his functional interpretation of HA.

3 The Yale Lecture: Gödel's Heuristics for the Functional Interpretation

Gödel lectured at Yale University on 15 April 1941 to a joint meeting of the Mathematics and Philosophy Clubs,[2] under the title "In what sense is intuitionistic logic constructive?" (C, in section 2). The text of this lecture is more informal and leisurely than the later *Dialectica* publication, and it explains (as the latter does not) the heuristic reasoning that led him to the functional interpretation. Again, he sets down three criteria for constructivity which are close to those of the 1933 lecture in Cambridge but differ in one essential respect:

> 1. All primitive (undefined) functions . . . must be calculable for any given arguments and all primitive relations must be decidable for any given arguments.
> 2. Existential assertions must have a meaning only as abbreviations for actual constructions . . . .
> 3. Universal propositions can be negated in the sense that a counterexample exists in the sense just described . . . . Therefore, leaving out abbreviations, universal propositions can't be negated at all. (Gödel 1995, p. 191)

The essential difference from the former criteria is that Gödel no longer requires universally quantified variables to range over totalities whose elements are generated by some finite procedure. Gödel calls a system "strictly constructive" or "finitistic" if it satisfies these three conditions, though he is troubled by the latter appellation. In fact, Gödel described later (in 1958) his use of functionals of finite type as a (novel) extension of finitistic methods. In any case, he says that intuitionism does not (on the face of it) meet these three conditions. Nevertheless, he is able to show that "in its application to definite mathematical systems intuitionistic logic can be reduced to finitistic systems" (in the above sense). This is illustrated by his interpretation of Heyting Arithmetic, HA, in a system Σ of functionals of finite type, which has variables ranging over natural numbers, functions of numbers, functions of functions (functionals), etc. Σ provides for two basic means of introducing specific functions and functionals: Explicit Definition and [Primitive] Recursive Definition, where in

[2] I learned of these auspices from John Dawson. Kreisel (1987, p. 104) erroneously states that this lecture was on the occasion of an honorary doctorate; Gödel was awarded an Honorary D. Litt. by Yale ten years later in June 1951.

$F(0) = G, F(x+1) = H(x, F(x))$ the values of $F(x)$ may lie in any given type. The axioms and rules of Σ allow for ordinary propositional calculus applied to decidable (quantifier-free) formulas, the rule of induction, and rules for equality and substitution. The "meaningful propositions" of Σ have the form $(\exists x_1, \ldots, x_n)(\forall y_1, \ldots, y_m)R(x_1, \ldots, x_n, y_1, \ldots, y_m)$ where R is quantifier-free and $x_1, \ldots, x_n, y_1, \ldots, y_m$ may be of arbitrary type. This comes with the understanding that one can assert such a statement in Σ only if specific instances $t_1, \ldots, t_n$ have been found such that $R(t_1, \ldots, t_n, y_1, \ldots, y_m)$ is established to hold for arbitrary $y_1, \ldots, y_m$. Gödel argues that requirements 2 and 3 above force one to limit the statements in this way, since propositional operations (in particular negation) may not be applied to universal statements, and existential quantification is only regarded as an abbreviation for successful instantiation.

Statements of Σ are indicated in the form $\exists x \forall y R(x, y)$, where x, y are understood to be sequences of variables of arbitrary type. The interpretation of intuitionistic logic and arithmetic in Σ is obtained by associating with each formula $A(z)$ of arithmetic, a formula $\exists x \forall y R(x, y, z)$ of Σ. This is done by induction on A; for simplicity, the parameters "z" are suppressed in the following. The most complicated association (and, in Gödel's words, "the most important") is with $A \to B$, where $\exists x \forall y R(x, y)$ is the interpretation of A and $\exists u \forall v S(u, v)$ is that of B. Gödel says that constructively

$$(1) \qquad \exists x \forall y R(x, y) \to \exists u \forall v S(u, v)$$

can only mean

$$(2) \qquad \forall x \exists u [\forall y R(x, y) \to \forall v S(u, v)].$$ [3]

This then is converted to the form

$$(3) \qquad \exists f \forall x [\forall y R(x, y) \to \forall v S(f(x), v)].$$

The problem then is how to interpret the implication in brackets; the simplest way, Gödel says, is to consider its contrapositive,

$$(4) \qquad \neg \forall v S(f(x), v) \to \neg \forall y R(x, y),$$

and here one would show how to convert any counterexample to the hypothesis into one for the conclusion, that is,

$$(5) \qquad \exists g_1 \forall v [\neg S(f(x), v) \to \neg R(x, g_1(v))].$$

Since the formulas R, S are decidable, classical propositional calculus then leads us from (3) via (4) and (5) to

$$(6) \qquad \exists f \forall x \exists g_1 \forall v [R(x, g_1(v)) \to S(f(x), v)]$$

and finally to

$$(7) \qquad \exists f, g \forall x, v [R(x, g(x, v)) \to S(f(x), v)]$$

as the interpretation of $A \rightarrow B$. Gödel then goes on to show how to interpret $\neg A, A \vee B, A \wedge B, \exists z A(z)$ and $\forall z A(z)$, given those for A and B. Treating $\neg A$ as $A \rightarrow (0 = 1)$ yields $\exists g \forall x \neg R(x, g(x))$ as its interpretation; the rest are treated in the obvious way and will be shown explicitly in the next section.

As examples, Gödel shows how the interpretation of $(A \rightarrow A)$ is provable in Σ and how the rule of modus ponens is preserved by the interpretation. Both of these use only the Explicit Definition principle. He says that the proofs for the other axioms and rules of HA are a little longer but quite straightforward. Gödel also remarks that his interpretation can be extended to other intuitionistic systems whose primitive functions (relations) are calculable (decidable). In addition, one obtains constructive consistency proofs of certain classical systems by first translating them into corresponding intuitionistic systems by the method of Gödel (1933) and then applying the functional interpretation. In particular, the system PA is thus reduced to Σ.

4 The *Dialectica* Paper

The title of Gödel's *Dialectica* paper, "Über eine noch nicht benützte Erweiterung des finiten Standpunktes,"[4] signals his principal foundational concern, developed in the discussion with which the paper begins. Here Hilbert's finitism is characterized as the mathematics of finite combinations of concretely representable and directly visualizable objects such as numbers and symbols. But, finitary mathematics is insufficient to establish the consistency of classical number theory, let alone of classical mathematics more generally. For that, continues Gödel, certain abstract notions are needed; one can retain the constructive component of the finitary standpoint while admitting such notions, as for example in intuitionistic logic. But the notion of computable function(al) of finite type, while also abstract, is more definite than the abstract notion of proof that underlies intuitionistic reasoning. The system T of the *Dialectica* paper [previously referred to as Σ in the 1941 Yale lecture] embodies directly evident principles for the functionals of finite type, and this extension of finitism may be used to prove the consistency of classical arithmetic.

As in the Yale lecture, the basic axioms and rules of T are only indicated; the functional interpretation is spelled out, but no details are given of the proof that HA is interpreted in T. In the supplemental footnotes to his 1972 version of the *Dialectica* paper, Gödel filled in various of those details. But he also endeavored there to strengthen the case for the foundational progress achieved by his interpretation, apparently without arriving at a formulation that he considered sufficiently convincing.

[4] Or, in the English translation in Gödel (1990), "On a hitherto unutilized extension of the finitary standpoint."

Gödel's functional interpretation was brought to the attention of the logic community in a lecture by Georg Kreisel at the Summer Institute in Symbolic Logic held at Cornell in 1957; of this, more below. From then on, a number of researchers worked out Gödel's interpretation in detail and extended it (as he had expected) to a variety of other systems. To my mind, the best introduction to Gödel's own work and the subsequent literature (up to 1990) is provided by Troelstra's introductory notes to (1958) and (1972) in volume 2 of the *Collected Works* (1990, pp. 217–241). For further study of the technicalities involved, Troelstra (1973, sec. III.5) is required reading. Kreisel (1987, pp. 104–120) provides a discursive assessment of Gödel's interpretation and some of its extensions.*

Due to limitations of space, I can only touch here on a few results that illustrate the kind of information that may be drawn from the *Dialectica* interpretation and which illustrate its adaptability to a variety of situations. We begin with setting down the interpretation in full and its first consequences, following the expositions by Troelstra just mentioned. Many of the necessary syntactic preliminaries found in those sources are omitted.

Let L^ω be the language of finite types over that of elementary number theory; in this, each term, and in particular each variable, has a specified type σ. Atomic formulas are supposed to be equations between terms of type 0. Equations between terms of higher type are supposed to be abbreviations for equality at all (variable) arguments (when driven down to type 0).** There are constants $K_{\sigma,\tau}$ and $S_{\rho,\sigma,\tau}$ of various types ρ, σ, τ as provided by the finite-typed combinatory calculus, satisfying the equations

$$(1) \qquad K_{\sigma,\tau}xy = x, \quad S_{\rho,\sigma,\tau}xyz = xz(yz)$$

(for x, y, z variables of appropriate type), which ensure closure under explicit definition. There are also, in the case specifically of Gödel's system T, *recursors* R_σ for each type σ satisfying

$$(2) \qquad R_\sigma xy0 = x, \quad R_\sigma xy(z') = y(R_\sigma xyz)z$$

with x of type σ, y of type $\sigma \to 0 \to \sigma$, and z of type 0. Other applications make use of other kinds of recursors in place of (2) and/or additional constants. QF denotes the class of quantifier-free formulas in L^ω, and for any class $\mathcal{F}$ of formulas, QF-$\mathcal{F}$ denotes the intersection of QF and $\mathcal{F}$.

Sequences of variables (which may be empty) are given by boldface lowercase or uppercase letters, namely, $\mathbf{x} = (x_1, \ldots, x_n)$ or $\mathbf{X} = (X_1, \ldots, X_n)$, where $n \geq 0$. When preceded by $\exists$ or $\forall$, $\mathbf{xy}$ is the concatenation of $\mathbf{x}$ and $\mathbf{y}$; within a formula, $\mathbf{Xy}$ indicates the result of applying each term of $\mathbf{X}$ to the sequence $\mathbf{y}$ (when this is well typed). With each formula ϕ of L^ω is associated its *Dialectica* (or *D*-) interpretation $\phi^D \equiv \exists\mathbf{x}\forall\mathbf{y}\phi_D(\mathbf{x}, \mathbf{y})$ where

*[A more recent exposition and survey of the applications and extensions of the interpretation is given in Avigad and Feferman (1998).]

**[Cf. ftn. 7 below.]

ϕ_D is QF; the free variables of ϕ will be exhibited only when necessary.[5] This is defined inductively as follows, using also $\psi^D \equiv \exists\mathbf{u}\forall\mathbf{v}\psi_D(\mathbf{x}, \mathbf{y})$:

(3) (i) For ϕ an atomic formula, $\mathbf{x}, \mathbf{y}$ are empty and $\phi^D \equiv \phi_D \equiv \phi$,

(ii) $(\phi \wedge \psi)^D \equiv \exists\mathbf{xu}\forall\mathbf{vy}(\phi_D \wedge \psi_D)$,

(iii) $(\phi \vee \psi)^D \equiv \exists z\mathbf{xu}\forall\mathbf{vy}[(z = 0 \wedge \phi_D) \vee (z = 1 \wedge \psi_D)]$ (z of type 0),

(iv) $(\phi \rightarrow \psi)^D \equiv \exists\mathbf{UY}\forall\mathbf{x}, \mathbf{v}[\phi_D(\mathbf{x}, \mathbf{Yxv}) \rightarrow \psi_D(\mathbf{Ux}, \mathbf{v})]$,

(v) $\forall z\phi(z))^D \equiv \exists\mathbf{X}\forall z, \mathbf{y}\ \ \phi_D(\mathbf{X}z, \mathbf{y}, z)$,

(vi) $(\exists z\phi(z))^D \equiv \exists z\mathbf{x}\forall\mathbf{y}\ \ \phi_D(\mathbf{x}, \mathbf{y}, z)$.

Gödel's motivation for (iv) has been described in the preceding section.

Negation of ϕ is regarded as defined by $\neg\phi \equiv (\phi \rightarrow 0 = 1)$. Then one obtains

(4) $$(\neg\phi)^D \equiv \exists\mathbf{Y}\forall\mathbf{x}\neg\phi_D(\mathbf{x}, \mathbf{Yx})$$

and so

(5) $$(\neg\neg\phi)^D \equiv \exists\mathbf{X}\forall\mathbf{Y}\neg\neg\phi_D(\mathbf{XY}, \mathbf{Y}(\mathbf{XY})).$$

This is important to take note of, since Gödel's 1933 double-negation or "negative" translation of classical into intuitionistic systems replaces $\exists z\psi(z)$ by $\neg\neg\exists z\psi(z)$ and $\psi \vee \theta$ by $\neg\neg(\psi \vee \theta)$. In particular, we have

(6) $$(\neg\neg\exists\mathbf{z}R(\mathbf{z}))^D \equiv \exists\mathbf{z}\neg\neg R(\mathbf{z}) \text{ for } R \in QF.$$

Let HA^ω be the system extending HA in L^ω by the combinatory axioms (1) and recursor axioms (2), with the induction scheme extended to arbitrary formulas of L^ω.[6] The quantifier-free part of QF-HA^ω is just another form of Gödel's system T [or his 1941 Σ];[7] in the following we shall use PR^ω as a more suggestive denotation for this theory of primitive recursive functionals of finite type. Gödel showed that the D-interpretation carries HA into PR^ω, in the sense that for every provable ϕ with $\phi^D \equiv \exists\mathbf{x}\forall\mathbf{y}\phi_D(\mathbf{x}, \mathbf{y})$ we can find a sequence of terms $\mathbf{t}$ (directly from a proof of ϕ) with $\phi_D(\mathbf{t}, \mathbf{y})$ provable in PR^ω. With hardly any additional work, one obtains

(7) $$\mathrm{HA}^\omega \vdash \phi \Rightarrow \mathrm{PR}^\omega \vdash \phi_D(\mathbf{t}, \mathbf{y}) \text{ for some } \mathbf{t}.$$

[5] While Gödel defined ϕ^D only for first-order ϕ, its extension to ϕ in L^ω is immediate.

[6] Troelstra (1990) uses "WE-HA^ω" for our HA^ω, where "WE" stands for "Weakly Extensional"; there is a slight risk of confusion since he uses "HA^ω" for a different system, but that is not needed here.

[7] Except that Gödel can be understood as assuming basic decidable equality relations at each type σ in 1958; this is the so-called *intensional* version of T (cf. Troelstra 1990, pp. 221–222).

The proof of (7) breaks into two parts. The first, quite general, part is to show that the axioms and rules of intuitionistic logic in L^ω are preserved under the D-interpretation (in the sense of (7)); this uses only the axioms (1) for explicit definition and the underlying logic of PR^ω. The second part is to show that the nonlogical axioms of HA^ω are preserved under the D-interpretation; here the main step is to use the recursor axioms (2) to verify the D-interpretation of the induction scheme. The latter involves a nice exercise in the application of the definition of $(\phi \to \psi)^D$.

It is also easy to verify that the D-interpretation preserves the Axiom of Choice, as given by the scheme

$$\text{AC} \qquad \forall x \exists y \phi(x,y) \to \exists f \forall x \phi(x, fx).$$

This principle is generally accepted by intuitionists. However, the following scheme is not (ordinarily), since it generalizes "Markov's principle":

$$\text{M}' \qquad \neg\forall \mathbf{v} R(\mathbf{v}) \to \exists \mathbf{v} \neg R(\mathbf{v}), \text{ for } R \in \text{QF}.$$

This principle figures in a slightly different analysis (than that given by Gödel in his 1941 lecture) of how one arrives at the D-interpretation of implication; cf. Troelstra (1990, p. 226). Another such schema which is problematic for intuitionists is a form of "independence of premise" principle:

$$\text{IP}' \qquad [\forall \mathbf{y} R \to \exists \mathbf{u} \forall \mathbf{v} S] \to \exists \mathbf{u}[\forall \mathbf{y} R \to \forall \mathbf{v} S], \qquad \text{for } R, S \in \text{QF}.$$

That was used in the step from (1) to (2) in the preceding section, again in the treatment of implication. Let $\widetilde{\text{HA}}^\omega = \text{HA}^\omega + \text{AC} + \text{M}' + \text{IP}'$; then (7) above can be strengthened to the following:

$$\text{(i)} \qquad \widetilde{\text{HA}}^\omega \vdash \phi \leftrightarrow \phi^D \text{ for any } \phi, \text{ and}$$

$$\text{(ii)} \qquad \widetilde{\text{HA}}^\omega \vdash \phi \Rightarrow \text{PR}^\omega \vdash \phi_D(\mathbf{t}, \mathbf{y}) \text{ for some } \mathbf{t}.$$

(See Troelstra 1990, p. 232.) The result (8) serves to simplify various verifications. As an immediate application of (8)(i) one has, by (6),

$$(8) \qquad \widetilde{\text{HA}}^\omega \vdash \neg\neg\exists \mathbf{z} R(\mathbf{z}) \leftrightarrow \exists \mathbf{z} R\mathbf{z} \text{ for } R \in \text{QF},$$

since R is decidable; of course, (9) also follows from M′.

5 Application of the D-Interpretation to Classical Systems

For $\phi \in L^\omega$ considered as a formula in a system with classical logic, let ϕ^- be its "negative" translation, obtained by prefixing every disjunctive or existential subformula of ϕ by double negation. Let PA^ω be HA^ω with classical logic, so PA^ω extends PA as HA^ω extends HA. Gödel (1933) showed that the negative translation sends PA into HA; this carries over immediately to the corresponding finite type extensions. In connection with the further application of the D-interpretation, Howard (1968, p. 115) observed the following useful strengthening of this result: let $\widetilde{\mathrm{PA}}^{\omega}$ be PA^{ω} together with all formulas of the form $\psi \leftrightarrow (\psi^{-})^{D}$ for $\psi \in \mathrm{L}^{\omega}$. Then

(1) $$\widetilde{\mathrm{PA}}^{\omega} \vdash \phi \text{ implies } \widetilde{\mathrm{HA}}^{\omega} \vdash (\phi^{-}), \text{ hence } \widetilde{\mathrm{HA}}^{\omega} \vdash (\phi^{-})^{D}.$$

Application of the *Dialectica* interpretation to classical systems involves calculating the effect of $\phi \mapsto (\phi^{-})^{D}$ for various ϕ.[8] Let S^{-} be the set of $(\theta)^{-}$ for $\theta \in S$. Since $(\ )^{-}$ preserves $\wedge$ and $\rightarrow$, we have

(2) $$\widetilde{\mathrm{PA}}^{\omega} + S \vdash \phi \Rightarrow \widetilde{\mathrm{HA}}^{\omega} + S^{-} \vdash \phi^{-}.$$

The easiest application of this observation is that

(3) $$\widetilde{\mathrm{PA}}^{\omega} + (\mathrm{QF}-\mathrm{AC}) \vdash \phi \Rightarrow \widetilde{\mathrm{HA}}^{\omega} \vdash \phi^{-}.$$

For, an instance $\forall \mathbf{x} \exists \mathbf{y} P(x, y) \rightarrow \exists f \forall \mathbf{x} P(\mathbf{x}, f\mathbf{x})$ of QF-AC, with $P \in \mathrm{QF}$, is sent into $\forall \mathbf{x} \neg\neg \exists \mathbf{y} P(\mathbf{x}, \mathbf{y}) \rightarrow \neg\neg \exists f \forall \mathbf{x} P(\mathbf{x}, f\mathbf{x})$. But the hypothesis is equivalent to $\forall \mathbf{x} \exists \mathbf{y} P(\mathbf{x}, \mathbf{y})$ by (9) of the preceding section and the conclusion then follows from AC in $\widetilde{\mathrm{HA}}^{\omega}$. As a corollary, we obtain, for $R \in \mathrm{QF}$,

(4) $$\widetilde{\mathrm{PA}}^{\omega} + (\mathrm{QF}-\mathrm{AC}) \vdash \forall \mathbf{x} \exists \mathbf{y} R(\mathbf{x}, \mathbf{y}) \Rightarrow \mathrm{PR}^{\omega} \vdash R(\mathbf{x}, \mathbf{tx}) \text{ for some } \mathbf{t}.$$

This shows among other things that every provably recursive function(al) of $\widetilde{\mathrm{PA}}^{\omega}$ + (QF-AC) is definable by a term of PR^{ω}. By formalizing in PA the model of PR^{ω} in HEO, the hereditarily extensional effective operations, this shows that

(5) *The provably recursive functions of* PA *are the same as those of* PA^{ω} + QF-AC, *and also the same as the functions of type* 1 *in* PR^{ω}.

It follows from (5) and Kreisel (1952) that the provably recursive functions of type 1 in PR^{ω} are exactly the functions definable by effective transfinite recursion on ordinals $< \epsilon_0$ (in its natural ordering). For more direct assignments of ordinals $< \epsilon_0$ to the terms of PR^{ω} giving the same result, see the references in Troelstra (1990, p. 238).

[8] Shoenfield (1967), pp. 219ff., provides an alternative interpretation which avoids passing through the negative translation. This associates with each ϕ a formula ϕ^{S} of the form $\forall \mathbf{x} \exists \mathbf{y} \phi_S(\mathbf{x}, \mathbf{y})$ with $\phi_S \in \mathrm{QF}$, such that if $\mathrm{PA} \vdash \phi$ then $\mathrm{PR}^{\omega} \vdash \phi_S(\mathbf{x}, \mathbf{tx})$ for some $\mathbf{t}$. However, this provides no real saving in work in general.

6 Functional Interpretation of Classical Analysis

The historical material described above shows that for Gödel the primary value of the functional interpretation was its use in reducing the consistency of HA to that of PR^{ω}, a system embodying notions which he thought to be closer to those of finitism than those underlying intuitionistic reasoning. He noted in his 1941 Yale lecture that this reduction leads via the negative translation to a proof of consistency of PA on the basis of PR^{ω} and, moreover, that if PA proves $\exists y R(y)$ with R in QF then $R(t)$ holds for some t in PR^{ω} (an evident precursor of (4) in the preceding section). Toward the end of the same lecture, Gödel speculated that

> It is perhaps not altogether hopeless to try to generalize these consistency proofs to analysis by means of functions of still higher (that is, transfinite) type. Future development will show if that is possible at all and in which sense the system necessary to accomplish this proof will be constructive. (Gödel 1995, p. 200)

Kreisel (1987, p. 104) reports that Gödel told him, very soon after their first meeting in 1955, about his functional interpretation work in the 1940s, "later incorporated in the so called *Dialectica* interpretation (with a total shift of emphasis)." Evidently Gödel misremembered: there is really no significant difference in emphasis, though the 1941 lecture mentions a few applications that are not contained in the 1958 *Dialectica* article.[9] Of Gödel's supposed change in emphasis, Kreisel (*loc. cit.*) goes on to say that Gödel "wanted to fill the superficially principal gap left by his negative translation . . . . He dropped the project after he learned of recursive realizability that Kleene found soon afterward." It is true that the association with each ϕ, provable in HA, of terms $\mathbf{t}$ in PR^{ω} such that $\phi_D(\mathbf{t}, \mathbf{y})$ holds, is akin to Kleene's (1945) original realizability interpretation, though, as Kreisel points out, neither one yields the existential definability property for HA.[10]

Be that as it may, given Kreisel's general interests in constructivity in the 1950s and his prior development of the no-counterexample functional interpretation for classical arithmetic, he was in an excellent position (perhaps uniquely so) to appreciate Gödel's accomplishment in this respect and to exploit it further. Before long he was spreading the word about Gödel's functional interpretation and, by 1957, arrived at an extension to a system for full second-order analysis. Kreisel first presented this work to a large

[9] It *is* a question of Gödel's memory, since he wrote Bernays in 1968, "In those days . . . I set no particular store by the philosophical aspect; rather, it was chiefly the mathematical result that was important to me, while now it is the other way around" (Troelstra 1990, p. 217).

[10] That property was established later by Kleene (1952), §82, using Γ-realizability (and, for that purpose, simplified still later to "slash" realizability).

audience of logicians at the Summer Institute for Symbolic Logic held at Cornell in June 1957, under the title "Gödel's interpretation of Heyting's Arithmetic," though most of his lecture was about his own extension of the interpretation to analysis. Later that same summer he lectured on this material at the Colloquium on Constructivity in Mathematics held in Amsterdam; his article for its proceedings appeared as Kreisel (1959). The formulation of analysis used there is a second-order system with full Comprehension Axiom scheme CA. In view of the preceding discussion, it is natural to extend this to the language L^ω and to infer CA from the (full) Axiom of Choice scheme AC. Now by (8)(i) of section 4 above we have

(1) $$\mathrm{PA}^\omega + \mathrm{AC} \vdash \phi \leftrightarrow (\phi^-)^D \text{ for each } \phi,$$

since ϕ is equivalent to ϕ^- with classical logic, and M′ and IP′ are classically valid. Another way of stating (1) using the notation of the preceding section, is that $\widetilde{\mathrm{PA}}^\omega$ is contained in PA^ω + AC.

One of the main results of Kreisel (1959) is that if ϕ is provable in analysis and $(\phi^-)^D \equiv \exists \mathbf{x} \forall \mathbf{y} R(\mathbf{x}, \mathbf{y})$ with $R \in$ QF, then $(\exists \mathbf{x} \in \mathcal{RC})(\forall \mathbf{y} \in \mathcal{C})$ $R(\mathbf{x}, \mathbf{y})$ holds, where $\mathcal{C}$ consists of the continuous functionals of finite type and $\mathcal{RC}$ of the recursively continuous functionals.[11] The classes $\mathcal{C}$ and $\mathcal{RC}$ had been developed independently by Kreisel and Kleene, with their respective treatments both published in the proceedings of the 1957 Amsterdam conference; however, only Kreisel pursued the connection with Gödel's functional interpretation.[12] A direct strengthening of Kreisel's result to the system in (1) is

(2) if PA^ω + AC $\vdash \phi$ and $(\phi^-)^D \equiv \exists \mathbf{x} \forall \mathbf{y} R(\mathbf{x}, \mathbf{y})$, then there exists $\mathbf{f} \in \mathcal{RC}$ such that for all $\mathbf{y}$ in $\mathcal{C}$, $R(\mathbf{f}, \mathbf{y})$ holds.

It was then natural to ask whether the class $\mathcal{RC}$ in (2) could be replaced by a class of functionals determined by schemata extending Gödel's PR^ω. This was pursued intensively by Clifford Spector during the year 1960–61 which he spent at the Institute for Advanced Study in Princeton, where he had frequent contact with Kreisel and also benefited from conversations with Gödel and Bernays. Spector succeeded in achieving this goal by the adjunction of a new scheme to PR^ω, called Bar Recursion:

$$\text{BR} \quad F(n,c) = \begin{cases} G\ (n, \langle c0, \ldots, c(n-1)\rangle) \text{ if } Y(\langle c0, \ldots, c(n-1)\rangle) < n, \text{ else} \\ H(\lambda a.F(n+1, \langle c0, \ldots, c(n-1), a\rangle), n, \langle c0, \ldots, c(n-1)\rangle). \end{cases}$$

[11] In fact, according to Kreisel (1959), ϕ is equivalent to $(\exists \mathbf{x} \in \mathcal{RC})(\forall \mathbf{y} \in \mathcal{C}) R(\mathbf{x}, \mathbf{y})$.

[12] There is a wealth of results and observations in Kreisel (1959). Among others, it is noted that the n.c.i. for PA follows from Gödel's functional interpretation. Kreisel also considered the status of the formulas $(\phi^-)^D$ in other classes of functionals of finite type including the hereditarily effective operations and Kleene's recursive functionals.

Here the values $c(k)$ may be at any type σ (with a of the same type). The scheme (BR) is justified for continuous Y, since $Y(c)$ will be determined by a finite amount of information $\langle c0, \ldots, c(n-1)\rangle$ about c and then for sufficiently large n one can make $Y(c) = Y(\langle c0, \ldots, c(n-1)\rangle) < n$. Given c, F "searches" for such an n in (BR). The terminology "Bar Recursion" was suggested by that of Brouwer's Bar Theorem, which uses an intuitionistically accepted form of the principle of transfinite induction on well-founded trees of natural numbers. This may be generalized to trees of objects of any type σ as a principle called Bar Induction (BI), and Bar Recursion is then a related principle of definition by recursion on such well-founded trees.

BR^ω is used here to denote the formal quantifier-free system extending PR^ω by adjunction of the defining equations (BR) at each type. The main result of Spector's work may then be formulated as follows:

(3) If $PA^\omega + AC \vdash \phi$ and $(\phi^-)^D \equiv \exists \mathbf{x} \forall \mathbf{y} R(\mathbf{x}, \mathbf{y})$ with $R \in QF$, then we can find terms $\mathbf{t}$ of BR^ω such that $BR^\omega \vdash R(\mathbf{t}, \mathbf{y})$.

This appeared in Spector (1962), brought to publication by Kreisel after Spector's tragic sudden death due to leukemia. Another proof of (3) was later given by William Howard (1968), which perhaps explains better the role of BR in Spector's interpretation of analysis. Howard showed that the scheme AC is implied classically by a subscheme of BI whose negative translation follows from $\widetilde{HA}^\omega$ + BI; he then used BR to realize the D-interpretation of BI. In other publications (cf. Troelstra 1990 for references), Howard made further substantial contributions to the analysis and use of the *Dialectica* interpretation for various systems.

While Spector's interpretation did not follow the lines of Gödel's speculation that one might be able to extend his functional interpretation to analysis by the use of transfinite types, it did bear transfinite features by its use of recursion on certain well-founded trees. The main foundational issue following Spector's result became the question as to whether it provided a constructive consistency proof of analysis. That it constituted a real advance in perspicuity over Kreisel's 1957 model in the continuous functionals was clear to all. But what was not evident was the intuitionistic acceptability of Bar Recursion; in this respect, Spector (1962, p. 2) reported varying degrees of opinion among Bernays, Gödel, Heyting, and Kreisel. Spector himself thought that its acceptability to the intuitionists was questionable, and that further work would be required to give it a "suitable foundation"; in this, Kreisel concurred.[13] It should be noted

[13]Spector had originally entitled his paper "Provably recursive functionals in analysis: A consistency proof by an extension of intuitionistic principles." At Gödel's suggestion (cf. Spector 1962, ftn. 1), "to avoid misunderstandings," this was changed to ". . . : A consistency proof by an extension of principles formulated in current intuitionistic mathematics."

that in this respect, the constructivity of intuitionism itself was taken for granted; this was not an attempt to reduce that to "more constructive" principles as had been Gödel's aim with his previous work.

The question of the constructivity of Bar Recursion was the main subject of a Seminar on the Foundations of Analysis, held at Stanford in the summer of 1963 under the leadership of Kreisel. Contributions to the seminar were prepared as a volume of reports, circulated to interested parties but never published per se, whose authors were G. Kreisel, W. A. Howard, W. W. Tait (with several parts each), J. Harrison, and R. J. Parikh. While much information was obtained in this seminar about classical and intuitionistic systems of analysis, theories of generalized inductive definitions, the functional interpretations, and classes of functionals of finite type, the conclusion about the main question was disappointing. In the words of Kreisel's introduction to volume 2 of these reports (p. i), in answer to the question: "*Can bar recursion of finite type be constructively justified?*"... "the answer is negative by a wide margin, since not even bar recursion of type 2 can be proved consistent."[14] That assessment seems unchanged to date.

7 The D-Interpretation as a General Proof-Theoretical Tool

Gödel's concentration on the consistency problem both in the development of his functional interpretation of arithmetic and in his special interest in Spector's extension of it to analysis was a direct continuation of the central concern of Hilbert's program, though with the recognized necessity of giving up adherence to strictly finitistic methods. But proof theory in general has had to broaden its concerns and methods. The main line of technique which has been developed from the original work of Herbrand and Gentzen uses normalization of derivations in some sense or other; modern extensions of these methods in many cases make heavy use of infinitary derivations. Except for some of the more traditional proof-theorists, establishing consistency of formal systems has receded as the main goal, to be replaced by a more general *reductive* program. This still takes foundational aims to be primary, but gives up hopes for any supposedly absolute foundation in favor of reductions of various systems to others recognizably more basic in terms of concepts and/or principles (cf. Feferman 1988 and 1993 [chapter 10 in this volume]). At the same time, proof-theoretical tools have been applied to obtain results of a more mathematical character that one may describe as *extractive*, namely, to draw explicit or computational information from proofs of statements of existential or universal-existential form. Under the

[14]On the basis of the species of representing functions of continuous functionals, which Kreisel then considered to be the only possible candidate which could be used to provide a justification (*op. cit.*, Vol. 1, p. 2 of the introduction).

latter one would count the use of proof theory in characterizing the *provably recursive functions* (or *functionals*, where that is appropriate) of a formal system. In practice, both reductive and extractive proof theory work hand in hand, and are related to assessments of proof-theoretic strength in one sense or another, most frequently as measured by the provably recursive well-orderings of a system.

Though the D-interpretation has not been applied as extensively as the Herbrand-Gentzen methods of syntactic transformation, it has proved to be a rather powerful and versatile tool with distinctive advantages, especially for extractive purposes. Applied to intuitionistic systems it takes care of the underlying logic once and for all, verifies the Axiom of Choice AC in all types, and interprets various forms of induction by suitably related forms of recursion. This then leads for such systems to a perspicuous mathematical characterization of the provably recursive functions and functionals. For application to classical systems, one must first apply the negative translation (again taken care of once and for all). Since the D-interpretation verifies Markov's principle even at higher types (principle M′ in section 4 above) at least the provably recursive functions and functionals are preserved, as well as QF-AC in all types and induction schemata. The main disadvantage, though, comes with the analysis of other statements whose negative translation may lead to a complicated D-interpretation; special tricks may have to be employed to handle these.

My own experience with the D-interpretation came toward the end of the 1960s, when I used it (among other things) to show that certain reductive results obtained by use of model-theoretic methods by Friedman for subsystems of classical analysis based on Σ^1_n-AC ($n = 1, 2, \ldots$) could be reobtained and strengthened (for example, to higher types) by use of the D-interpretation. The trick here was to reduce Σ^1_n-AC to QF-AC by the introduction of Skolem functions which are eventually eliminated (cf. Feferman 1977, §8, and Feferman 1987, pp. 466ff.). I subsequently came to the conclusion that Herbrand-Gentzen methods worked just as well and were simpler; evidence for this was given in Feferman and Sieg (1981) and Feferman and Jäger (1983). In more recent years, however, I have returned to a more positive view of the merits of the D-interpretation, and want to encourage people to learn how it has worked so far and to get a sense of its potentialities.

Thus, in conclusion, I want to mention some results which show how the D-interpretation may be stretched down to fragments of arithmetic and analysis, that is, in the direction opposite to the stretch upward to full analysis described in the preceding section. An early instance of such, and first to be mentioned here, is Parsons' result (1970) that the subsystems Σ^0_1-IA and Π^0_2-IR of PA are both conservative over PRA (Primitive Recursive Arithmetic) for Π^0_2 statements, and hence have exactly the primitive recursive functions as their provably recursive functions. This was established by a D-interpretation into a subclass of the primitive recursive functionals with restricted recursors. Following later model-theoretic work of Friedman which showed that Σ_1^0-IA + Δ_1^0-CA + WKL is conservative over PRA, Sieg (1985) showed how to obtain the same result by Herbrand-Gentzen methods, and also strengthened it by replacing Δ_1^0-CA by Σ_1^0-AC. (WKL is the so-called Weak König's Lemma, that is, for subtrees of $2^{<\omega}$.) A key step in Sieg's treatment of WKL was the use of a majorization argument due to Howard (1973), originally applied to Gödel's PR^ω functionals for quite different purposes. In unpublished notes (Feferman 1990a) I returned to the D-interpretation and showed how this majorization argument could also be applied to systems based on Kalmar-elementary arithmetic as well as Buss' feasible arithmetic S_2^1 (Buss 1986), when QF-AC is allowed in all types and WKL is adjoined. I learned then that Cook and Urquhart had already applied the D-interpretation to S_2^1 in 1988 in order to recapture Buss' characterization of its provably recursive functions (in a suitably modified sense) as the polynomial time computable functions; their work appeared finally in Cook and Urquhart (1993). However, my result for the adjunction of WKL strengthened both that and the result of Ferreira (1988) for S_2^1 + Δ_1^0-CA + WKL [published subsequently in Ferreira (1994)]. In the meantime, Sieg has independently pursued Herbrand-Gentzen methods to obtain all the same results (cf. Sieg 1991). Thus, to a considerable extent, for these purposes it is a matter of taste which approach to prefer. In any case, I strongly recommend reading the earlier sources referred to in sections 4–6 above [as well as Avigad and Feferman (1998)], which already demonstrate the considerable elasticity of Gödel's functional interpretation.

[Since writing the article from which this chapter is drawn, a gap in my argument for the D-interpretation of Σ_1^0-IA + Σ_1^0-AC + WKL in the unpublished notes mentioned above was brought to my attention by Ulrich Kohlenbach. However, in his dissertation (1990) he had independently found a reduction of this system to PRA by a variant combination of functional interpretation with majorization; see Kohlenbach (1992a, especially pp. 1260 ff.). That argument can also be applied to the reduction of S_2^1 + Σ_1^0-AC + WKL indicated above. Kohlenbach has subsequently obtained a number of further extensions of his methods in his papers (1992, 1993, 1993a, 1994) with interesting applications to mathematical analysis.]

Part V

COUNTABLY REDUCIBLE MATHEMATICS

12

Infinity in Mathematics: Is Cantor Necessary? (Conclusion)

Is the Cantorian Transfinite Necessary for Finitary Mathematics? Gödel's Doctrine

By Gödel's doctrine I mean the view first enunciated in footnote 48a of Gödel (1931) that the "true reason" for the incompleteness phenomena is that "the formation of ever higher types can be continued into the transfinite," both in systems explicitly using types and in systems of set theory such as ZF for which the (cumulative) type structure is implicit in the axioms. For, as Gödel says, the "undecidable propositions constructed here become decidable whenever appropriate higher types are added." Since the undecidable propositions are of finitary character, Gödel's doctrine says in effect that the unlimited transfinite iteration of the power-set operation is necessary to account for finitary mathematics.

In order to discuss Gödel's doctrine in detail, we shall have to introduce various classifications of statements and formulas. In modern logical symbolism, statements of the form $\forall n(F(n) = 0)$ with F primitive recursive are in the class Π^0_1 and their negations $\exists n(F(n) \neq 0)$ in the class Σ^0_1. More generally $\Pi^0_k(\Sigma^0_k)$ is used for the class of statements having a primitive recursive matrix preceded by k alternating quantifiers of which the first is a universal (existential) quantifier, and in which all variables range over the natural numbers. The same classification is applied to formulas with free variables and to the relations which such formulas define in $\mathbf{N}$. The negation of a formula in $\Pi^0_k(\Sigma^0_k)$ is equivalent to one in $\Sigma^0_k(\Pi^0_k)$. A relation is said to be in class Δ^0_k if it is in both Π^0_k and Σ^0_k. The class Σ^0_1 is identical

This chapter first appeared as the last third of "Infinity in mathematics: Is Cantor necessary?" in G. Toraldo di Francia (ed.), *L'infinito nella scienza* (*Infinity in Science*), Istituto della Enciclopedia Italiana, Rome (1987), 151–209 (Feferman 1987a); the first two-thirds of that article appears as chapter 2 in this volume. The combined pieces are reprinted here with the kind permission of the publisher. There are some minor corrections and additions.

with that of the recursively enumerable relations, and thus, by a theorem of Post, Δ^0_1 consists exactly of the recursive relations. Π^0_∞ is used for the union of the classes $\Pi^0_k (k < \omega)$; every arithmetical formula is equivalent to one in Π^0_∞.

When second-order quantifiers with variables $X, Y, \ldots$ ranging over sets of natural numbers are introduced, a formula is said to be in $\Pi^1_k(\Sigma^1_k)$ form if it has an arithmetical matrix preceded by k alternating second-order quantifiers, beginning with a universal (existential) quantifier. The negation of a formula in $\Pi^1_k(\Sigma^1_k)$ is equivalent to one in $\Sigma^1_k(\Pi^1_k)$. Again, Δ^1_k is used for $\Pi^1_k \cap \Sigma^1_k$. In some cases it is preferable to use variables ranging over functions of natural numbers rather than over sets. The resulting classifications are essentially the same, using the equivalence of sets with their characteristic functions and of functions with their graphs (sets of ordered pairs coded as numbers using a primitive recursive pairing function $\langle n, m\rangle$).

If $\mathcal{F}$ is a class of formulas, the *comprehension axiom schema* (CA) for formulas in $\mathcal{F}$ is

$$(\mathcal{F}\text{-CA}) \qquad \exists X \forall n[n \in X \leftrightarrow P(n)],$$

where $P \in \mathcal{F}$ and P may have free variables (first or second order) besides "n." To express comprehension for properties definable by formulas both in $\mathcal{F}$ and in complementary form, we use

$$(\Delta_{\mathcal{F}}\text{-CA}) \qquad \forall n[P(n) \leftrightarrow Q(n)] \rightarrow \exists X \forall n[n \in X \leftrightarrow P(n)]$$

where $P \in \mathcal{F}$ and $(\sim Q) \in \mathcal{F}$ (up to logical equivalence). In particular, (Δ^1_1-CA) is used for ($\Delta_{\Pi^1_1}$-CA). In should be noted that (Π^0_∞-CA) follows from (Δ^1_1-CA).

Gödel's doctrine is illustrated in the case of the extension of the first-order system PA of Peano arithmetic to a second-order language and axioms. Basically, one introduces axioms which are sufficiently strong to define the formal notion of truth for the first-order language and to establish the formal statement expressing that every statement provable in PA is true; this theory thus proves the consistency statement Con_{PA} which, as we have seen, is in Π^0_1 form.

Such an argument will be carried out in an extension of PA_2, that is, second-order PA, by suitable comprehension axioms. Here PA_2 has the same axioms as PA but with the induction scheme extended to apply to all second-order formulas $P(n)$. It is usual to take ($\mathcal{F}$-CA) as an abbreviation for the system $\mathrm{PA}_2 + (\mathcal{F}\text{-CA})$. Incidentally, we shall deal later on with systems with restricted induction: $(\mathrm{PA}_2)\restriction$ has the induction schema replaced by the second-order axiom

$$\forall X[0 \in X \wedge \forall n(n \in X \rightarrow (n+1) \in X) \rightarrow \forall n(n \in X)],$$

and in general if S is any system with the full induction schema, S$\restriction$ is the corresponding system with the schema replaced by this axiom. ($\mathcal{F}$-CA)$\restriction$

thus abbreviates $(\text{PA}_2)\restriction + (\mathcal{F}\text{-}CA)$; this system includes PA as long as $\mathcal{F}$ includes Π^0_∞. For any system S, $\text{S} \vdash A$ means that A is provable in S.

For the argument indicated above, the obvious step is to define the set of (Gödel numbers of) true statements as the smallest set satisfying certain arithmetical closure conditions; this requires (Π^1_1-CA). However, one can make do with less, namely, (Δ^1_1-CA). For one first defines a formula $Tr_1(X, k)$ which expresses that X satisfies the conditions to be the truth-set for the set St_k of statements of logical complexity $\leq k$. Then it is easily proved using (Π^0_∞-CA) that

$$(1) \qquad \forall X \forall k[Tr_1(X,k) \rightarrow \exists Y(Tr_1(Y, k+1) \wedge Y \cap St_k = X)].$$

Then $(\Pi^0_\infty - \text{CA}) \vdash \forall k \exists! X Tr_1(X,k)$ by induction on k. Finally, one can define $Tr_1(a)$ in the two equivalent forms

$$St(a) \wedge \exists k \exists X[c(a) \leq k \wedge Tr_1(X,k) \wedge a \in X], \text{ and}$$
$$St(a) \wedge \forall k \forall X[c(a) \leq k \wedge Tr_1(X,k) \rightarrow a \in X]$$

where $c(a)$ = complexity of a. Thus

$$(2) \qquad (\Delta^1_1 - \text{CA}) \vdash \exists X \forall a[a \in X \leftrightarrow Tr_1(a)].$$

With the existence of the truth-set thus established, one can go on to prove by induction in (Δ^1_1-CA) that every numerical instance of every formula provable in PA is true. Hence

$$(3) \qquad (\Delta^1_1 - \text{CA}) \vdash \forall a[St(a) \wedge Prov_{\text{PA}}(a) \rightarrow Tr_1(a)].$$

Since $\neg Tr_1(\ulcorner 0 = 1 \urcorner)$, it follows that

$$(4) \qquad (\Delta^1_1 - \text{CA}) \vdash Con_{\text{PA}}.$$

As we have seen, Con_{PA} is a true Π^0_1 statement not provable in PA assuming PA is consistent, so in this case adjunction of higher types leads to new results with finitary content.

This argument is paradigmatic: for any system S whose axioms are intuitively true, we can define a formal extension S* of S using quantification at a next higher level and axioms analogous to full induction plus Δ^1_1-comprehension, for which we can establish

$$(5) \qquad \text{S}^* \vdash Con_{\text{S}}.$$

In the case of systems of set theory, we proceed in a variant manner. Most reasonable theories of sets S have natural models in the cumulative hierarchy; that is, for suitable α satisfying a special property $K(\alpha)$, the structure $(V_\alpha, \in \cap V_\alpha)$ is a model of S. For example, for S = ZFC and $K(\alpha) = Inacc(\alpha)$ (the property that α is an inaccessible cardinal), we have that V_α is a model of S whenever $K(\alpha)$ holds. For such S in general let $\text{S}^* = \text{S}$

$+ \exists\alpha K(\alpha)$. Then in $S^\star$ one can define the notion of truth in the structure $(V_\alpha, \in \cap V_\alpha)$ and prove that this is a model of S whenever $K(\alpha)$; hence (5) also holds in this situation. Thus Gödel's doctrine is verified in these cases by adjoining a "new type" V_α in the form of an assumption $\exists\alpha K(\alpha)$. (Here the implicit adjunction of new types is far more wasteful than the direct extension of a system by the adjunction of one higher type, since from the axioms of ZFC, once we have an ordinal α we have many ordinals β larger than α and thence the corresponding new type levels V_β.)

In this formal sense, then, Gödel's doctrine can be verified: by adding higher types to a system S either explicitly or implicitly we are able to prove new Π^0_1 statements, and by iterating this process transfinitely, we are led to ever stronger such statements. Of course this assumes that at each stage along the way, the system S is consistent. That will be the case if we start out with a system true for the natural numbers, and we accept that the iteration of the power-set operation leading to higher types has a well-determined meaning so that at each stage, the full comprehension axiom is validated (using the language available at that stage). In particular, both full induction and the analogue of the Δ^1_1-comprehension axiom schema will be true under that assumption and so the extension from S to $S^\star$ as above will preserve correctness and hence consistency. An analogous style of reasoning can be applied to the set-theoretical case.

However, Gödel's doctrine can be challenged when it is read as asserting that the platonistic view of the determinateness of the power-set operation and its iteration through all the ordinals is *necessary* for the derivation of *previously* undecidable but true Π^0_1 statements. Consider the situation at the outset: assuming one grants the correctness in an informal sense of PA under its intended interpretation in the structure of natural numbers, would not one immediately accept a predicate $Tr_1(x)$ with the appropriate closure axioms for truth in the first-order language, without regarding it as necessary to *define* this predicate in second-order terms? If so, extension of PA by these axioms (with induction extended to the new formulas) suffices to prove Con_{PA}. Call this extension $Tr_1(\mathrm{PA})$; it is a first-order theory which serves to decide the undecidable proposition constructed for PA. So why is the extension to a higher type necessary? To be sure, $Tr_1(\mathrm{PA})$ is again subject to the incompleteness results. But the same line of reasoning, which begins with an informal judgment of the correctness of PA under its intended interpretation and a recognition of the meaningfulness of the predicate $Tr_1(x)$ for the language of PA, can be repeated: one recognizes the correctness of $Tr_1(\mathrm{PA})$ and the meaningfulness of a truth predicate $Tr_2(x)$ for the language of $Tr_1(\mathrm{PA})$, and then the correctness of new axioms involving Tr_2 by means of which we can prove $Con_{Tr_1(\mathrm{PA})}$, and so on.

According to this line of thinking, it is equally reasonable to replace Gödel's doctrine by a variant doctrine, according to which the "true reason" for the incompleteness phenomena is that the truth definition for the language of a formal system is not expressible in that language (Tarski's theorem) and that by adjunction of the notion of truth with suitable axioms, the undecidable statements produced by Gödel become undecidable. In fact, even less will do: one need merely adjoin to S the formal *reflection scheme*

(6) $$Prov_S(\ulcorner A \urcorner) \to A, \text{ for each sentence } A,$$

as a means of expressing faith in the correctness of S without any new predicates at all. For, if one takes the specific instance of (6) with $A = (0 = 1)$, the consistency statement Con_S is a consequence of S. In other words, for $S^\star = S + \{Prov_S(\ulcorner A \urcorner) \to A : A$ any sentence of $S\}$, we have $S^\star \vdash Con_S$. So still another variant of Gödel's doctrine is that the "true reason" for the incompleteness phenomena is that though a formal system S may be informally recognized to be correct, we must adjoin formal expression of that recognition by means of a reflection principle in order to decide Gödel's undecidable statements. Either way, one has a reasonable alternative to Gödel's doctrine—the arguments for which are certainly no less persuasive than for his—and which does *not* claim the necessity to accept higher type notions *à la* Cantorian set theory.

These variant forms of Gödel's doctrine, which lead to the consideration of certain uniform extension procedures $S \mapsto S^\star$, may also be iterated into the transfinite. But here, if one is not to accept the transfinite in the Cantorian sense, ordinals must be understood and treated in a more constructive way. That has been carried out in the subject of *transfinite recursive progressions of formal systems*, initiated by Turing (1939) and continued in Feferman (1962) and, for predicative extension procedures, in Feferman (1964). The conclusion is that the incompleteness phenomena do not, by themselves, force one to accept the "formation of ever higher types . . . into the transfinite," that is, the transfinite iteration of the power-set operation. Application of Occam's razor would cut the discussion at that.

However, the platonist will still insist on a rejoinder: notwithstanding this counterargument to Gödel's doctrine, the systems of analysis (= full second-order CA) and of type theory and, further on, of ZFC and ZFC $+\forall\alpha\exists\beta(Inacc(\beta) \land \beta > \alpha)$ etc., are in fact *all true*, and for each of these S the consistency statement will be unprovable in S. Moreover, these statements Con_S go, in strength, far beyond those of the piddling extensions produced above by iterating formal reflection principles or truth definitions. So, according to this line, very strong set-theoretical principles are needed to decide statements of finitary character.

But isn't this argument begging the question? Sure, whatever leads one to accept "correctness" of such S will, with one little additional step, lead one to accept the consistency statement for S. But, as Brouwer repeatedly (and Gödel himself) emphasized, consistency does not by itself ensure correctness. While consistency of such systems may be plausible, the plausibility of such can (as I indicated at the end of chapter 2) rest on quite different grounds than the assumption of an underlying set-theoretical reality in its platonistic sense. In other words, unless one is *already* a set-theoretical platonist, Gödel's doctrine does not by itself provide compelling reasons to embrace the Cantorian transfinite.[1]

There remains the question: What, if any, relevance do the incompleteness phenomena have to finite combinatorial mathematics in the everyday sense of the word? There are a number of specific results in recent years which it is claimed establish that relevance. We turn next to a review of those results.

Undecidable Diophantine Problems

The results here, and the question of their relevance to everyday mathematics just raised, are comparatively easy to discuss. Diophantine equations are those of the form $p(x_1, \ldots, x_n) = q(x_1, \ldots, x_n)$ where p, q are polynomials in one or more variables with coefficients in $\mathbf{Z}$ (the integers). A Diophantine problem concerns whether there exists an integer solution of such an equation, that is, whether $(\exists x_1, \ldots, x_n \in \mathbf{Z})p(x_1, \ldots, x_n) = q(x_1, \ldots, x_n)$. Hilbert's Tenth Problem in his 1900 list called for an effective method to determine whether or not a Diophantine equation is solvable, that is, has any integer solutions [see chapter 1 in this volume for an introduction to Hilbert's Tenth Problem]. There is a closely related group of problems for solutions in $\mathbf{N}$, and for solutions in $\mathbf{Q}$. Specific such questions were first considered by the Greek mathematician Diophantus (third century A.D.); after a long lapse, the subject was revived by Pierre de Fermat in the seventeenth century and has been a staple of number theory ever since. Many specific Diophantine problems have so far resisted attack.* [Also it follows from the work described below that the (still unproved) Goldbach Conjecture, according to which every even integer greater than two is a sum of two primes, reduces to a Diophantine problem. It happens that Gödel frequently referred to his undecidable propositions as being of "Goldbach type."]

The first step toward a negative solution of Hilbert's Tenth Problem was made by Gödel in his 1931 incompleteness paper. He showed there that every primitive recursive function $F(x)$ is arithmetically definable, and hence can be put in the form

$$(1) \qquad F(x) = y \leftrightarrow Q_1 z_1 \ldots Q_n z_n R(x, y, z_1, \ldots, z_n),$$

[1]The viewpoint here concerning independence results for arithmetical statements should be compared with that of Isaacson (1987). He argues that the truths expressible in the first-order language of arithmetic which are not derivable in PA "contain essentially higher-order, or infinitary concepts." This is opposite to the position taken here, that one may be led to accept such results without assuming higher-type notions in the platonist sense. However, as regards the finite combinatorial independence results to be discussed in the next subsection, Isaacson and I agree that they are obtained by a transmutation of essentially metamathematical statements coded in the language of arithmetic.

*[At the time I wrote the article from which this chapter was drawn, the example given of the most famous such problem was Fermat's Last Theorem. This was subsequently settled by A. Wiles by very advanced methods; see chapter 1 in this volume.]

where each Q_i is the universal ($\forall$) or existential quantifier ($\exists$), the variables z_i range over $\mathbf{N}$, and R is built up by $\wedge, \vee$, and $\neg$ from polynomial equations. It follows that every Π_1^0 statement $\forall x F(x) = 0$ is likewise definable in that form. Gödel concluded from (1) that for each (ω-)consistent formal system S containing a sufficient amount of number theory, there are arithmetical propositions A which cannot be decided by S, that is, such that neither S $\vdash A$ nor S $\vdash \neg A$.

In the later development of the theory of effective computability at the hands of Church, Kleene, Turing, Post, and others, attention shifted from individual problems A undecidable relative to given formal systems S, to effectively undecidable problems for subsets C of $\mathbf{N}$, that is, for which there is no effective method to determine, given an arbitrary x in $\mathbf{N}$, whether or not $x \in C$. By Church's thesis, the effectively decidable sets C are exactly those which have a general recursive characteristic function, and then these are exactly the same as the Δ_1^0 sets. Church gave examples of recursively enumerable sets which are not recursive, in other words, which are in the class Σ_1^0 but not in Π_1^0.

Beginning in the 1950s, considerable progress was made on Hilbert's Tenth Problem through the work of M. Davis, H. Putnam, and J. Robinson. They succeeded in showing that if *variable* exponents are permitted, every Σ_1^0 set C is definable in the form

$$(2) \qquad x \in C \leftrightarrow \exists z_1, \ldots, z_n[p(x, z_1, \ldots, z_n) = q(x, z_1, \ldots, z_n)],$$

where p, q are *exponential* integer polynomials. Finally, in 1970, J. Matiyasevich succeeded in establishing a result of the same form for *ordinary* integer polynomials; this finally showed the answer to Hilbert's Tenth Problem to be negative. Subsequently, Matiyasevich and Robinson succeeded in showing that it suffices to take $n \leq 13$ to represent as in (2) every recursively enumerable set with ordinary p, q. This and a number of other results concerning Hilbert's Tenth Problem are reviewed in Davis, Matiyasevich, and Robinson (1976). A few years later, Matiyasevich managed to reduce the number of variables in the representation (2) of recursively enumerable sets from 13 to 9 for ordinary polynomials, and even further to 3 variables for exponential polynomials.[2]

Now, returning to undecidable propositions, it follows by Gödel's results that for suitable consistent systems S of arithmetic, there are true Π_1^0 statements A of the form

$$(3) \qquad A = \forall x, z_1, \ldots, x_n[p(x, z_1, \ldots, z_n) \neq q(x, z_1, \ldots, z_n)]$$

which are not provable in S; by formalizing the work described above, this can be reduced to $n \leq 9$ for ordinary integer polynomials p, q.

[2]For references to this further work see *Mathematical Reviews*, 81f: 03055 [and Matiyasevich (1993)].

Impressive as these results are, the problems they concern are still very distant from the bread-and-butter Diophantine problems of everyday number theory ("everyday" over the last three hundred years), because the number of variables is so much larger than is considered in such problems, and the complexities of the polynomials involved are so great (according to various measures of complexity). There is no evidence that these undecidability and incompleteness results have any relevance to the classic unsettled problems that have challenged generations of number theorists. [Besides the example of the Goldbach Conjecture mentioned above, it has been shown by Davis, Matiyasevich, and Robinson (1976) that the Riemann Hypothesis—which is one of the most important outstanding problems in number theory—reduces to a Diophantine problem. However, the numbers of variables and degrees of the polynomials required for this might be rather large.] Though the chain from the original undecidable propositions to the undecidable propositions of the form (3) is a long and technically complicated one (so that the point of departure recedes into the background), it remains the case that the problems thus shown to be undecidable were not of any prior mathematical interest—they have simply been derived in a step-by-step process from problems "cooked up" to demonstrate the incompleteness of formal systems. As a further reinforcement to the view here that the results on undecidable Diophantine problems are irrelevant to everyday mathematical concerns is that the character of these problems is insensitive to the formal system considered. If one takes A in (3) to be equivalent to Con_{PA}, there is no way to tell it apart from an A taken equivalent to Con_{ZFC} or from one which is equivalent to the consistency (with ZFC) of the existence of measurable cardinals. This suggests that the mathematical content of the resulting individual Diophantine problems simply has nothing to do with the internal mathematical content of the formal systems from which their independence is established.

How much different it would be if one showed that some outstanding number-theoretic Π_1^0 statement A [such as the Goldbach Conjecture or the Riemann , Hypothesis] is not provable in PA, or even more strikingly, in ZFC—and thus demonstrated why it's so difficult to prove it (if true)! There is nothing in the work on undecidable diophantine problems to suggest that one is anywhere near obtaining such a result—or anything comparable—nor that further efforts in chopping n down in (2) or (3) is going to get one any closer to achieving such a result. The situation at present is thus analogous to that concerning transcendental numbers with which we began this story in chapter 2; Liouville and Cantor showed how to construct (explicitly or implicitly) transcendental numbers, but mathematicians wanted to know whether the numbers π and e are transcendental. For those results the work of Liouville and Cantor was useless. All they did was demonstrate that the effort to show specific numbers to be transcendental numbers was not a waste of time, since such numbers do after all exist. Working number theorists want to know about the *truth* of specific number-theoretic problems which can be put in diophantine form. It is highly unlikely that one will simultaneously demonstrate the unprovability of, say, the Riemann Hypothesis RH from certain S *and* the truth of RH, as was the case with the nonprovable propositions produced by Gödel. But mathematicians would no doubt sit up and take notice if *merely* unprovability were established. All that the chain of work from Gödel to Matiyasevich permits one to say currently is that efforts to obtain such independence results for propositions of prior mathematical interest are not necessarily a waste of time.

Independence Results for Statements of Finite Combinatorial Character

Starting in the mid 1970s, a number of interesting independence results have been obtained with respect to a wide range of formal systems for statements which are prima facie relevant to finite combinatorial mathematics in its everyday sense. The expository article Simpson (1987a) provides an excellent introduction to this area of work, and has been quite useful to me in the following. Many of the results are "finitizations" P_{fin} of strong infinitary propositions P, where P implies P_{fin}. In some cases we even have P equivalent to P_{fin}. The classic example of this sort is *König's Infinity Lemma*, according to which if T is a finitely branching infinite tree then T has an infinite branch and (as is obvious) conversely. For T represented as a collection of finite sequences in N, the statement P that T has an infinite branch is Σ^1_1, while for T finitely branching the equivalent statement P_{fin} that it contains infinitely many nodes is Π^0_2, and even Π^0_1 for branching with fixed bounds.

As Simpson points out, König in his 1927 paper "Über eine Schlussweise aus dem Endlichen ins Unendliche" thought of his lemma as a method of obtaining infinitary results from finitary ones. Indeed, one of the most striking early applications of KL (König's Lemma) was Gödel's use of it in his proof of the completeness theorem for first order predicate logic: if a sentence A is consistent in that logic then it has a model.[3] The hypothesis of consistency is Π^0_1, but this is used to build a binary branching infinite tree whose nodes correspond to sequences of statements or their negations consistent with S. In this case we obtain a Σ^1_1 conclusion from a Π^0_1 hypothesis. The argument requires classical logic and can be carried out formally within PA, since the model constructed is arithmetically definable. (This situation is in a way a vindication of Hilbert's view discussed in chapter 2 that the completed infinite is already implicit in the use of the Law of Excluded Middle when combined with quantifier logic.)

Now, to continue with Simpson's point, the newer results discussed here proceed in the opposite direction. Starting with P which is independent

[3]Gödel did not explicitly acknowledge the use of König's Lemma; cf. the discussion in Gödel (1986), pp. 53–54, of his 1929–1930 work on completeness.

of a formal system S, the effort is to obtain a finitary consequence P_{fin} of P which is still independent of S. The first striking example of such was provided by the theorem of Paris and Harrington (1977), which is a modified form of the famous combinatorial theorem due to F. P. Ramsey in 1930. Ramsey's theorem concern partitions of the set of all k-element subsets of a set X, that is,

(1) $$[X]^k = C_1 \cup \ldots \cup C_\ell$$

where $[X]^k = \{Y : Y \subseteq X \text{ and } card(Y) = k\}$ and $C_1, \ldots, C_\ell$ are pairwise disjoint. We deal here only with denumerable X and may assume $X \subseteq \mathbf{N}$. The *Infinite Ramsey Theorem* P is the statement that if X is infinite then for any k and partition of $[X]^k$ as in (1), there exists an $i \leq \ell$ and an infinite subset of Y of X such that

(2) $$[Y]^k \subseteq C_i.$$

In the case $k = \ell = 2$ this is interpreted as saying that any infinite graph whose edges (that is, members of $[X]^2$) are colored by one of two colors (C_1 or C_2), there exists an infinite subgraph all of whose edges have the same color. Ramsey proved a finitary form P_{fin} of P which is as follows:

(3) for any k, ℓ, and m there exists n such that for any X with $card(X) = n$ and partition $[X]^k = C_1 \cup \ldots \cup C_\ell$, there exists $i \leq \ell$ and $Y \subseteq X$ such that $card(Y) \geq m$ and $[Y]^k \subseteq C_i$.

This statement can be seen to be a consequence of P by use of KL; if (3) is assumed false, then a finitely branching tree T can be constructed such that an infinite branch X through it violates the conclusion of the Infinite Ramsey Theorem. It turns out that this *Finite Ramsey Theorem* (3) can be proved in Peano Arithmetic PA. The surprising result found by Paris and Harrington is that a slightly modified form $P^{\star}_{fin}$ of (3) is independent of PA. This *Modified Finite Ramsey Theorem* differs from (3) only by addition, to the conclusion, of the requirement that $card(Y) \geq min(Y)$, where $min(Y)$ is the least element of Y. The argument that P implies $P^{\star}_{fin}$ can again be carried out by contradiction and use of KL (though, as pointed out by Simpson, this use of KL is inessential).

Since the work of Paris and Harrington, a number of other results of finite combinatory character have been shown to be independent of PA. One of the simplest is a theorem due to Goodstein concerning the effect of shifting bases in the representations of natural numbers to various bases b, which has been shown by Kirby and Paris to be independent of PA.[4]

It should be emphasized at this point that the statements P_{fin} thus shown independent of PA are recognized to be true by infinitary methods

[4]See Simpson (1987a) for details.

which go beyond *PA*. They can in fact be proved in the second-order system (Π^0_∞-*CA*) based on the arithmetical comprehension axiom with full induction. Moreover, P_{fin} is equivalent to the 1-consistency of PA (a notion intermediate between ω-consistency and consistency). Incidentally (as is discussed in the next and final section), the system $(\Pi^0_\infty\text{-CA})\restriction$ with restricted induction proves the same arithmetical statements as PA.[5]

The system (Π^0_∞-CA) is *predicatively justified*, since the second-order variables can be interpreted as ranging over the arithmetically definable subsets of **N**. Schütte and I both analyzed in formal terms the transfinite iteration of the formal process of introducing sets by definitions referring only to previously determined collections of sets, where the iteration extends only to those ordinals corresponding to previously recognized well-orderings. The limit of these is a certain countable ordinal Γ_0.[6] This characterization of *predicativity* yields a sequence of *ramified systems of analysis* R_α each of which is predicatively justified for $\alpha < \Gamma_0$, but not for $\alpha = \Gamma_0$. In my 1964 paper and in several subsequent papers I produced a variety of single unramified systems S which are proof-theoretically of the same strength as $\bigcup R_\alpha[\alpha < \Gamma_\alpha]$ and hence are *locally predicative*.[7] Another such system of strength Γ_0 has been introduced and utilized by Harvey Friedman, and denoted ATR_0 by him; in our notation this is denoted $ATR\restriction$, since it uses restricted induction on **N**. The locally predicative system ($ATR\restriction$) is itself a relatively weak subsystem of the patently *impredicative* system $(\Pi^1_1\text{-CA})\restriction$.

The point of all this here is that Friedman found a finite combinatorial version P_{fin} of an existing infinite combinatorial theorem P known as *Kruskal's Theorem*, such that P_{fin} is independent of $ATR\restriction$ and hence cannot be proved by predicative methods. In this case Kruskal's proposition P concerns a certain (relatively simple) embeddability relation $\leq$ between finite trees; it says that the collection of such trees is "well-quasi-ordered" under $\leq$, that is,

(4) for any infinite sequence $T_1, T_2, \ldots, T_n, \ldots$ of finite trees there exist i, j with $i < j$ and $T_i \leq T_j$.

(In other words, there is no infinite descending sequence in this collection under the relation $<^*$ defined by $T <^* T' \leftrightarrow T' \not\leq T$). Note that (4) is equivalent to a Π^1_1-statement. Now the following is Friedman's finite form P_{fin} of Kruskal's theorem:

(5) For any k there exists n such that for any finite sequence $T_1, T_2, \ldots, T_n$ of trees with $card(T_i) \leq k \cdot i$ for all $i \leq n$, there exist i, j with $i < j$ and T_i embeddable in T_j.

---

[5]See Simpson (1987a) for details.

[6]Cf. Feferman (1964).

[7]See Feferman (1968a) for references.

This P_{fin} is a consequence of Kruskal's P; its proof goes by assuming (5) false and (once again) applying KL in a suitable way. Friedman showed that the statement (5) implies the 1-consistency of ATR$\restriction$ and hence is not predicatively provable; it is, however, provable in the system $(\Pi^1_1\text{-CA})\restriction$. Friedman also found an extended finite form of Kruskal's theorem which is independent of $(\Pi^1_1\text{-CA})\restriction$ though provable in $(\Pi^1_1\text{-CA})$ + BI, where BI is a principle described below. This makes use of labeled finite trees and a stronger embeddability relation. Details of the proofs of these independence results are in Simpson (1985).

At the time of writing the strongest subsystem of full second-order analysis $(\Pi^1_\infty\text{-CA})$ for which a finite combinatory independence result has been found is the system $(\Pi^1_1\text{-CA})$ + BI, where BI is the scheme expressing that transfinite induction can be carried out (with respect to any second-order formula) along any well-founded ordering relation in **N**. This result is due to Buchholz (1987); the statement shown independent concerns "hydra" games which involve specified alternations of amputation and regeneration of limbs in finite trees.

With reference to our original question as to the relevance of such independence results to finite combinatorial mathematics, I would say the following. None of the statements P_{fin} thus shown to be independent were previously considered per se in ordinary combinatorial work. It is a matter of pure speculation whether they would naturally have arisen for consideration in the normal course of such work, had they not come out of the efforts by logicians to establish the mathematical relevance of finitary results. At any rate, all these statements are at first sight close enough to the kinds of results which combinatorists *have* been dealing with and, conceivably, *could* have dealt with, so that there is at least a prima facie case for their relevance. Taking that for granted for the moment, what conclusions are to be drawn as to the significance of this work?

The most favorable interpretation to be placed on these results is that they tend to support Gödel's doctrine. The argument goes somewhat as follows. The system $(\Pi^1_1\text{-CA})\restriction$ is based on the impredicative comprehension axiom:

(6) $$\exists X \forall n[n \in X \leftrightarrow (\forall Y)A(n,Y)],$$

where A is arithmetical. Now (the argument continues), this system is justified only if one assumes that the power set $\mathcal{P}(\mathbf{N})$ exists as a fixed definite totality. Hence, according to this line of argument, it is necessary to make platonistic assumptions concerning the existence of uncountable totalities in order to derive finite combinatorial truths P_{fin} such as those, indicated above, independent of $(\Pi^1_1\text{-CA})\restriction$. In other words, according to this line of thought, at least the first stage of the Cantorian transfinite is necessary for everyday combinatorial mathematics.

Now there is an immediate counterargument to this particular effort to support Gödel's doctrine. Namely, although the infinitary statements P from which the finitary P_{fin} are derived as consequences clearly make use of impredicative principles, it turns out that the statements P_{fin} only require the (1-)consistency of such principles. And, as we have stressed above, one cannot argue from consistency (or even 1-consistency) to existence of the concepts in question, at least in their supposed "standard" intended sense.

There is a second line of response to the above argument, which makes use of a body of proof-theoretical work on reductions of the subsystems of analysis in question to constructive theories, in particular to various intuitionistic systems ID^i_α of iterated inductive definitions. ("ID" abbreviates "inductive definition," "i" is for "intuitionistic logic," and the ordinal subscript "α" measures the extent of the iteration.) This work is reported in Buchholz, Feferman, Pohlers, and Sieg (1981), which brings together a series of earlier contributions by the individual authors. In particular, it is demonstrated there that the system $(\Pi^1_1\text{-CA})\restriction$ is finitarily reducible to $\mathrm{ID}^i_{<\omega} = \bigcup_{n<\omega} \mathrm{ID}^i_n$, and that $(\Pi^1_1\text{-CA}) + \mathrm{BI}$ is finitarily reducible to ID^i_ω (in both cases preserving Π^0_2 sentences). Now the justification for the systems ID^i_α (for low α) is quite opposite to that for the classical subsystems of analysis which have been shown reducible to them. The systems ID^i_α concern specific countable sets (the recursive number classes) which are inductively generated "from below"; moreover, since only intuitionistic logic is used, it may be claimed there is no implicit appeal to the completed infinite in them. (This last is particularly clear for ID^i_1, where it turns out that Friedman's first finite form of Kruskal's theorem can be proved, but calls for argument in the case of ID^i_α when $\alpha \geq 2$.) At any rate, what the above reductions demonstrate is that the classical systems in question have an *alternative constructive justification* which does not require anything like belief in a preexisting totality of subsets of $\mathbf{N}$. Moreover, the same kind of alternative constructive justification has been given for far stronger theories than $(\Pi^1_1\text{-CA}) + \mathrm{BI}$, namely, for the system $(\Delta^1_2\text{-CA})$ in the work of Buchholz, Feferman, Pohlers, and Sieg cited above, and for $(\Delta^1_2\text{-CA}) + \mathrm{BI}$ in further work of Jäger and Pohlers.[8] At present, none of the independence results for finite combinatorial systems of the kind described above come close to exhausting the strength of systems for which we thus have a completely opposite (nonplatonist) constructive justification.

In the preceding discussion I have, for the sake of argument, taken for granted that the independent statements produced are indeed relevant to everyday combinatorial mathematics, even though they do not solve problems of prior mathematical interest. But even that may be challenged: it seems to me that as one passes from statements of the sort provided by Paris-Harrington and Kirby-Paris for PA, through those provided by Friedman for $\mathrm{ATR}\restriction$ and $(\Pi^1_1\text{-CA})\restriction$, on to those provided by Buchholz for $(\Pi^1_1\text{-CA}) + \mathrm{BI}$, the connection to practice becomes more and more tenuous. Viewed as an observer from the sidelines, there is a nagging feeling that the statements still have a "cooked up" look. This already appears with the use

[8] See Buchholz et al. (1981) for a general review of this work and further references.

in the Paris-Harrington statement of the condition $card(Y) \geq min(Y)$, and in Friedman's finite form of Kruskal's theorem of the condition $card(T_i) \leq k \cdot i$. Examination of the details of the argument in Simpson (1985) for the independence of the latter from ATR$\restriction$ makes clear how the imposition of the linear condition on growth allows one to pass from that statement (with the aid of König's Lemma) to the well-foundedness with respect to a wide-enough class of (possible) descending sequences to carry through a proof-theoretical demonstration of the 1-consistency of ATR$\restriction$. Note that the Paris-Harrington statement, which is most clearly relevant to results of established mathematical interest (finitary forms of Ramsey's theorem), does nothing to support Gödel's doctrine, since it can be proved in a mild, predicatively justified, extension of PA. But, in my view, the further one moves to support Gödel's doctrine by independence results for increasingly stronger systems, the less convincing is the case for the relevance of such.[9]

This section would not be complete without some discussion of the leap that has been taken in Friedman (1986). The effort there is to produce finite combinatorial statements P_{fin} independent of relatively strong systems of set theory, and in particular of ZFC + $\{$"(there exists an n-Mahlo cardinal)"$\}_{n<\omega}$. While the statement thus produced[10] does make use only of concepts from ordinary finitary mathematics, it is extremely remote from any ordinary mathematical reading. (Indeed, so much so that it would be unrewarding to reproduce the statement here—the reader must see for himself in Friedman's paper.) Here, I am evidently in complete disagreement with Friedman that his statement is "readily understandable." But Friedman himself goes on to say that this statement "is still not quite as simple or as natural as we would like [and] we are continuing to strive for an improved form."* What Friedman believes his result in this instance accomplishes is to "open up, for the first time, the realistic possibility, if not probability, that strong abstract set theory will prove to play an essential role in a variety of more standard finite mathematical contexts." If that case can be made then, according to Friedman, "this would open up a foundational crisis of nearly unprecedented magnitude since we seem to have no way of convincing ourselves of the correctness or consistency of such set theoretic principles short of faith in our very uneasy intuition about them."[11] In other words, Friedman is hesitatingly plumping for some form of Gödel's doctrine.

[9]In personal communications to me, both Friedman and Simpson have taken issue with the views expressed in this paragraph, particularly with respect to the comparison of the finite form of Kruskal's theorem given in the text and the Paris-Harrington statement (PH). I agree with them that the former statement is more readily understandable (or visualizable) than the latter, and also that cardinality restrictions on the sizes of finite trees are more natural than the ad hoc cardinality condition appearing in PH. However, this does not affect my overall conclusion at the end of the paragraph, which also concerns much stronger statements.

[10]Friedman (1986), proposition VIII.

*[His most recent efforts in that direction are presented in Friedman (1998).]

[11]The quotations are from Friedman (1986), pp. 92–93.

My general arguments against that doctrine in the last section, except one, apply equally to this specific case. Once more there is a *petitio principii*: would we believe Friedman's independent finitary statement to be true if we did not believe in the correctness of ZFC + $\forall n$ ("there exists an n-Mahlo cardinal") in some independent set-theoretical reality? So how does it justify the latter to demonstrate independence of the former from a slightly weaker system? Since the statement is clearly fabricated to do a job, and the evidence for its truth begs the question, the necessary use of such strong systems for everyday finitary mathematics has yet to be demonstrated. As I read Friedman's statements above, he agrees that the case has yet to be clinched; clearly, we differ on how much farther one will have to go in order to do that decisively, if at all. But my overall conclusion (announced in the introduction to chapter 2) about the independence results discussed in this section is even stronger: the case remains to be established that any use of the Cantorian transfinite beyond $\aleph_0$ is necessary for the mathematics of the finite in the everyday sense of the word.*

Countable Foundations for Applicable Mathematics

In this final section,** I shall explain the basis for the second claim announced in the introduction to chapter 2, that higher set theory is dispensable in scientifically applicable mathematics. The argument for this follows the trail blazed by Weyl in his 1918 monograph *Das Kontinuum*, but goes much farther by making use of modern logical tools. First of all, the work sketched by Weyl can be formalized in $(\Pi^0_\infty\text{-CA})\!\restriction$, which is a *conservative extension* of PA; that is, for any first order A, if $(\Pi^0_\infty\text{-CA})\!\restriction \vdash A$ then PA $\vdash A$. This conservation result can be proved very simply by use of Gödel's completeness theorem, as follows. If PA $\nvdash A$ then there is a model M of PA $+ \{\neg A\}$; one then expands M to a model M' of $(\Pi^0_\infty\text{-CA})\!\restriction + \{\neg A\}$ by taking the range of the set variables to be the collection of all first-order definable subsets of M. There are also standard proof-theoretical techniques which may be used to prove this conservation result by strictly finitary means. Thus, the portion of classical mathematical analysis that can be formalized in $(\Pi^0_\infty\text{-CA})\!\restriction$, following Weyl's plan, rests on first-order Peano Arithmetic as a foundation. Since the general notion of real number is defined in the wider system, the conservation result shows that such uses of the uncountable in classical analysis can be eliminated.

Though Weyl did not himself carry his plan very far, the direction he set for it was clear enough to see how to formalize substantially all of the classical analysis of piecewise continuous functions of a real variable developed through the nineteenth century (and relatedly for complex analysis).

*[For a more recent discussion of the issues in this and the preceding sections, see Feferman (1998).]

**[The material of this section overlaps that of chapters 13 and 14 in this volume, whose writing it preceded.]

Since this part of analysis already includes a considerable portion of scientifically applicable mathematics, the execution of Weyl's program would go far toward substantiating the claim made above. However, one would also have to be able to provide a countable foundation for significant portions of twentieth-century analysis and accessory parts of mathematics in algebra and topology in order to establish the claim in full. Among other things one would want to be able to deal with the *Lebesgue integral* and more general theories of integration, as well as with a variety of *function spaces* and more general abstract spaces (*Banach spaces, Hilbert spaces,* etc.), and the operators on them, which form the subject matter of *functional analysis.* One way in which this might be done is to deal only with countably presented mathematical objects (functions, sets, structures, spaces), each of which is determined by a countable amount of information and hence can be represented by a subset of **N**. While this approach is possible in principle, it is not very natural. It is preferable to find a system S for which one can proceed in a direct way from a body of mathematical notions and theorems to the formalization of such in the system, and such that S can be given a countable foundation (as $(\Pi^0_\infty\text{-CA})\restriction$ is justified by PA). However, a system of this type would have to incorporate higher type notions. For, as soon as we wish to speak about *functions* of real numbers in a direct way in general, we must move to *third-order* concepts; then direct discussion of *functionals* (such as various linear operators on functions) would require dealing with (prima facie) *fourth-order* concepts, and so on.

Thus the aim would be to find a *finite type theory* S which can be given a countable foundation, say by proof-theoretical reduction to PA, and in which scientifically applicable twentieth-century mathematics can be directly formalized. The first part of this aim was realized in two ways by systems presented in Feferman (1977) and Takeuti (1978). But these systems are still not as convenient as one would like for the second part of the aim. The reason is that types are *fixed* syntactic objects in these systems, so there is no simple direct means to handle *subtypes* given by separating with respect to properties having variable parameters. A second reason for wanting *variable* types is that, following modern abstract mathematics, one wants to speak of arbitrary structures of any given algebraic, topological, or analytic kind and there is no natural sense in which these should be restricted to range over a fixed type. For example, one would like to speak of *all* groups (or all linear spaces, or all Banach spaces, etc.) and not just of all groups (etc.) within a given type. What is thus required is a formal system in which the types themselves may be *variable* and in which subtypes can be freely separated by suitable definitions. Such a system would approach set theory in appearance. Indeed, Friedman (1980) provided a theory ALPO which is contained in ZFC and is conceptually rich, yet is a conservative extension of PA. However, Friedman's system is semi-intuitionistic (in the sense that classical logic is applied in it only to number-theoretic equations).

Since the use of classical logic pervades modern mathematics, in my view it is preferable to have a formal system with variable types using classical logic in full and having a countable foundation, say by reduction to PA. Such a theory has been described in Feferman (1985a)[12]; it was there denoted Res-VT + (μ) but is denoted $\text{VT}_\mu\restriction$ in the following ("VT" abbreviates "Variable Types" and "μ" is for the unbounded minimum operator described below). While the formalism employs set variables, it differs from set theory in that the notion of function is treated as a primitive and not reduced to the notion of set. This conceptual separation is characteristic of certain approaches to constructive mathematics. At the same time it accords with the highly asymmetric use that working mathematicians make of functions and sets (in general, in practice, functions are not nearly as complicated as sets).

It is not possible here to go into details of the system $\text{VT}_\mu\restriction$, but the following sketches its main features. Besides class variables (or "type variables") X, Y, Z . . . there are class terms $U, V, W, \ldots$ built up from class constants and variables by the following operations: (i) if U, V are class terms then so also are $U \times V$ and $(U \to V)$; and (ii) if U is a class term and A is a bounded formula (defined below), then $\{x \in U : A\}$ is a class term. Individual terms are all introduced in context: for any class term U (which may contain variables) there is a list of individual variables $x^U, y^U, z^U, \ldots$ of type U; further terms are built from these and individual constants by pairing and projections (corresponding to $U \times V$), and by function application and abstraction (corresponding to $U \to V$). Thus we have the full means of a typed λ-calculus but with variable types. Unlike ordinary type theory, the formalism permits equations $s = t$ between terms of arbitrary type. Formulas are built from atomic formulas $(s = t)$ by $\neg, \wedge, \vee, \rightarrow$, and universal and existential quantification with respect to both class variables and (relative to any class term U), individual variables $\exists x^U(\ldots)$ and $\forall x^U(\ldots)$.[13] Bounded formulas are those containing no quantified class variables. (The preceding description of the syntax requires a simultaneous definition of class terms, individual terms and formulas). One defines $(t \in U)$ as $\exists x^U(t = x)$ and $\{x \in U : A\} = \{x^U : A\}, (\exists x \in U)A = \exists x^U A$, and $(\forall x \in U)A = \forall x^U A$.

The basic axioms of $\text{VT}_\mu\restriction$ are the usual ones for abstraction and application, pairing and projections, and separation. The further axioms concern $\mathbf{N}$ (a constant symbol for the set of natural numbers). This comes equipped with constants for 0 and the successor function $sc(x) = x'$, and a primitive recursor $r_\mathbf{N}$ which gives for each $a \in \mathbf{N}$ and $g \in \mathbf{N} \times \mathbf{N} \to \mathbf{N}$ a function $f = r_\mathbf{N}(a, g) \in (\mathbf{N} \to \mathbf{N})$ with $f(0) = a \wedge \forall n \in \mathbf{N}[f(n') = g(n, f(n))]$. From this we can derive primitive recursion with parameters drawn from

[12]This paper was presented to a meeting in Bogotá, Colombia, in 1981. The exceptional delay in publication of the proceedings of the meeting accounts for its date. [An improved system W for this purpose is described in chapters 13 and 14.]

[13]The system in Feferman (1985a) did not employ quantified class variables, using such only for pure statements of generality. The present extension is conservative over that system.

any classes. The axiom of induction for $\mathbf{N}$ is given in a form restricted to classes defined by characteristic functions:

(1) $f(0) = 0 \wedge \forall x \in \mathbf{N}(f(x) = 0 \rightarrow f(x') = 0) \rightarrow \forall x \in \mathbf{N}(f(x) = 0).$

Finally, there is an axiom for a function $\mu \in (\mathbf{N} \rightarrow \mathbf{N}) \rightarrow \mathbf{N}$ with

(2) $\mu(f) = \underset{x \in \mathbf{N}}{min}[f(x) = 0]$ if $(\exists x \in \mathbf{N})[f(x) = 0]$, otherwise 0;

μ is called the *unbounded minimum operator*. Using it we can define an operator $E_{\mathbf{N}}$ with $E_{\mathbf{N}}(f) = 0 \leftrightarrow (\exists x \in \mathbf{N})(f(x) = 0)$.

The main mathematical result obtained for this system is that

(3) $\mathrm{VT}_{\mu}\restriction$ is a conservative extension of PA.

The proof of (3) described in Feferman (1985a) proceeds by a series of reduction steps—first from variable types to constant types, then by elimination of subtypes, and finally by reduction to the system Res-$\hat{\mathrm{Z}}^{\omega} + (\mu)$ of Feferman (1977); it was there shown how to reduce the latter system to PA by proof-theoretical means. All of this argument is finitary; furthermore, conservation with respect to arithmetical statements is preserved at each step.*

While (3) shows that $\mathrm{VT}_{\mu}\restriction$ has a countable foundation, many sets which are ordinarily thought of as being uncountable can be defined to exist in the system. To prepare the way, $\mathbf{Z}$ is defined as usual in terms of $\mathbf{N} \times \mathbf{N}$, and $\mathbf{Q}$ in terms of $\mathbf{Z} \times (\mathbf{Z} - \{0\})$. Then the class $\mathbf{R}$ of real numbers is defined to be the class of all Cauchy sequences of rationals, that is, all functions from $\mathbf{N}$ to $\mathbf{Q}$ satisfying the Cauchy convergence criterion; an equality relation $=_{\mathbf{R}}$ is introduced to tell when these represent the same real number. The class $Fun_{\mathbf{R}}$ of all real functions is then defined to be the subclass of $(\mathbf{R} \rightarrow \mathbf{R})$ consisting of all f which preserve $=_{\mathbf{R}}$. We can then proceed to define the class of all functionals on these and so on. Of course the complex numbers $\mathbf{C}$ are defined as usual from $\mathbf{R}$ via $\mathbf{R} \times \mathbf{R}$. Since n-tupling can be explained in terms of pairing, one has all finite-dimensional spaces $\mathbf{R}^n$ and $\mathbf{C}^n$; one also has Cantor space and Baire space via $\mathbf{N} \rightarrow \{0, 1\}$ and $\mathbf{N} \rightarrow \mathbf{N}$, respectively. All of these spaces may be shown to be *locally sequentially compact*; that is, every bounded sequence contains a convergent subsequence. (The proof uses König's Lemma, which in turn is proved by a combination of primitive recursion and the $E_{\mathbf{N}}$ operator.) The *Cauchy completeness* of these spaces follows. More generally one deals with arbitrary *complete separable metric spaces*, for which fundamental results like the Baire Category Theorem and the Contraction Mapping Theorem (existence of a fixed point) can be proved.

The space $\mathbf{R}$ as dealt with in $\mathrm{VT}_{\mu}\restriction$ cannot be proved complete in the sense that the l.u.b. (or g.l.b.) axioms holds for arbitrary *sets* of reals; one

*[The reduction to PA of the improved system W referred to in the postscript to ftn. 12 above is carried out in Feferman and Jäger (1993) and (1996).]

only has that for arbitrary *sequences* of reals. Since the l.u.b. (or g.l.b.) axiom for sets of reals is applied ubiquitously in analysis, one must in each case see whether such an application can be replaced by use of the corresponding principle for sequences. Often that is very easy but in some cases it requires more care; finally, there are certain cases in actual analysis where this replacement simply cannot be made. One such is when Lebesgue measure $meas(X)$ of a set $X \subseteq R$ is defined in terms of its outer measure $meas^\star(X)$, which is the g.l.b. of $\{meas(Y) : X \subseteq Y \text{ and } Y \text{ is open}\}$ (where for open Y, $meas(Y)$ is the sum of the lengths of the components of Y). Failing this, we cannot deal with nonmeasurable sets. But for applications it is only necessary to deal with *measurable sets* X, and for these a direct definition of measure is possible in terms of sequential open covers of X and of the complement of X. Similarly *measurable functions* can be dealt with directly in terms of limits (a.e.) of sequences of step functions. By this means we can represent the positive theory of Lebesgue measurable sets and functions (and similarly for more general theories of measure and integration).

Finally, one can develop substantial portions of functional analysis in $\mathrm{VT}_\mu \restriction$ for linear operators on separable Banach and Hilbert spaces, including the principal results for the spectral theory of compact self-adjoint operators. I have verified that usable forms of the Riesz Representation Theorem, Hahn-Banach Theorem, Uniform Boundedness Theorem, and the Open Mapping Theorem can all be proved in $\mathrm{VT}_\mu \restriction$. It seems, then, that all of *applicable classical and modern analysis* can be developed in this conservative extension of PA.

There is further detailed evidence which can be brought from another corner to support this claim. Namely, in his famous 1967 work, Errett Bishop gave a new constructive redevelopment of classical and modern analysis which he carried out in great detail for a number of fundamental results (many of which could be considered test cases). Bishop said that each of his theorems provides a constructive substitute $A^\star$ for a classical theorem A. The relationship is that A implies $A^\star$ and that $A^\star$ implies A by use of classical logic. In fact, according to Bishop the use of classical logic can be reduced in each case to that of what he called the Limited Principle of Omniscience (LPO), namely, that

$$\forall n \in \mathbf{N}(f(n) = 0) \vee \exists n \in \mathbf{N}(f(n) \neq 0) \tag{4}$$

holds for each $f \in (\mathbf{N} \to \mathbf{N})$.[14] Now Friedman had found in 1977 a subsystem of intuitionistic ZF in which all of Bishop's constructive analysis (with a later form of his theory of measure) can be formalized, and which is conservative over HA. In my paper "Constructive theories of functions and classes"[15] I provided an alternative constructive theory of functions

[14]This explains Friedman's use, in his 1980 paper, of "ALPO" as an abbreviation for "Analysis with the Limited Principle of Omniscience."

[15]Feferman (1979).

and classes T_0, containing a subtheory $EM_0\restriction$ in which all of this work of Bishop can be formalized. I showed that $EM_0\restriction$ with classical logic is conservative over PA and Beeson showed its conservation over HA in intuitionistic logic.[16] The crucial point for the present purposes is that when classical logic is added, one regains from Bishop's work all the classical theorems A for which he found constructive substitutes $A^\star$, in a theory conservative over PA. (The system $EM_0\restriction$ is related in certain ways to $VT\restriction$, without the μ operator; addition of the latter builds in a strong form of LPO.) Actually as far as the claim here is concerned, the detour via Bishop's work takes one through unnecessary technical complications, since those are necessitated only by his insistence on constructivity. But at least the foregoing provides another way of giving massive detailed support for the main claim of this section.

Still further evidence for that has been gathered in the program of "Reverse Mathematics" initiated by Friedman and pursued by Simpson and others, which shows that even much weaker theories, conservative over PRA (Primitive Recursive Arithmetic), suffice for the development of significant portions of analysis and algebra.[17]

Nothing here is meant to suggest that a system S which makes prima facie use of uncountable sets of various kinds *must* be reduced to PA in order to show that higher set theory is eliminable in the applications of S. Any predicatively reducible system is a candidate for that and even, from a certain point of view (suggested above), are certain (prima facie) impredicative theories of iterated inductive definitions. Moreover, there is no question that in the current practice of mathematics there are a number of results in which the use of transfinite set theory is essential. While these results are far from the applications of mathematics, further efforts should be made to see whether any of them might lead to a crucial test case of the claim made here.

Independently of such detailed work which puts into question the necessity of higher set theory for everyday mathematics,* I am convinced that the platonism which underlies Cantorian set theory is utterly unsatisfactory as a philosophy of our subject, despite the apparent coherence of current set-theoretical conceptions and methods. To echo Weyl, *platonism is the medieval metaphysics of mathematics*; surely we can do better.

---

[16] Feferman (1979), pp. 218–219.

[17] See Simpson (1985a), for an introduction and references [and, now, Simpson (1998)].

*[In this respect, cf. also the following chapters 13 and 14.]

13

Weyl Vindicated: *Das Kontinuum* Seventy Years Later

In his book *Das Kontinuum* (1918), Hermann Weyl initiated a program for the arithmetical foundation of mathematics. This was overshadowed by Hilbert's program for finitist foundations via formal consistency proofs and Brouwer's program for the "intuitionistic" constructive redevelopment of mathematics, both of which eventually met serious problems. Modern logical work has made it possible to give considerable substance to Weyl's program, showing it to be surprisingly viable for scientifically applicable mathematics.

1 Weyl's Career

Hermann Weyl is widely noted as one of the outstanding mathematicians of this century, with deep contributions to many fields. A brief review of his career helps understand the nature of his excursions into the foundations of mathematics.[1]

Hermann Weyl was born 9 November 1885 in Schleswig-Holstein. His university studies in Münich and Göttingen were completed in 1908 with a doctorate under the direction of David Hilbert at the University of Göttingen. Weyl became a Privatdozent in 1910 and stayed in Göttingen until 1913, when he accepted a professorship at the Eidgenössische Technische Hochschule (Federal Institute of Technology) in Zürich. Overlapping with him for one year as a colleague was Albert Einstein, with whom he became friendly. Einstein was then laying the ground for the theory of

"Weyl vindicated: *Das Kontinuum* seventy years later" originally appeared in C. Cellucci and G. Sambin (eds.), *Temi e prospettive della logica e della scienza contemporanee*, vol. 1 (1988), 59–93 (Feferman 1988a). It is reprinted here with the kind permission of the publisher, CLUEB (Cooperative Libraria Universitaria Editrice, Bologna). Some minor corrections and additions have been made to bring it up-to-date.

[1]For biographical material I have drawn on the obituaries of Weyl by C. Chevalley and A. Weil (1957), and by M. H. A. Newman (1957), and from the book by C. Reid on Hilbert (1970), in which Weyl figures prominently.

general relativity, a subject to which Weyl himself was to contribute before long.

Except for visits to Göttingen and Princeton, Weyl was to remain in Zürich until 1930. Around 1923, Hilbert attempted to get Weyl to return to Göttingen as the successor of Felix Klein. The Mathematical Institute in Göttingen was then at its apogee as one of the most exciting mathematical research centers in the world. Weyl was naturally strongly attracted by the opportunity, but he was very happy with his situation in Zürich and eventually turned down the offer. In 1930, though, he finally agreed to accept the position as Hilbert's successor, despite his concerns about the unstable political situation in Germany. His fears were well justified: the advent of the Nazi regime in 1933 brought devastating changes to the Institute in Göttingen, which Weyl and his colleagues were unable to stem. At the urging of Einstein, Weyl accepted an appointment in that year as professor at the newly established Institute for Advanced Study in Princeton, which he was to retain until his retirement in 1951. In the following period until his death on 9 December 1955, Weyl spent half of each year in Princeton and half at his old institute in Zürich.

In his mathematics, Weyl was largely imbued with Hilbert's spirit and interests, and he was almost as broad and deep as his teacher. He made many contributions to number theory, algebra, geometry, analysis, mathematical physics, logic, and the philosophy of mathematics and of science. Weyl's *Gesammelte Abhandlungen* (1968), in four thick volumes, and his various books and monographs[2] testify to his keen intelligence, breadth of interests, and creative originality. Weyl was also very cultured outside of mathematics; he was familiar with the classics and could draw with confidence and style on his extensive knowledge of art, literature, history, and philosophy.

2 Weyl's Foundational Contributions

Prior to his major work *Das Kontinuum* (1918), which is our main concern here, Weyl made one original contribution to the foundation of mathematics, namely his 1910 paper on the definition of fundamental mathematical concepts.[3] This provided a means to replace the vague idea of "definite property" in Zermelo's *Aussonderungsaxiom* in his theory of sets by a more precisely defined notion, as explained in section 4 below. Weyl's procedure was also adapted to his 1918 work, which will be described in more detail in section 5 below when we reconstruct the axiomatics of *Das Kontinuum* in modern terms. Incidently, other solutions to the problem of explaining the concept of "definite property" in set theory were later offered both by

---

[2]The bibliography of Weyl here is not intended to be comprehensive.

[3]See also the section on Weyl in my article "Infinity in mathematics" [reproduced in chapter 2 of this volume].

Fraenkel and by Skolem in 1922, the latter being the one that is now commonly accepted; Weyl's solution largely anticipated that due to Skolem.[4]

After 1918, Weyl made no further original contributions to the foundations of mathematics, but he maintained a great interest in it and wrote and lectured periodically on foundational questions until the end of his life. Of the 167 items in Weyl's collected works (1968), I count a dozen on this subject. One should also add the posthumous publication (1985) of a lecture on this subject given toward the end of his life. Finally, and by no means least, Weyl devoted a major part of the book *Philosophy of Mathematics and Natural Science* (1949) to the foundations of mathematics.[5]

At least from 1917 on, Weyl was critical of the set-theoretical foundations of mathematics as developed from Cantor's theory of transfinite entities. The problem of the paradoxes which had appeared with the unrestricted application of Cantor's ideas dominated foundational discussions well through the 1920s and beyond. Zermelo's 1908 axiomatization of a restricted concept of set led with Fraenkel's axiom of replacement to a mathematically adequate theory of transfinite cardinals and ordinals in which the paradoxes were (prima facie) blocked in a simple way. This should, perhaps, have calmed fears about the inherence of contradictions in set theory. But Zermelo did not give a guiding informal interpretation of his axiom system in (what has come to be called) the cumulative hierarchy until 1930, and this idea was not widely appreciated until the 1950s. Even to this day, the possibility of a contradiction in set theory is raised by some critics as a reason to pursue alternative foundational schemes.

It was partly out of concern for the paradoxes that Weyl called the set-theoretical foundations of mathematics a house built to an essential extent on sand (1918, preface). But there was a more basic disagreement in viewpoint as to the fundamental nature of mathematics: set theory requires a platonistic conception of the subject matter of mathematics as a universe of abstract entities existing independently of human conceptions, constructions, and definitions, in direct opposition to what Weyl espoused. A more specific delineation of Weyl's viewpoint in *Das Kontinuum* is provided in the next section. It suffices for the moment to say that in 1918, Weyl accepted the natural number system as a basic conception and insisted on working only with sets and functions given by explicit or recursive definitions (following the sequence of natural numbers). Moreover, he regarded statements formulated in terms of these notions as having a definite truth value, true or false, and thus accepted classical logic (first-order predicate calculus) as basic.

In the years directly following the publication of *Das Kontinuum*, Weyl became familiar with the more trenchant criticism of set-theoretical platonism (and axiomatic formalism) by L. E. J. Brouwer, whose own "intu-

[4] For a comparison of the three, see the discussion on p. 285 of van Heijenoort (1967).

[5] This was a considerable expansion of an encyclopedia article he wrote in the 1920s.

itionistic" program was based on the view of mathematics as being entirely about (the possibilities of) human constructions. Thus Brouwer was willing to ascribe the truth value "true" to a mathematical statement only if it had been established or proved by such a construction, and the truth value "false" only if it had been correspondingly disproved. The Law of Excluded Middle was necessarily rejected on these grounds, forcing a complete revamping of basic logical reasoning in mathematics. In addition, Brouwer accepted no "completed" infinite totalities, the existence of which would make sense only on set-theoretical platonistic grounds. In particular, according to Brouwer the entire theory of real numbers had to be reconsidered, with each real number now to be represented by a "potentially" infinite sequence of rational numbers proceeding by a sequence of arbitrary "free" choices; this in turn required a radical redevelopment of mathematical analysis.

Brouwer also argued against Hilbert's view, which was to treat mathematics as given by axiomatic systems. Hilbert's idea was that the consistency of such systems would guarantee the "existence" of the corresponding concepts; for example, the consistency of (something like) Peano's axioms would assure the existence of the structure of natural numbers, and the consistency of (something like) Dedekind's axioms would assure that of the structure of real numbers. Hilbert's program aimed to develop systematic methods for establishing the consistency of stronger and stronger systems, presupposing only the most elementary, surely-to-be-granted, "finitistic" concepts and methods. Hopefully, this was to lead to proofs of consistency of systems of axiomatic set theory such as given by Zermelo, and thereby justify the use of substantial tracts of abstract nonconstructive mathematics as well as their bread-and-butter consequences.[6] However, Brouwer criticized Hilbert's program on the grounds that consistency was not sufficient to assure correctness (according to some basic intuitive conception).[7]

Not long after publishing *Das Kontinuum*, Weyl became a convert to Brouwerian intuitionistic constructivism. In one of his lectures on Brouw-linebreak er's ideas in 1920 he said, "I now give up my own attempt and join Brouwer."[8] During the following years he continued to champion Brouwerian intuitionism and to criticize Hilbert's program, much to Hilbert's annoyance. However, he gradually became pessimistic about the prospects for the Brouwerian revolution. Thus in his late summing up (1949), he says:

> Mathematics with Brouwer gains its highest intuitive clarity. He succeeds in developing the beginnings of analysis in a nat-

[6]See the section on Hilbert in Feferman (1987a) [chapter 2 in this volume] for a more detailed and careful presentation of Hilbert's program.

[7]As is well known, Gödel's theorems on the unprovability of the consistency of a system by no stronger (consistent) means raised a more patent obstacle to Hilbert's program.

[8]See Weyl (1968), vol. 2, p. 158.

> ural manner, all the time preserving the contact with intuition much more closely than had been done before. It cannot be denied, however, that in advancing to higher and more general theories the inapplicability of the simple laws of classical logic eventually results in an almost unbearable awkwardness. And the mathematician watches with pain the larger part of his towering edifice which he believed to be built of concrete blocks dissolve into mist before his eyes. (Weyl 1949, p. 54)

On the other hand, he continued Brouwer's criticism of Hilbertian formalism: "Hilbert's mathematics may be a pretty game with formulas . . . but what bearing does it have on cognition, since its formulas admittedly have no material meaning by virtue of which they could express intuitive truths?" (*op. cit.*, p. 61). But, he says, a consistency proof of axiomatic arithmetic "would vindicate the standpoint taken by the author in *Das Kontinuum*, that one may safely treat the sequence of natural numbers as a closed sequence of objects" (*op. cit.*, p. 60).

In Weyl's lecture "Axiomatic versus constructive procedures in mathematics," given late in his life and published posthumously (1985), Brouwer is not even mentioned, though Weyl expresses his lingering sympathy for the Brouwerian approach: "Indeed my own heart draws me to the side of constructivism. Thus it cost me some effort to follow the opposite direction, putting axiomatics before construction, but justice seemed to require this from me" (Weyl 1985, p. 38). Here Weyl employs "constructivism" in a wide sense, embracing both a genetic conception of basic structures and basic constructive procedures; he further employs "axiomatics" in a wide sense of the approach to mathematics exemplified by abstract twentieth-century algebra and analysis, and not in the narrow sense of Hilbert's program. In the struggle or tension between these two approaches, Weyl urges resisting the adoption of "one of these two views as the genuine primordial way of mathematical thinking to which the other merely plays a subservient role" (*loc. cit.*).

Though wavering to the end between genetic constructivism and the abstract axiomatic approach, Weyl remained constant throughout the years in his criticism of set-theoretical platonism and its assumption of closed transfinite totalities:

> The leap into the beyond occurs when the sequence of numbers that is never complete but remains open toward the infinite is made into a closed aggregate of objects existing in themselves. Giving the numbers the status of ideal objects becomes dangerous only when this is done. . . . The vindication of this transcendental point of view forms the central issue of the violent dispute . . . over the foundations of mathematics. (Weyl 1949, p. 38)

In particular, he criticizes the assumption, for any infinite set, of the closed totality of all subsets of that set (*op. cit.*, p. 49).

In *Das Kontinuum*, only one closed infinite totality is assumed, namely, that of the set $\mathbf{N} = \{0, 1, 2, \dots\}$ of natural numbers. The definiteness of this concept is reflected in the assumption of classical logic with its consequence, for each number-theoretic property $P(n)$, that either $P(n)$ is true for all n in $\mathbf{N}$ or $P(n)$ is false for some n in $\mathbf{N}$, that is, in logical symbolism, $\forall n P(n) \vee \exists n \neg P(n)$. Weyl did not seem to be aware of the simple proof-theoretic reduction of classical logic to intuitionistic logic accomplished in 1933 by Gödel (and, independently, Gentzen), carried over to axiomatic systems of number theory.[9] In consequence of this reduction, even the natural numbers can be conceived of as a "potentially infinite" totality. Though Weyl may have been equivocal on this point of underlying logic, either way he clung to the definiteness of the natural number concept, which he calls "the simplest example of what I like to call an *a priori* surveyable range of variability" (Weyl 1985, p. 13).

3 Definitionism, Predicativity, and Arithmetical Mathematics

In *Das Kontinuum*, Weyl to a large extent accepted Henri Poincaré's definitionist philosophy of mathematics, which featured the following general ideas:[10] (i) the conception of the natural number sequence is an irreducible minimum for (abstract) mathematical thought, and the associated principles of proof and definition by induction are to be accepted as basic;[11] (ii) all other mathematical concepts, in particular those of sets and functions, are to be introduced by explicit definition; (iii) there are no completed infinite totalities; and (iv) great care must be taken to exclude apparent definitions which seem to single out an object from a presumed completed totality of objects, by essential reference (either explicit or implicit) to that totality. Poincaré did not elaborate these ideas in detail; in particular, he was silent about the choice of underlying symbolism for definitions and of underlying logic for proofs.

Improper (apparent) definitions of an object A which essentially assume the existence of a completed infinite totality S containing A as member are said to be *impredicative*, while proper ones are said to be *predicative*. Thus Poincaré's definitionist philosophy of mathematics is also called that of *predicativity*. The idea (iv) was embodied in Poincaré's *vicious circle*

[9] See the remark quoted above, from Weyl (1949), p. 60.

[10] See, for example, chapter 4 of Poincaré (1963) (a translation of *Dernières Pensées*). Poincaré's ideas about geometry are of a different character. [Cf. also Mooij (1966) and Folina (1992)].

[11] Poincaré argued that all attempts to define the natural numbers in terms of some more basic notions (such as in the "logicist" program of Frege and Russell) involve a *petitio principii*, either explicitly or implicitly. [Cf., in this respect, Goldfarb 1988.]

principle, which banned the use of impredicative definitions, and thence of the assumption of totalities S underlying such definitions. Poincaré was led to this principle by his analysis of the paradoxes, which he viewed as implicitly admitting such totalities as the set S of all sets; that is, $S = \{x|x \text{ is a set}\}$ with $S \in S$, and thence the paradoxical $A = \{x|x \in S \wedge x \notin x\}$ with (supposedly) $A \in S$.

The exact meaning and acceptability of the vicious circle principle have been the subject of much discussion, which will not be reviewed here.[12] Our concern is rather to see its effect on the formulation of specific axiomatic systems, namely, Zermelo's theory of sets, Russell's (ramified) theory of types, and Weyl's system in *Das Kontinuum.* In the case of Zermelo Set Theory Z (and likewise its extension ZF by Fraenkel), the vicious circle principle is accepted only in a minimal way, namely, by excluding the existence of a set of all sets. Zermelo (and Fraenkel) balanced this by positive set-existence axioms designed (with the aid of the Axiom of Choice) to justify Cantor's theory of transfinite ordinals and cardinals. But, as we shall see in a moment, Z still permitted the use of impredicative definitions, in fact in the very set-theoretical foundations of the real number system. Thus Zermelo by no means accepted the vicious circle principle as a general guide, only enough of it to block (as it seems) the paradoxes of set theory.

Among other things, the axioms of Z guarantee the existence of the infinite set $2^{\mathbf{N}}$ and, for each set S, of the totality $\mathcal{P}(S)$ of arbitrary subsets of $S : \mathcal{P}(S) = \{X|X \subseteq S\}$. In the foundations of analysis one builds in succession the rational numbers $\mathbf{Q}$ and the real numbers $\mathbf{R}$. $\mathbf{Q}$, being denumerable, is in 1–1 correspondence with $\mathbf{N}$; in symbols $\mathbf{Q} \sim \mathbf{N}$. Following Dedekind, $\mathbf{R}$ may be characterized as being isomorphic to the set $\mathcal{D}(\mathbf{Q})$ of all (Dedekind-)sections in $\mathbf{Q}$, that is, of all nonempty sets X of $\mathbf{Q}$ which are closed to the left. It is not hard to show that $\mathcal{D}(\mathbf{Q}) \sim \mathcal{P}(\mathbf{Q}) \sim \mathcal{P}(\mathbf{N})$.[13] Now the essential statement about $\mathbf{R}$ is its *completeness*, according to which $\mathcal{D}(\mathbf{R})$ gives no new numbers, or equivalently that every nonempty subset S of $\mathbf{R}$ which is bounded above has a *least upper bound* (l.u.b.), or dually that every non-empty subset S of $\mathbf{R}$ which is bounded below has a *greatest lower bound* (g.l.b.). With S identified as a subset of $\mathcal{D}(\mathbf{Q})$, the l.u.b. and g.l.b. of S are simply, respectively, the union $U = \bigcup X[X \in S]$ and intersection $I = \bigcap X[X \in S]$ of all sets X in S. Under the correspondence $\mathcal{D}(\mathbf{Q}) \sim \mathcal{P}(\mathbf{N})$, we have $S \sim S'$ with $S' \subseteq \{X|X \subseteq \mathbf{N}\}$ and U, I correspond to U', I' defined by $U' = \{n|\exists X(X \in S' \wedge n \in X)\}$ and $I' = \{n|\forall X(X \in S' \Rightarrow n \in X)\}$. But the latter *may be* impredicative definitions of subsets of $\mathbf{N}$ and thus subject to ban if one accepts the vicious circle principle in full.

[12]See, for example, Gödel (1944) and Chihara (1973). [Cf. also the introductory note by C. Parsons to Gödel 1944 in Gödel 1990.]

[13]With the set-theoretical notation $B^A = \{f|f \text{ is a function from } A \text{ into } B\} = \{f|f : A \to B\}$, we have $\mathcal{P}(S) = \{0,1\}^S$ and Card $(\mathcal{P}(S)) = 2^{\mathrm{Card}(S)}$ for the corresponding cardinal numbers. Thus Card($\mathbf{R}$)=$2^{\mathrm{Card}(\mathbf{N})} = 2^{\aleph_0}$. Cantor's characterization of $\mathbf{R}$ as the set of equivalence classes of all Cauchy sequences in $\mathbf{Q}$ leads to $\mathbf{R} \sim \mathbf{Q}^{\mathbf{N}} \sim \mathbf{N}^{\mathbf{N}}$ and Card($\mathbf{R}$) $= \aleph_0^{\aleph_0}$, which again reduces to Card($\mathbf{R}$) $= 2^{\aleph_0}$.

Logically speaking, the purported definitions of the sets in question take the form $\{n|\exists X\phi(n,X)\}$ and $\{n|\forall X\psi(n,X)\}$, respectively, where "n" ranges over $\mathbf{N}$ and "X" over $\mathcal{P}(\mathbf{N})$; the problem concerning the impredicativity of these is the same, since the complement of $\{n|\exists X\phi(n,X)\}$ is $\{n|\forall X\neg\phi(n,X)\}$. Now with $A = \{n|\forall X\psi(n,X)\}$, in order to determine, given n, whether or not $n \in A$, we need to "check" all possible subsets X of $\mathbf{N}$ to see whether or not $\psi(n,X)$, but one of the sets which must be considered in this process is A itself, and to know whether $\psi(n,A)$ holds we shall in general have to know exactly what members A has. Thus we are defining A in $\mathcal{P}(\mathbf{N})$ in terms of the totality $\mathcal{P}(\mathbf{N})$ without, in general, having any other way of singling out A, and we are in a vicious circle. This sort of definition is, however, justified by the axioms of Z, but it becomes problematic when the vicious circle principle is incorporated more fully into an axiomatic system, as was done by Russell and Weyl.[14]

In Russell's Ramified Theory of Types RTT, the vicious circle principle is employed in a twofold way via a syntactic scheme of *types* and *levels*, indexed by natural numbers. Variables $X_0, Y_0, \ldots$ are used for objects of type 0, and $X_i^{(j)}, Y_i^{(j)}, \ldots$ for objects of type $i > 0$ and level j. Intuitively, these range over S_0 and $S_i^{(j)}$, respectively, where for $i > 0$, $S_i = \bigcup_j S_i^{(j)}$ and we have $S_{i+1} \subseteq \mathcal{P}(S_i)$. At the base, S_0 is supposed to be some set of individuals; for the applied theory we may interpret S_0 as $\mathbf{N}$.[15] When the levels j are ignored, we just have the *simple theory of types*, with only atomic formulas of the form $X_i \in Z_{i+1}$ and $X_i = Y_i$ being admitted as sensible. (Thus it is senseless in Russell's system to ask whether a set X contains itself as a member.) Now the levels are supposed to correspond to *stages of definition*, where a set $X_i^{(j)}$ is supposed to be defined only in terms of individuals or of sets in $Y_\ell^{(k)}$ for some $k < j$, and $\ell \leq i$, possibly by quantification over $S_\ell^{(k)}$. Intuitively speaking, sets are conceived to be introduced in stages by successive definitions, where at each stage we are only allowed to refer to the totality of previously defined sets. The set existence (comprehension) axioms of RTT correspond directly to this idea. But, in addition to these, Russell felt compelled to add his own problematic *Axiom of Reducibility*, which completely vitiated the purpose of ramification by levels.

In order to explain the issues involved, we simplify Russell's theory to a ramified theory of objects $n, m, \ldots$ of type 0 and $X^{(j)}, Y^{(j)}, \ldots$ of type 1, assuming Peano's axioms of natural numbers for the type 0 objects. Call a formula ϕ in this language *suitable for level j* if it is built up by the propositional connectives and quantifiers from numerical atomic formulas

[14]There is no problem from a platonist point of view, since $\mathcal{P}(\mathbf{N})$ is supposed to exist independently of any means of definitions, and for any formula ψ, $\{n|\forall X\psi(n,X)\}$ determines a definite subset A of $\mathbf{N}$, whether or not we can tell *which* subset that is.

[15]Actually, Russell sought to define $\mathbf{N}$ in his system without assuming it at the outset. We shall consider his theory as applied over $\mathbf{N}$.

and membership atomic formulas of the form $n \in Y^{(k)}$ with $k \leq j$ but where all *bound* variables $Y^{(k)}$ have $k < j$. Then the comprehension axiom

$$\exists X^{(j)} \forall n [n \in X^{(j)} \iff \phi(n)]$$

is taken as basic for each ϕ suitable for level j. Now Russell's Axiom of Reducibility in this setting takes the form

$$\forall X^{(j)} \exists Y^{(0)} \forall n [n \in X^{(j)} \iff n \in Y^{(0)}].$$

In other words, it says that every set of higher level is coextensive with one of the lowest level. The resulting theory is then equivalent to one without *any* level distinctions whatever, since they all collapse to one: we have variables $X, Y, \ldots$ for sets of natural numbers, and for any well-formed formula ϕ in this symbolism (which may involve quantifiers "$\forall Y$," "$\exists Y$" unrestrictedly), we have the comprehension axiom

$$\exists X \forall n [n \in X \iff \phi(n)].$$

The reason that Russell felt compelled to introduce his Axiom of Reducibility was simply pragmatic: he saw no way of providing a workable foundation for analysis in the ramified system without it. For, in RTT, reals correspond to subsets of **N** (via **Q** $\sim$ **N**) at various levels, so that one can only speak of real numbers of level 0, real numbers of level 1, etc., but not of real numbers in general. Furthermore, sets of reals must also be at a specific level; for example, a set S of reals may be represented as $\{X^{(j)} | \phi(X^{(j)})\}$ where ϕ is suitable of level j. But then the l.u.b. and the g.l.b. of S will be represented, respectively, by

$$\{n | \exists X^{(j)} [\phi(X^{(j)}) \wedge n \in X^{(j)}]\} \text{ and } \{n | \forall X^{(j)} [\phi(X^{(j)}) \Longrightarrow n \in X^{(j)}]\},$$

which are sets of level $j + 1$. By adjoining the Axiom of Reducibility, Russell could succeed in formally obtaining the l.u.b. and g.l.b. axioms already at level 0, since no level gives any more (extensionally speaking).

Russell acknowledged the opportunistic character of the Axiom of Reducibility, but saw no alternative to adopting it. Weyl, however, found this completely unacceptable, and said that it created a gulf between the two of them (Weyl 1918, p. 36). As he put it vividly thirty years later,

> Russell, in order to extricate himself from the affair, causes reason to commit harikari, by postulating the above assertion [the Axiom of Reducibility] in spite of its lack of support by any evidence. [But] in a little book *Das Kontinuum*, published in 1918, I have tried to draw the honest consequence and constructed a field of real numbers of the first level, within which the most important operations of analysis can be carried out. (Weyl 1949, p. 50)

In other words, Weyl saw that he could remain faithful to the tenets of predicativity by confining himself entirely to **N** and subsets of **N** of lowest level (and thereby not requiring any level distinctions), while at the same time accomplishing substantial portions of classical analysis. This meant sacrificing the l.u.b. principle for *sets* of reals, but Weyl recognized that the l.u.b. principle for *sequences* of reals could still be derived at level 0, and that this is sufficient for most applications. This development, which we shall examine below, may be considered to be a form of *arithmetical mathematics*, since the sets of level 0 are just those which are arithmetically definable. And, as the quotation from 1949 shows, Weyl never did give up his "own attempt."

4 Weyl's Definition of Definable Relations (1910)

In his 1910 paper on definition principles for relations, Weyl gave a procedure for generating all relations explicitly definable from given ones, and he incorporated a variant of this in his 1918 system. It is of interest to compare the two approaches, each of which anticipates much later developments.

In the 1910 paper, Weyl assumes given some basic relations of one or more arguments, between objects in a basic domain D. If R is a k-ary such relation, then $R(a_1, \dots, a_k)$ expresses that the relation R holds between $(a_1, \dots, a_k)$. Weyl defines five operations (or definition principles) leading from given relations R to new ones R', as follows.

(1) *Permutation.* If R is k-ary and π is a permutation of $\{1, \dots, k\}$ then $R' = R_\pi$ is given by $R'(a_1, \dots, a_k)$ if $R(a_{\pi(1)}, \dots, a_{\pi(k)})$.

(2) *Negation.* If R is k-ary then $R' = Rn$ is given by $R'(a_1, \dots, a_k)$ if not $R(a_1, \dots, a_k)$.

(3) *Expansion.* If R is k-ary then $R' = R+$ is $(k+1)$-ary, given by $R'(a_1, \dots, a_{k+1})$ if $R(a_1, \dots, a_k)$.

(4) *Projection.* If R is $(k+1)$-ary then $R' = R-$ is k-ary given by $R'(a_1, \dots, a_k)$ if there exists x in D with $R(a_1, \dots, a_k, x)$.

(5) *Conjunction.* If R, S are k-ary then $R' = \left.\begin{matrix} R \\ S \end{matrix}\right\}$ is k-ary given by

$R'(a_1, \dots, a_k)$ if $R(a_1, \dots, a_k)$ and $S(a_1, \dots, a_k)$.

The class of relations explicitly definable from given relations $R_1, \dots, R_m$ is then defined to consist of all relations generated from $R_1, \dots, R_m$ by repeated application (in any order) of the operations (1)–(5). In more modern symbolism, we might write $\bar{R}$ for Rn, $R \times D$ for $R+$, $\exists R$ for $R-$, and $R \cap S$ for $\left.\begin{matrix} R \\ S \end{matrix}\right\}$. It is easily seen that if equality is among the basic relations, then the relations definable from $R_1, \dots, R_m$ in Weyl's sense are exactly those which are definable in the language of the first-order predicate calculus with equality over the structure $(D, R_1, \dots, R_m)$.[16]

Weyl gives two applications of his notion of definability, one in geometry and the other in set theory. In the latter case, he conceives of D as the domain of all sets and takes $=$ and $\in$ as the basic relations. Then he proposes to make Zermelo's *Aussonderungsaxiom* precise by asserting, for each $(k+1)$-ary definable relation R and each $(k+1)$-tuple of sets $a, a_1, \dots, a_k$, the existence of the set $b = \{x \in a | R(x, a_1, \dots, a_k)\}$.

Despite the equivalence of Weyl's notion of definability with that given in modern semantical terms noted above, Weyl's viewpoint in 1910 is clearly mathematical rather than logical, in the sense that he has no language under consideration and is attending only to the actual relations generated over a fixed mathematical structure. Indeed, in connection with his treatment of Zermelo's axioms Z he even suggests that one ought to add a "completeness" axiom to Z (à la Hilbert's "completeness" axiom for geometry), to the effect that $(D, =, \in)$ is the largest structure satisfying Z.[17] Thus in 1910, he still has an informal conception of axiomatics in mind.

Weyl's purely mathematical explanation of the class of definable relations over a structure anticipates that given by Tarski (for the special case of real numbers, but clearly applicable to arbitrary structures) in 1931.[18] But it also leads directly to a formalism for representing all definable relations from given $R_1, \dots, R_m$, by using symbols for the operations (1)–(5) above. There would be in such a formalism a compound relation symbol for each definable relation. If one allows the number k of arguments of a relation to be 0, then also all first-order propositions about the underlying structure are expressible in such a symbolism. In this way, one may set up the language of *first-order predicate calculus without variables*, a project which was carried out much later by Quine in 1936 and by others since then.[19] Finally, there is a direct relationship between Weyl's operations generating the definable relations between sets and the class construction axioms used by Gödel in his 1940 theory of sets and classes; the main difference is that Gödel replaces the permutation principle (1) by finitely many instances (in a straightforward way).

---

[16]However, when equality is not included we lack the means in (1)–(5) of going from $R(a_1, a_2)$ to $R(a, a)$.

[17]See ftn. 4, p. 304 of Weyl (1968) Vol. I. Note also his remark on the difference between the denumerability of the class of all *concepts* (relations) and the nondenumerability of the class D of all *objects* (sets).

[18]See the translation "On definable sets of real numbers" in the collection of papers Tarski (1956), pp. 110–142.

[19]See Quine (1963) for the history. Note that setting up the language is only the first step; one needs to set up a suitable system of axioms and rules of inference in this language.

5 *Das Kontinuum* Reexamined

In the following, I shall give a tour of the contents of Weyl (1918), though where possible I shall recast this in modern terms. It is not my purpose to give a detailed step-by-step analysis of this monograph, though I think that doing so would be of historical value.[20] All chapter, section, and page references are to Weyl (1918).

There are two chapters, the first of which, chapter I, presents a theory of sets and functions within Weyl's predicativist position. It in turn is divided into two parts, the first of which he calls logical (§§1–3) and the second mathematical (§§4–8). Chapter II treats the real number system and the beginnings of a development of analysis in this predicative framework.

Section I.1 concerns properties and relations and existence, but the basic idea is that of a *judgment* ("Urteil"), that is, an *assertion of a state of affairs*, which may be *true* or *false*. Only meaningful statements express judgments, according to which a state of affairs *holds*. A *property judgment* is one that asserts that a certain object has a certain property. Not every formally grammatical sentence is a meaningful statement; for example, the ascription to an object of a property not appropriate to it is (said to be) meaningless. Weyl uses the heterological paradox to bring out the dangers of using formally grammatical but meaningless sentences. In the following, I shall use "statement" for Weyl's "judgment."

After the property statements, Weyl takes up *relation statements*, which express that a relation holds between specified objects. Examples of these are "point A lies between point B and point C" and "5 follows 4." These are represented by using *statement schemata*, for example, "x follows y" in the latter case. These contain *indeterminates* "x," "y,' . . . , that is, (symbolic) *variables* in modern terminology. Definite statements are obtained from statement schemata by substituting names of specific objects of the appropriate kind for the variables. Properties may be considered as relations of one argument, and thus property statements are special cases of relation statements.

A body of meaningful statements is obtained by starting with one or more basic domains of objects which are given directly to the intuition (for example, "point in space" or "number"), and with certain basic relations between these (for example, "lies between" or "follows"), and proceeds by applying *principles of statement combination* as described in I.2. One of the most important of these for mathematics is the principle which allows the assertion of the *existence* of objects with specified properties, for example, "there exists an x such that $P(x)$," "there exists an x such that $R(x, a)$," etc. Here, always, one must refer existence to a prior domain of objects; that is, the variable is to range over one of the given basic domains.

---

[20] A beginning for such has been made in the notes to the Italian language edition of Weyl (1918) translated by A. B. Veit Riccioli (Weyl 1977, pp. 169–182). See also the English translation by S. Pollard and T. Bole (Weyl 1987).

It is the approach in I.2 which is to be compared with that of Weyl (1910) described in the preceding section. Here Weyl definitely shifts over to a formal language close to that of the first-order predicate calculus (with equality) in the modern sense. For the relation schemata corresponding to the basic assumed relations, Weyl takes *atomic formulas* $R(x_1, \ldots, x_k)$ where $x_1, \ldots, x_k$ are distinct variables, including the *equality formulas* $x = y$ where x, y are distinct variables (of the same sort). Then *formulas* $U(x_1, \ldots, x_k)$ (with $k \geq 0$) in general are obtained from these by repeated application (in any order) of the operations: (1) negation, (2) identification of variables, (3) conjunction, (4) disjunction, (5) substitution of an object (that is, constant symbol) for a variable, and (6) existential quantification. Actually, Weyl does not describe these in full generality, being content to illustrate each case, but the intent is always clear. Thus, for example, (2) is illustrated by the passage from $U(x_1, x_2)$ to $V(x) = U(x, x)$.

The symbols Weyl uses for (1), (3), and (4) are $^-$, $\cdot$, and $+$, respectively, and in their place, we shall take $\neg$, $\wedge$, and $\vee$. Since classical logic is to be used, disjunction is dispensable in favor of negation and conjunction (as in (1910)). Actually, (2) is also dispensable, using equality formulas and existential quantification (cf. ftn. 16 above). In the case of the operation (6), Weyl makes an unfortunate notational choice: existential quantification with respect to a variable in a formula R is indicated by replacing that variable by "*" in R. Thus, for example, given $R(x, y, z)$, he writes $R(x, y, ^*)$ for what we would write as $\exists z R(x, y, z)$ and $R(^*, y, ^*)$ for what we would write either as $\exists x \exists z R(x, y, z)$ or $\exists z \exists x R(x, y, z)$. Weyl points out a possible ambiguity (p. 6): when $R(x, y) \wedge S(x, y) = T(x, y)$, we should not conclude that $R(x, ^*) \wedge S(x, ^*) = T(x, ^*)$, for (in modern terms) the lefthand side is to be read as $\exists y R(x, y) \wedge \exists y S(x, y)$, which is not equivalent to $\exists y T(x, y)$, that is, $\exists y[R(x, y) \wedge S(x, y)]$. Another possible ambiguity arises with $\neg R(x, ^*)$: is this $\neg \exists y R(x, y)$ or $\exists y \neg R(x, y)$? To resolve such ambiguities, Weyl introduces a *new* (atomic) formula symbol each time an operation is applied; thus where we would write $\neg \exists y \neg R(x, y)$ (or $\forall y R(x, y)$), he forms a succession: $S(x, y) = \neg R(x, y), T(x) = S(x, ^*)$ and $U(x) = \neg T(x)$. Finally, we may reach formulas with no free variables, as for example, $V = U(^*)$ in the preceding, which expresses $\exists x \forall y R(x, y)$.

The mathematical part of chapter I begins in I.4 with the treatment of sets, which are simply explained as being the extensions of definable properties: to every property E on a basic domain D corresponds a set (E) such that for a in D we have

$$a \in (E) \text{ if and only if } E(x) \text{ is true for } x = a.$$

Two sets $(E), (E')$ are identified: $(E) = (E')$, when the properties of E and E' hold of exactly the same elements in D. More generally, relations $R(x_1, \ldots, x_n)$ determine multidimensional sets. The one- and multidimensional sets over an original domain of objects constitute a new domain of "ideal" objects, the formation of which Weyl calls the *mathematical process*. Weyl describes this as the logical [predicative] conception of sets as opposed to the objective [platonistic] conception. Furthermore, under his conception there is no notion of set independent of given basic domains and relations. Though Weyl finds the set-theoretical attempts to found the notion of natural number interesting, he is convinced that this proceeds the wrong way around—rather that set theory must already be based on the intuition of iteration and the natural number sequence.

This leads in I.5 to a discussion of the system of natural numbers [which are positive integers in Weyl's account], treated to begin with as a structure consisting of the domain $\mathbf{N}^+ = \{1, 2, 3, \dots\}$ together with the successor relation $F(x, y)$ [y immediately follows x], where each x has a unique y with $F(x, y)$; 1 is defined as the unique element without a successor. The axioms for the natural numbers are not presented explicitly, but Weyl says that the *principle of complete induction* rests on the intuition that every natural number is reachable from 1 by repeated application of the passage from a number to its successor.[21]

In connection with the natural number sequence Weyl remarks on Cantor's notion of *denumerability*, which when combined with the notion of *definability* apparently gives rise to a paradox, namely, that of Richard on the definable real numbers. The paradox carries over to definable sets of natural numbers as follows: the class $\mathcal{D}$ of all such is evidently denumerable, but one can then define a set S of numbers not in $\mathcal{D}$ (by diagonalization), so that $\mathcal{D}$ does *not* consist of all definable sets after all. Weyl's way out is to note that under his precise explanation of definable set, the enumeration of $\mathcal{D}$ is given externally, but cannot be given by a set in $\mathcal{D}$ itself; thus S is not definable in the precise initial sense used to describe $\mathcal{D}$. In other words, there is a shift in the meaning of "definability" in the course of the argument for the paradox. However, Weyl does not face up to the fact that the enumeration of $\mathcal{D}$ *ought* to be accepted as a legitimate definition once $\mathcal{D}$ is accepted, and that this is perfectly reasonable under a definitionist philosophy, even if it means that there is no one privileged concept of definability. Lying behind this is his decision to deal only with definable sets of level 0 (first level) in order to avoid the troubles of ramification.

That issue is taken up next in I.6; Weyl there posits a general "operation domain" consisting of one or more basic categories of objects together with basic properties and relations. The category of natural numbers with the successor relation F is called the *absolute* operation domain and is always assumed to be included in the given operation domain. Now by the "mathematical process" of I.4 one can form a new object category consisting of the definable sets (of any number of arguments), that is, the sets of first level. By iterating the mathematical process we may move on to introduce sets of the second and higher levels. One is then tempted to ignore the level distinctions, that is, (essentially) to assume the Axiom of Reducibility. But Weyl points out that this will lead to impredicative definitions in analy-

[21]The usual axiom that the successor operation is one-to-one is not mentioned.

sis, and hence to a violation of the vicious circle principle (as explained in section 3 above). On the other hand, he says that analysis *with* level distinctions is artificial and unworkable ("künstlich und unbrauchbar," p. 23). Weyl concludes that the only natural procedure is to restrict attention to definable sets of the first level, even if this means (as we saw in section 3) giving up the general l.u.b. principle for the real number system.

The second half of I.6 is devoted to the question of how functions of real numbers are to be treated under the identification of the latter with Dedekind sections, which are simply certain kinds of sets of rationals. More generally, given a basic domain D, how is one to define functions from individuals $x, y, \ldots$ and/or sets $X, Y, \ldots$ in D to new such sets $U, V, \ldots$? The answer is simply that the set which is to be the function value has to be defined explicitly by some property or relation in the function arguments as variables. For example, if we want to define a function $U = \Phi(x)$, we must provide a relation $R_\Phi(u, x)$ with $U = \{u : R_\Phi(u, x)\}$. If U is to be a function of a set X, $U = \Psi(X)$, then we must provide a relation $R_\Psi(u, X)$ with $U = \{u : R_\Psi(u, X)\}$. In the latter case R_Ψ must be built by the basic definition principles (1)–(6) of I.2, but where now the membership relation $x \in X$ is included among the basic relations; it is essential that in the buildup of R_Ψ, quantification is still applied only to variables ranging over one of the original basic domains and not to set variables, that is, not over the new "ideal" domain of sets, for otherwise one is led to impredicative definitions. To indicate the distinction between "dependent" and "independent" variables, Weyl separates them by a vertical bar. Thus in general,

$$R_\Phi(u_1, \ldots, u_k \mid x_1, \ldots, x_\ell, X_1, \ldots, X_m)$$

denotes a definable relation in all the displayed variables which is used to represent a function $U = \Phi(x_1, \ldots, x_\ell, X_1, \ldots, X_m)$ from individuals $x_1, \ldots, x_\ell$ and (possibly multi-)sets $X_1, \ldots, X_m$ to a k-ary set U given by

$$U = \{(u_1, \ldots, u_k) : R_\Phi(u_1, \ldots, u_k \mid x_1, \ldots, x_\ell, X_1, \ldots, X_m)\}.$$

Thus the variables $u_1, \ldots, u_k$ serve to mediate between the independent variables $x_1, \ldots, x_\ell, X_1, \ldots, X_m$ and the actual dependent variable U. Now it is easily seen that if each X_i is definable in the original sense over the operation domain then so also is U, simply by substituting the defining properties for the X_i in R_Φ. In other words, $\mathcal{D}$ is closed under this kind of functional definition.

The closure of set functions under *substitution* and *iteration* is then taken up in Weyl's framework in I.7. For substitution the matter is quite straightforward: say, given set-valued functions $\Phi(x)$ and $\Psi(x, X)$, we can then obtain $\Theta(x) = \Psi(x, \Phi(x))$ taking

$$R_\Theta(u|x) = R_\Psi(u|x, \hat{v} R_\Phi(v|x)),$$

which indicates that in $R_\Psi(u|x, X)$ we substitute for each occurrence $v \in X$ the relation $R_\Phi(v|x)$. We can similarly define $\Psi(x, \Phi(x))$ from $R_\Phi(v|X)$

and $R_\Psi(u|x, X)$. The principle of substitution for set-valued functions is numbered (7), in continuation of the previous list (1)–(6) of definition principles.

Matters are not so clear with Weyl's *Principle of Iteration* (8). For one thing, this is presented in three different forms, and it is not explained how to subsume them under a single form or principle of definition for functions. More seriously, as we shall see, even accepting the first (also the simplest) of these principles conflicts with Weyl's decision to stick to sets of lowest level in the ramified hierarchy. This first form serves to define a function $\Phi^*(n, X)$ from given $\Phi(X)$ by[22]

$$\text{(8a)} \qquad \begin{cases} \Phi^*(1, X) & = \Phi(X) \\ \Phi^*(n+1, X) & = \Phi^*(n, \Phi(X)). \end{cases}$$

Thus $\Phi^*(n, X) = \Phi^n(X)$, so that (8a) gives iteration in the usual sense—though ordinarily one defines $\Phi^*(n+1, X)$ instead as $\Phi(\Phi^*(n, X))$. The second form of "iteration" Weyl accepts is the extension of (8a) to definition by simultaneous recursion, as follows, where $\Phi(X, Y)$ and $\Psi(X, Y)$ are given functions and $\Phi^*(n, X, Y), \Psi^*(n, X, Y)$ are determined by

$$\text{(8b)} \begin{cases} \Phi^*(1, X, Y) & = \Phi(X, Y) \\ \Phi^*(n+1, X, Y) & = \Phi^*(n, \Phi(X, Y), \Psi(X, Y)) \\ \Psi^*(1, X, Y) & = \Psi(X, Y) \\ \Psi^*(n+1, X, Y) & = \Psi^*(n, \Phi(X, Y), \Psi(X, Y)). \end{cases}$$

The third form is obtained by a modification of (8a) for the "iteration" of a function $\Phi(n, X)$; this is given by

$$\text{(8c)} \qquad \begin{cases} \Phi^*(1, X) & = \Phi(1, X) \\ \Phi^*(n+1, X) & = \Phi^*(n, \Phi(n+1, X)). \end{cases}$$

In this case we have, for example, $\Phi^*(3, X) = \Phi(1, \Phi(2, \Phi(3, X)))$. This should be contrasted with the usual form of primitive recursion where, given Ψ and Φ, we may define $\Theta(n, X)$ by

$$\text{(8c)}' \qquad \begin{cases} \Theta(1, X) & = \Psi(X) \\ \Theta(n+1, X) & = \Phi(n+1, \Theta(n, X)). \end{cases}$$

In particular, if one takes $\Psi(X) = \Phi(1, X)$ we have $\Theta(3, X) = \Phi(3, \Phi(2, \Phi(1, X)))$. It is not difficult, though, to show in the presence of other principles of explicit definition and primitive recursive definition for

[22]Here and in (8b), (8c), I am rewriting Weyl's relational description of functions following the explanations above. In (8a), X and $\Phi(X)$ should be multisets of the same arity; similar conditions apply to (8b), (8c) below.

$\mathbf{N}^+$-(natural-number) valued functions that each of $(8c), (8c)'$ can be derived from the other.

Curiously, Weyl does not give a direct treatment of recursively defined $\mathbf{N}^+$-valued functions, but extracts them from set-valued recursion in chapter II; this is described below. But for the time being we shall assume that all primitive recursive functions and relations on $\mathbf{N}^+ = \{1, 2, 3 \dots\}$ can be obtained by Weyl's principles. It is well known from the work of Kleene that there is a function $\Phi(X) = X'$, called the *jump* of X such that (i) Φ is definable from a relation of the form $R_\Phi(m|X) = \exists k T(m, k, X)$, where T is a certain primitive recursive relation (in X); (ii) every predicate of the form $\exists k S(m, k, X)$ with S primitive recursive (in X) is primitive recursively definable from $\Phi(X)$; and (iii) moreover, the same holds for every predicate of the form $\forall k S'(m, k, X)$. Note that (iii) follows from (ii) simply by closure of primitive recursive relations under negation. Now it follows from (i)–(iii) that for each relation $U(m, X)$ of the form

$$U(m, X) = (Q_1 k_1)(Q_2 k_2) \dots (Q_n k_n) S(m, k_1, \dots, k_n, X)$$

with each Q_i an existential or universal quantifier, we have U primitive recursive in $\Phi^n(X)$. It follows that every relation arithmetical in X is primitive recursive in the function Φ^* given by (8a), where $\Phi^*(n, X) = \Phi^n(X)$. Hence the relation

$$R_{\Phi^*}(p|n, X)$$

is not arithmetical in X by the usual diagonalization argument.[23] In particular, if we start with $X = \mathbf{N}$, this gives a relation definable by the iteration principle (8) which is not arithmetical, hence not definable at level 0. This shows that Weyl's principle (8) is in conflict with his plan to work only with sets definable at the lowest level. Clearly, Weyl was not aware of this problem with his principles.[24]

We now have two options in the reconstruction of Weyl's program: either (α) restrict the iteration principles for ($\mathbf{N}$-valued and) set-valued functions in such a way that they do not lead out of level 0 definability, or (β) accept the iteration principles in full, and show that they are predicatively justified. In case (α), one must show that the development of analysis in Weyl's program can still be carried out under the restricted principles. In fact that *is* the case, and hence we shall concentrate our attention on option (α) in the remainder of this chapter; however, (β) is also a reasonable option, as will be explained later on.

Weyl's chapter I concludes in I.8 with a recapitulation of the leading ideas, summary of the defining principles (1)–(8) for relations and functions, and reassertion of the necessity on philosophical grounds to pursue

---

[23] For details, cf., for example, Rogers (1967), chapters 13, 14.

[24] The prima facie incoherence of Weyl's principles at this juncture was also observed by Giuseppe Longo, in personal conversation.

restricted predicative foundation of analysis (1918, pp. 34–38). Here we find him reiterating his rejection of the Dedekind-Cantor-Zermelo (philosophically platonistic) program of set-theoretical foundations,[25] and his agreement with Poincaré on the opposed need for definitionist (predicative) foundations, with the vicious circle principle as a crucial test for acceptability of proper definition principles. In this respect, Russell's ad hoc assumption of the Axiom of Reducibility is seen as a violation of the vicious circle principle, despite the predicative formalism of ramified type theory, "welche Kluft mich trotz allem noch von Russell trennt," and thus the necessity for Weyl to follow the "engeren Verfahren" of restricting to sets of level 0. Finally, Weyl allies himself with Poincaré on the primacy of the natural number concept for abstract mathematical thought: "die *Vorstellung der Iteration, der naturlichen Zahlenreihe, ein letztes Fundament des mathematischen Denkens* ist" (italics in the original; 1918, p. 37).

Chapter II of *Das Kontinuum* is devoted to setting up the real number system (the continuum) and the initial development of the foundations of analysis on the basis of the principles in chapter I. We shall be much briefer here, since this part is superseded by later developments of arithmetical mathematics as will be described below.

The operations $+$ and $\cdot$ on natural numbers are defined by iteration of set functions as follows. For X a set of pairs of natural numbers, let $R(p, m|X)$ be the relation $\exists q[(q, m) \in X \wedge F(q, p)]$, where $F(q, p)$ means $p = q + 1$. Then take $\Phi(X) = \{(p, m) : R(p, m|X)\}$. Thus

$$\Phi^n(X) = \{(p, m) : \exists q[(q, m) \in X \wedge p = q + n]\}.$$

In particular, if I is the equality relation on N we have

$$\Phi^n(I) = \{(p, m) : p = m + n\}.$$

Hence we can define the relation $m + n = p$ by $(p, m) \in \Phi^n(I)$. The relation $m \cdot n = p$ is defined similarly. Basic properties of $+, \cdot$, and $<$ are then derived. Finally II.1 finished with a treatment of cardinal numbers of finite subsets X of an arbitrary set D, via the recursively defined relations $\bar{\bar{X}} \geq n$.

Fractions m/n for $m, n \in \mathbf{N}$ are treated in II.2 as pairs with the usual equivalence relation, so that m/n is sometimes regarded as a set α of pairs (m', n') or as a representative pair (m, n) in α. In order to close under subtraction, the rational numbers are defined as pairs (α, β) of fractions, representing $(\alpha - \beta)$ again with the usual equivalence relation.[26] Thus a rational number is treated either as a set of pairs of sets (α, β), or as a set of 4-tuples $(m, n; p, q)$ (representing $m/n - p/q$), or as a representative such 4-tuple.

[25] Though Weyl remarks that he began his foundational work in 1910 with the clarification of Zermelo's "definite properties" in the formulation of the *Aussonderungsaxiom*.

[26] Warning: Weyl writes $\alpha \div \beta$ for $\alpha - \beta$.

Next, the real numbers are introduced in II.3 as lower Dedekind sections in the set of rational numbers, and hence may be treated as sets X of 4-tuples $(m_1, n_1; m_2, n_2)$ in N. In general, then, functions on and to the real numbers get treated as functions on and to certain sets of 4-tuples in N. Thus for example, the operation $X + Y$ is given as the function

$$\Phi(X,Y) = \{(m_1, n_1; m_2, n_2) : S(m_1, n_1; m_2, n_2 | X, Y)\},$$

where $S(m_1, n_1; m_2, n_2 | X, Y)$ holds if and only if there exists $(p_1, q_1; p_2, q_2)$ in X and $(r_1, s_1; r_2, s_2)$ in Y such that

$$\left(\frac{m_1}{n_1} - \frac{m_2}{n_2}\right) = \left(\frac{p_1}{q_1} - \frac{p_2}{q_2}\right) + \left(\frac{r_1}{s_1} - \frac{r_2}{s_2}\right).$$

This eventually gets reduced to an arithmetical relation over N. The operation of multiplication of reals is dealt with in a similar fashion. Section II.3 concludes with a sketchy treatment of real polynomials, algebraic numbers, and complex numbers.

We shall here break off following Weyl further step by step. The topics and results he takes up in succession starting with II.4 are (1) sequences of real numbers, (2) the Cauchy convergence principle, (3) the l.u.b. and g.l.b. principle for bounded sequences of reals, (4) the Heine-Borel theorem for coverings of $[0, 1]$ by sequences of intervals, (5) infinite series and power series, (6) the elementary functions (defined by series), (7) continuous functions, (8) the mean-value theorem for such, (9) attainment of maxima and minima for continuous functions on a closed interval, (10) uniform continuity of such, and (11) differentiation and integration of continuous functions. It is asserted that the l.u.b. principle for *sets* of reals fails, as does the Heine-Borel theorem for covering *sets* of intervals, but no actual counterexamples are given.[27] Finally, it is stated that the extension of integration to wider classes of functions via measure theory comes out less simply, but no indications are given as to how this is to be approached, if at all (1918, p. 65). After an informal discussion of the intuitive continuum versus the mathematical continuum, and of measurement, the monograph concludes with a quick sketch of the treatment of curves and surfaces.

Weyl's last words in *Das Kontinuum* are:

> Hier brechen wir unsere Entwicklung ab. Wir haben gesehen, dass sich auf unseren Prinzipien sehr wohl eine Analysis aufbauen lässt und haben diesen Aufbau in seinen ersten Stadien vollzogen . . . jene abstrakten Schemata, welche uns die Mathematik liefert, [sind] erforderlich, um exakte Wissenschaft solcher Gegenstandsgebiete zu ermöglichen, in denen Kontinua eine Rolle spielen.

[27] Such would required *nondefinability* results, thus going outside of Weyl's framework.

In other words: arithmetical (predicative) analysis suffices for scientific applications. Weyl must have been especially aware of the need for his theory to meet at least this requirement, since in the same year he published his famous exposition of relativity theory, *Raum, Zeit, Materie* (1918a).

6 The Axiom Systems of *Das Kontinuum* Reconstructed

In this section we begin the process of reworking the axiom system of *Das Kontinuum* in terms which meet the modern requirements of formalization while remaining faithful as far as possible to the underlying ideas. As we have seen, one must deal with the basic problem met in I.7 of Weyl's monograph that acceptance of the principle of iteration of set-valued functions leads one outside the intended arithmetical (level 0) interpretation, though not outside the more general predicative interpretation. That suggested two options (α) and (β); under the latter one would incorporate all the definition principles in question, while under the former one would restrict them so as to be able to retain the strictly arithmetical interpretation, (hopefully) without losing the applications to mathematical analysis of Weyl's chapter II. In this section I shall formulate a theory $\mathrm{K}^{(\alpha)}$ which follows option (α). Then in the next section, I shall describe a much more flexible extension W of $\mathrm{K}^{(\alpha)}$ which maintains the arithmetical interpretation, but in which a much greater body of analysis and other parts of mathematics can be formalized more readily. Finally, it will be indicated there how a theory $\mathrm{K}^{(\beta)}$ following option (β) can be embedded in a simple extension of W. The theory $\mathrm{K}^{(\alpha)}$ is second order, while $\mathrm{K}^{(\beta)}$ is third order; W itself permits one to deal with objects of all finite orders.

The language $\mathcal{L}(\mathrm{K}^{(\alpha)})$ has as its first-order variables the *numerical variables* $n, m, p, q, \ldots$, and as its second-order variables, *relation variables* $a, b, c, \ldots$ and *function variables* $f, g, h, \ldots$ of any number of arguments. There are also *relation constants* and *function constants* of various numbers of arguments, to be specified. The intended range of the numerical variables is the set $\mathbf{N}$ of natural numbers, here taken to include zero; there is thus also a numerical constant symbol 0. The successor function is also represented by a function symbol *sc*. As explained above, Weyl defined in II.1 the specific numerical operations $+$ and $\cdot$ in terms of suitable third-order relations obtained by iteration. Since that is not available in $\mathrm{K}^{(\alpha)}$, we shall introduce $+$ and $\cdot$, and in fact all further primitive recursive operations, directly in $\mathrm{K}^{(\alpha)}$ as function constants with recursive defining conditions as axioms. This simply incorporates into $\mathrm{K}^{(\alpha)}$ the principle of iteration for numerical-valued functions rather than set-valued functions. Relation constants will also be introduced to represent primitive recursive relations. However, the axioms will allow us in addition to assert the existence of relations and functions defined more generally by arbitrary arithmetical conditions, and indeed their closure under such conditions.

The (numerical) *terms* $t, t_1, \ldots$ of $\mathcal{L}(\mathrm{K}^{(\alpha)})$ are generated as follows:

(i) each numerical variable is a term;

(ii) 0 is a term;

(iii) if f is a k-ary function variable or constant and $t_1, \ldots, t_k$ are terms then $f(t_1, \ldots, t_k)$ is a term.

The set of *variables* of a term is defined as usual; then $t[n_1, \ldots, n_k, f_1, \ldots, f_\ell]$ represents a term t all of whose variables are among those shown.

The *atomic formulas* are those of the form $t_1 = t_2$ and $a(t_1, \ldots, t_k)$ where a is a k-ary relation variable or constant; to avoid confusion, we write $(t_1, \ldots, t_k) \in a$ for $a(t_1, \ldots, t_k)$.

The *formulas* $\phi, \psi, \theta, \ldots$ of $\mathcal{L}(\mathrm{K}^{(\alpha)})$ are generated as follows:

(i) every atomic formula is a formula;

(ii) if ϕ, ψ are formulas then so also are $\neg\phi$ and $\phi \vee \psi$;

(iii) if ϕ is a formula then so also are $\exists n\phi, \exists a\phi$ and $\exists f\phi$, where "n," "a," "f" are respectively any numerical, relation, or function variable.

The *free variables* of a formula are defined as usual; a *closed formula* is one without any free variables. $\phi[n_1, \ldots, n_k, a_1, \ldots, a_j, f_1 \ldots, f_\ell]$ represents a formula all of whose free variables are among those displayed. To simplify notation, we shall use underlined letters $\underline{n}, \underline{a}, \underline{f}$ to denote the lists of variable in ϕ and then write $\phi[\underline{n}, \underline{a}, \underline{f}]$ for ϕ. Variants of this notation below will be obvious.

The logical operations $\wedge, \Rightarrow, \Leftrightarrow, \forall$ are defined by $\phi \wedge \psi = \neg(\neg\phi \vee \neg\psi), (\phi \Rightarrow \psi) = (\neg\phi \vee \psi), (\phi \Leftrightarrow \psi) = (\phi \Rightarrow \psi) \wedge (\psi \Rightarrow \phi), \forall n\phi = \neg(\exists n)\neg\phi$, and similarly for $\forall a\phi, \forall f\phi$.

It is not clear what position Weyl would take on quantification over relations and functions as part of the language, since they are certainly to be excluded in defining conditions of relations and functions. A formula ϕ is said to *elementary*, or *arithmetical*, if it contains no bound second-order (relation or function) variables, though it may contain such variables in free positions.

The *logic of* $\mathrm{K}^{(\alpha)}$ is the *classical* predicate calculus for a many-sorted language, with usual axioms and rules of inference, including those for $=$ between numerical terms. The reader may take his favorite system for this.

The (nonlogical) *axioms of* $\mathrm{K}^{(\alpha)}$ are as follows, where we write t' for $sc(t)$.

I. *Zero and successor.*

(i) $n' \neq 0$

(ii) $n' = m' \Rightarrow n = m$.

II. *Induction* $\quad 0 \in a \wedge \forall n(n \in a \Rightarrow n' \in a) \Rightarrow \forall n(n \in a)$.

III. *Explicit function definition* (λ-*abstraction*). For each term $t[\underline{n}]$ there is a constant function symbol f(denoted $\lambda\underline{n} \cdot t[\underline{n}]$) with the axiom

$$f(\underline{n}) = t[\underline{n}].$$

IV. *Definition by primitive recursion.* For each k-ary constant g and $(k+2)$-ary constant h there is a $(k+1)$-ary constant function symbol f(denoted $Rc_{\mathbf{N}}(g,h)$) with the axioms

$$f(0, \underline{m}) = g(\underline{m})$$
$$f(n', \underline{m}) = h(n, \underline{m}, f(n, \underline{m})).$$

V. *Explicit relation definition.* For each function constant f we have a relation constant a with axiom

$$\forall \underline{n}(\underline{n} \in a \Leftrightarrow f(\underline{n}) = 0).$$

VI. *Elementary comprehension axioms.*

(i) For each arithmetical formula $\phi[\underline{n}, \underline{m}, \underline{b}, \underline{g}]$ we have the axiom

$$\exists a \forall \underline{n}\{\underline{n} \in a \Leftrightarrow \phi[\underline{n}, \underline{m}, \underline{b}, \underline{g}]\}.$$

(ii) For each arithmetical $\phi[\underline{n}, p, \underline{m}, \underline{b}, \underline{g}]$ we have the axiom

$$\forall \underline{n} \exists ! p \phi[\underline{n}, p, \underline{m}, \underline{b}, \underline{g}] \Rightarrow \exists f \forall \underline{n}, p\{f(\underline{n}) = p \Leftrightarrow \phi[\underline{n}, p, \underline{m}, \underline{b}, \underline{g}]\}.$$

This completes the description of the axiom system $\mathrm{K}^{(\alpha)}$.

Remarks

1. It is not clear whether Weyl would have admitted the *full scheme of induction*

$$\phi[0] \wedge \forall n(\phi[n] \Rightarrow \phi[n']) \Rightarrow \forall n \phi[n]$$

where $\phi[n]$ (or $\phi[n, \ldots]$) is an *arbitrary* formula of $\mathcal{L}(\mathrm{K}^{(\alpha)})$. This would not change having an arithmetical model of the theory, but it would affect proof-theoretical strength drastically (see below).

2. We follow standard current terminology in referring to axioms such as VI(i) and (ii) as *comprehension axioms* (CA) for relations and functions.

3. From VI(i) we can obtain $\forall f \exists a \forall \underline{n}[\underline{n} \in a \Leftrightarrow f(\underline{n}) = 0]$, and in this sense relations are dispensable in terms of (characteristic) functions.

4. From VI(ii) we can obtain

$$\forall a\{\forall \underline{n}\exists! p(\underline{n}, p) \in a\} \Rightarrow \exists f \forall \underline{n}, p\{f(\underline{n}) = p \Leftrightarrow (\underline{n}, p) \in a\},$$

   and in this sense functions are dispensable in terms of their graphs as relations.

5. All primitive recursive functions are introduced successively in $K^{(\alpha)}$ using λ-abstraction (III) and definition by primitive recursion (IV). Then the primitive recursive relations are obtained from V.

6. Equality of relations of the same number of arguments is defined by

$$a_1 = a_2 \Leftrightarrow \forall \underline{n}(\underline{n} \in a_1 \Leftrightarrow \underline{n} \in a_2).$$

   Then the relation a asserted to exist in VI (i) is unique up to equality; this relation is denoted $\{\underline{n} : \phi[\underline{n}, \underline{m}, \underline{b}, \underline{g}]\}$.

7. Equality of functions of the same number of arguments is defined by

$$f_1 = f_2 \Leftrightarrow \forall \underline{n}(f_1(\underline{n}) = f_2(\underline{n})).$$

8. We do not introduce a symbol for the function f asserted to exist in VI(ii), though it is also unique up to equality.

It is familiar that we can define primitive recursive pairing functions and their inverses, and so we can represent k-tuples of numbers by individual numbers. Then we need only take a comprehension axiom for sets in place of VI(i). If we follow this procedure and that of Remark 4 above, then we can simplify the system $K^{(\alpha)}$ to the following: the only *second-order variables* are *set variables* (unary relations), and the axiom VI is replaced by

$(\star)$ $$\exists a \forall n(n \in a \Leftrightarrow \phi[n, \underline{m}, \underline{b}]).$$ [28]

Then function variables are introduced by convention to range over graphs of many-one relations (as in Remark 4). The resulting simplified version of $K^{(\alpha)}$ is denoted variously in the literature as Restricted-(Π^0_∞-CA)$\restriction$ or (Π^0_∞-CA)$\restriction$ or $(\mathrm{ACA})_0$. The principle $(\star)$ itself is called the *Arithmetical Comprehension Axiom*, denoted (ACA) or (Π^0_∞-CA), where Π^0_∞ denotes the class of arithmetical formulas.[29] The modifiers "restricted" or "$\restriction$" or subscript "0" are used to indicate that the induction principle is taken in its form II as a single axiom for sets, and not as the full scheme in remark 1 above. In the following, "Restricted" is abbreviated to "Res."

[28] However, we still retain the primitive recursive function and relation constants introduced by axioms III–V.

[29] Since one can obtain $(\star)$ from a weaker principle (Π^0_1-CA), these systems are also denoted (Res-Π^0_1-CA) or (Π^0_1- CA)$\restriction$.

7 Strength and Limitations of Weyl's System(s)

The first-order consequences of $\mathrm{K}^{(\alpha)}$ can be axiomatized as follows: drop the comprehension axioms VI, and replace the induction axiom II by the full first-order induction scheme

$$\phi[0, \underline{m}] \wedge \forall n\{\phi[n, \underline{m}] \Rightarrow \phi[n', \underline{m}]\} \Rightarrow \forall n \phi[n, \underline{m}],$$

that is, where ϕ is any formula without second-order variables (free or bound). The resulting system is a form of first-order *Peano Arithmetic* (PA). Clearly PA is contained in $\mathrm{K}^{(\alpha)}$. We are thus asserting the following:

Metatheorem. *$K^{(\alpha)}$ is a conservative extension of PA; that is, if $K^{(\alpha)}$ proves a first-order formula ϕ, then PA proves ϕ.*

Proof. There is a well-known simple standard model-theoretic argument to show that Res-(Π°_{∞}-CA) is conservative over PA, by Gödel's completeness theorem for the (many-sorted) predicate calculus (see, for example, Feferman 1977, §5.5, p. 940). Namely, if ϕ is a closed first-order formula not provable from PA, then PA + $\{\neg\phi\}$ has a model $\mathcal{M} = \langle M, 0, sc, \ldots\rangle$. This is expanded to a model $\mathcal{M}^*$ of Res-(Π°_{∞}-CA) + $\{\neg\phi\}$ by taking the set variables to range over all subsets of M definable in $\mathcal{M}$ (by a first-order formula) from parameters in M. For a finitistic proof-theoretic argument, see, for example, Feferman and Sieg (1981, p. 112). Both proofs are readily extended to $\mathrm{K}^{(\alpha)}$ in place of Res-(Π^0_∞-CA).

Note that the model-theoretic argument shows that $\mathrm{K}^{(\alpha)}$ has a model in which the relation (function) variables range over the arithmetically definable relations (functions) in the standard structure $\mathcal{N} = \langle \mathbf{N}, 0, sc, \ldots\rangle$. But the same holds true for $\mathrm{K}^{(\alpha)}$ + Full Induction Scheme, which is no longer conservative over PA (cf. for example, Feferman 1977, §6.3.3, pp. 946–947). Thus, the above conservation result provides a much stronger sense in which $\mathrm{K}^{(\alpha)}$ has an arithmetical interpretation.

One often reads that the axiom system on which *Das Kontinuum* is based is that of Res-(Π^0_∞-CA), or an equivalent version (such as $\mathrm{K}^{(\alpha)}$). Indeed, this is what I had said on various occasions until looking more closely at the monograph for the purposes of the present essay. It is more accurate to say that it formalizes that part of Weyl's system which meets his aim of a purely arithmetical interpretation, though the system as a whole (reconstructed in something like $\mathrm{K}^{(\beta)}$, below) is stronger than Res-(Π^0_∞-CA).

In addition, Weyl's development of analysis in his chapter II appears to require means going beyond $\mathrm{K}^{(\alpha)}$ or Res-(Π^0_∞-CA), since we do not have set functions as objects of discourse. However, the same part of analysis as treated by Weyl *can* be developed in this restricted system by some simple modifications. First of all, the introduction of the rationals ($\mathbf{Q}$), and then of the real numbers ($\mathbf{R}$) as lower Dedekind sections in $\mathbf{Q}$, goes much as before. Let $\mathbf{Q} = \{r_n : n \in \mathbf{N}\}$ be a standard enumeration of the rationals. Using this, properties of subsets b of $\mathbf{Q}$ are reduced directly to properties of the corresponding sets $a = \tilde{b}$ of $\mathbf{N}$, where $n \in a \Leftrightarrow r_n \in b$. Now sequences $\langle a_n \rangle_{n \in \mathbf{N}}$ of subsets a_n of $\mathbf{N}$ are treated as binary relations $a \subseteq \mathbf{N}^2$ with $m \in a_n \Leftrightarrow (m,n) \in a$. Thus, for example, to prove the g.l.b. property for sequences $b = \langle b_n \rangle_{n \in \mathbf{N}}$ of real numbers, we simply want to establish the existence of $\bigcap b_n[n \in \mathbf{N}]$, and hence of $\bigcap a_n[n \in \mathbf{N}]$ for $a_n = \tilde{b}_n$. This is given by

$$m \in \bigcap a_n[n \in \mathbf{N}] \Leftrightarrow \forall n[(m,n) \in a],$$

where $a = \langle a_n \rangle_{n \in \mathbf{N}}$, and so the g.l.b. of the sequence exists by the arithmetical comprehension axiom. In much the same way we recover the other forms of sequential completeness and compactness for $\mathbf{R}$, obtained by Weyl in his II.4. It is only when we pass to functions $F : \mathbf{R} \to \mathbf{R}$ of real numbers that something different must be done, since this requires talking about set-valued functions $F(a) = b$ for (Dedekind) sets $a, b \subseteq \mathbf{Q}$. However, there is no problem if we restrict attention to *continuous functions* F, since such functions are determined by their values on $\mathbf{Q}$. One first considers $F_1 = F \restriction \mathbf{Q} : \mathbf{Q} \to \mathbf{R}$ and then passes to $F_1^* = \{(n,m) : r_m \in F_1(r_n)\}$, a binary relation in $\mathbf{N}$. Hence sets F_1^* of pairs in $\mathbf{N}$ can be used as surrogates of continuous functions F on $\mathbf{R}$. With this modification, the rest of the development of analysis sketched by Weyl can be carried out just as well in $\mathrm{K}^{(\alpha)}$ as in $\mathrm{K}^{(\beta)}$. This includes the theory of differentiation and integration of continuous functions together with applications of the theory to the classical functions given by power series representations. In this way, the option (α) can be completely realized; that is, not only does the restricted system $\mathrm{K}^{(\alpha)}$ have a direct arithmetical interpretation (both semantically and proof-theoretically), but it also serves to account for the same body of classical mathematics as claimed by Weyl for his full system in *Das Kontinuum.*

The real limitations of $\mathrm{K}^{(\alpha)}$ emerge when we want to treat much wider classes of functions including very discontinuous ones such as those measurable in the sense of Lebesgue, and when we want to pass on to modern (functional) analysis as exemplified by the use of various function spaces and (functional) operators on them. These are limitations just as well of $\mathrm{K}^{(\beta)}$ for we should want to deal at least with subclasses X of the class of "all" functions $f : \mathbf{R} \to \mathbf{R}$, and then functionals $F : X \to \mathbf{R}$, which takes us to the fourth order in type theory (the real numbers being second order, real functions third order, and functionals fourth order). But in the spirit of modern abstract analysis, we should want to be able to form more generally for any two spaces X, Y a space of "all" functions $f : X \to Y$. Doing so requires an axiomatic theory for all finite types.

The question then is whether we can develop a theory of finite types which has an arithmetical interpretation, if not directly then at least proof-theoretically (by being conservative over PA), and in which much of modern analysis can also be formalized. First examples of such theories were provided by Feferman (1977, §8.7) and Takeuti (1978). But in these theories, the types $A, B, \ldots$ are syntactically fixed, whereas for abstract analysis one wants to treat them as *variable types* (or *classes*) $X, Y, \ldots$. Moreover, it is natural to demand that with each class X and suitable property ϕ we should be able to form the subclass $\{x \in X | \phi[x, \ldots]\}$. In usual type theories such as those mentioned, it is not possible to form subtypes directly, and it becomes very awkward to account for their use. In the next section I present a theory of "variable" finite types, conservative over PA, which does have these kinds of flexibility and in which a substantial portion of classical and modern analysis can be directly formalized.

8 A Theory of Flexible Finite Types for Weyl's Program

The theory W presented in this section is a simplification and still more flexible version of the system of variable finite types introduced in Feferman (1985).[30] It is a two-sorted theory, where now the individual variables $a, b, c, \ldots, x, y, z$ range over a universe containing numbers, sets, functions, functionals, etc., and the variables $A, B, C, \ldots, X, Y, Z$ range over *types*, or *classes* as we shall call them here. Function objects are in general partial, and so for convenience we make use of a logic in which terms may be partially defined.

The *individual terms* $(s, t, \ldots)$ of $\mathcal{L}(\mathrm{W})$ are generated as follows:

(i) each individual variable is an individual term;

(ii) the constants $0, sc, \mu$ are individual terms;

(iii) if s, t, s_1, t_1 are individual terms then so also are $\langle s, t\rangle$, $P_1(s), P_2(s), D(s_1, s_2, t_1, t_2), Rc_{\mathbf{N}}(s, t)$, and $s(t)$;

(iv) if t is an individual term and S is a class term then $(\lambda x \in S)t$ is an individual term.

The *class terms* $(S, T, \ldots)$ of $\mathcal{L}(\mathrm{W})$ are generated as follows:

(i) each class variable is a class term;

(ii) the constant $\mathbf{N}$ is a class term;

(iii) if S, T are class terms then so also are $S \times T$ and $S \overset{\sim}{\rightarrow} T$;

(iv) if S is a class term and ϕ is a formula then $\{x \in S | \phi\}$ is a class term.

[30] The adjective "flexible" in place of "variable" was suggested to me by Gerhard Jäger. The designation "W" for this system is in honor of Weyl.

The *atomic formulas* are those of the form $s = t, t\downarrow$ and $t \in S$, where s, t are individual terms and S is a class term.

The *formulas* $(\phi, \psi, \ldots)$ of $\mathcal{L}(W)$ are generated as follows:

(i) every atomic formula is a formula;

(ii) if ϕ, ψ are formulas then so also are $\neg\phi$ and $(\phi \vee \psi)$;

(iii) if ϕ is a formula then so also are $(\exists x)\phi$ and $(\exists X)\phi$ where "x" ("X") is an individual (class) variable.

Note that the above is a simultaneous inductive definition of the individual terms, class terms, and formulas of $\mathcal{L}$(W).

The remaining logical operations are defined by $(\phi \wedge \psi) = \neg(\neg\phi \vee \neg\psi), (\phi \Rightarrow \psi) = (\neg\phi \vee \psi), (\phi \Leftrightarrow \psi) = (\phi \Rightarrow \psi) \wedge (\psi \Rightarrow \phi), (\forall x)\phi = \neg(\exists x)\neg\phi$, and $(\forall X)\phi = \neg(\forall X)\neg\phi$.[31] We also put $(\exists x \in S)\phi = (\exists x)(x \in S \wedge \phi)$ and $(\forall x \in S)\phi = \forall x(x \in S \Rightarrow \phi)$ where "x" is not free in S; these are called *bounded quantifiers*. The formulas built up from atomic formulas by $\neg, \wedge, \vee, \Rightarrow$ and only bounded quantifiers $(\forall x \in S)(\ldots), (\exists x \in S)(\ldots)$ are called the *bounded predicative formulas*.[32] We shall make use of the subclass construction operation $\{x \in S|\phi\}$ *only* in the case that ϕ is such a formula.

In the axioms, we shall reserve "m," "n," "p," (with or with subscripts) as variables ranging over $\mathbf{N}$; that is, for example $(\exists n)\phi$ is written for $(\exists n \in \mathbf{N})\phi$ and $(\forall n)\phi$ for $(\forall n \in \mathbf{N})\phi$. We put $n' = sc(n)$ and $1 = 0'$; then we take $\{0, 1\} = \{x \in \mathbf{N}|x = 0 \vee x = 1\}$. The variables "$f$," "$g$," "$h$" will be used primarily to represent partial functions, and $f(x)\downarrow$ means that x is in the domain of f. Then $(S \tilde{\rightarrow} T)$ will represent the class of partial functions which satisfy $(\forall x)(x \in S \wedge f(x)\downarrow \Rightarrow f(x) \in T)$. Thus we define $(S \rightarrow T) = \{f \in (S \tilde{\rightarrow} T)|(\forall x \in S)f(x)\downarrow\}$, and $\mathcal{P}(S) = (S \rightarrow \{0, 1\})$. In other words, $\mathcal{P}(S)$ is the class of *characteristic functions on* S, which we shall also call the *class of subsets* of S. We shall use the variables "a," "b," "c," . . . primarily to range over *subsets* of a given class. These must be distinguished from the *subclasses* (or *subtypes*) of a given class: we write $S' \subseteq S$ if $\forall x(x \in S' \Rightarrow x \in S)$; subclasses need not have characteristic functions. Finally, we write $(x \in a)$ for $a(x) = 0$ when $a \in \mathcal{P}(S)$ and $x \in S$; actually, we should use a different membership symbol for sets, but there will be no confusion in using the same "$\in$."

[31] These definitions are justified by the assumption of classical logic in W. Systems related to W may be used for purposes of formalizing constructive mathematics, in which case all of $\neg, \wedge, \vee, \Rightarrow, \forall, \exists$ should be taken as basic and the logic taken to be intuitionistic.

[32] Except for the use here of atomic formulas $(t\downarrow)$, these correspond exactly to the formulas in the language of Feferman (1985), since there we have no quantification with respect to class variables and all individual variables are typed $x^S, y^S, \ldots$.

With free variables of terms and formulas defined as usual, we write $\sigma[x_1, \ldots, x_k]$ for a term or formula σ all of whose free variables are among $x_1, \ldots, x_k$. When giving attention to some one variable "x," we write $\sigma[x, \ldots]$ for σ; then $\sigma[t, \ldots]$ is to be the result of substituting t for all free occurrences of x in σ, assuming there are no collisions with bound variables.

The *logic of partial terms* LPT which underlies W is adapted from Beeson (1985, pp. 97–99). First of all, each object exists; that is, $(\forall x)(x\downarrow)$. For the term-building operations such as pairing, projection, etc., we posit that if the compound term is defined then each subterm is defined, for example, $\langle s, t\rangle\downarrow \Rightarrow s\downarrow \wedge t\downarrow$, $P_i(s)\downarrow \Rightarrow s\downarrow$, etc. However, this does not apply to term-building by abstraction operators; in fact $(\lambda x \in S)t[x, \ldots]\downarrow$ will always be true (since it always names a partial operation on S), but $t[x \ldots]$ may be undefined for particular (or even all) x in S. LPT also includes axioms for equality, such as $s = t \Rightarrow s\downarrow \wedge t\downarrow$, and $s \simeq t \wedge \phi[s, \ldots] \Rightarrow \phi[t, \ldots]$, where $s \simeq t$ means $[s\downarrow \vee t\downarrow \Rightarrow s = t]$. In LPT, if an atomic relation holds between terms, then each of them is defined; in the case of W this tells us that $(t \in S) \Rightarrow t\downarrow$. Note that we do not apply "$\downarrow$" to class terms, since, intuitively speaking, all class terms denote. The crucial modification of ordinary classical predicate calculus in LPT is in the instantiation axioms:

$$\forall x\phi[x, \ldots] \wedge t\downarrow \Rightarrow \phi[t, \ldots], \quad \phi[t, \ldots] \wedge t\downarrow \Rightarrow \exists x\phi[x, \ldots].$$

But for class terms we have, unrestrictedly,

$$\forall X\phi[X, \ldots] \Rightarrow \phi[S, \ldots] \text{ and } \phi[S, \ldots] \Rightarrow \exists x\phi[X, \ldots].$$

In stating the axioms of W all unbound variables are regarded as being closed universally; for example, we may write an axiom in the form $\psi[X, \ldots]$ where $\forall X \ldots \psi[X, \ldots]$ is intended. By the preceding we then have that $\psi[S, \ldots]$ is derivable for any class term S.

We can now state the *Axioms of W* (I–VII):

I. *Separation classes.* For each bounded predicative ϕ,

$$y \in \{x \in X | \phi[x, \ldots]\} \Leftrightarrow y \in X \wedge \phi[x, \ldots].$$

II. *Function classes, application, abstraction.*

(i) $f \in (X \stackrel{\sim}{\rightarrow} Y) \Leftrightarrow (\forall x \in X)[f(x)\downarrow \Rightarrow f(x) \in Y]$.

(ii) $(\lambda x \in X)t[x, \ldots]\downarrow$.

(iii) $u \in X \Rightarrow (\lambda x \in X)t[x, \ldots](u) \simeq t[u, \ldots]$.

III. *Pairing, projections, Cartesian products.*

(i) $(\forall x, y)[\langle x, y\rangle\downarrow] \wedge \forall z[P_1(z)\downarrow \wedge P_2(z)\downarrow]$.

(ii) $P_1(\langle x, y\rangle) = x \wedge P_2(\langle x, y\rangle) = y$.

(iii) $x \in (X \times Y) \Leftrightarrow (\exists x \in X)(\exists y \in Y)[z = \langle x, y\rangle]$.

IV. *Definition by cases.*

(i) $\forall x, y, z, w[D(x, y, z, w)\downarrow]$.

(ii) $[x = y \Rightarrow D(x, y, z, w) = z] \wedge [x \neq y \Rightarrow D(x, y, z, w) = w]$.

V. *Natural numbers, induction and recursion.*

(i) $0 \in \mathbf{N} \wedge sc \in (\mathbf{N} \to \mathbf{N})$.

(ii) $n' \neq 0 \wedge (n' = m' \Rightarrow n = m)$.

(iii) $(\forall a \in \mathcal{P}(\mathbf{N}))[0 \in a \wedge \forall n(n \in a \Rightarrow n' \in a) \Rightarrow \forall n(n \in a)]$.

(iv) $f = Rc_{\mathbf{N}}(g, h) \wedge g \in (X \to \mathbf{N}) \wedge h \in (\mathbf{N} \times X \times \mathbf{N} \to \mathbf{N}) \Rightarrow$
$f \in (\mathbf{N} \times X \to \mathbf{N}) \wedge$
$(\forall n)(\forall x \in X)[f(0, x) = g(x) \wedge f(n', x) = h(n, x, f(n, x))]$.

VI. *Search operator on* **N**.

(i) $\mu \in \mathcal{P}(\mathbf{N}) \to \mathbf{N}$.

(ii) $(\exists n)(n \in a) \Rightarrow \mu(a) \in a$.

To see the working and possible modifications of some of the axioms a little better, the following points should be noted.

1. The theory W is not extensional, either for functions or classes, in keeping with the definitionist interpretation. However, extensionality for both is consistent with W, simply by using the obvious set-theoretical interpretation. On the other hand, extensionality is inessential for the mathematical applications of W.

2. It may be that for $f \in (X \tilde{\to} Y)$ we have $f(x)\downarrow$ for some $x \notin X$. Similarly, as we have defined $f \in (X \to Y) \Leftrightarrow f \in (X \tilde{\to} Y) \wedge (\forall x \in X) f(x)\downarrow$, the domain of f might properly include X. Put in other terms, if $f \in (X \to Y)$ and $X_1 \subseteq X$ then $f \in (X_1 \to Y)$.

3. For applications, it is sufficient in place of D to take definition by Boolean cases, that is, a function $D_B(x, u, w)$ with $D_B(0, z, w) = z$, $D_B(1, z, w) = w$; $D_B(x, z, w)$ is definable as $D(x, 0, z, w)$.

4. Implicitly in V(iv) we are writing $f(u, x)$ for $f(\langle u, x\rangle), h(u, x, y)$ for $h(\langle u, x, y\rangle)$ where, say, $\langle u, x, y\rangle = \langle\langle u, x\rangle, y\rangle$ and $U \times X \times Y = (U \times X) \times Y$.

5. In V(iii) we have taken the weakest of three levels of induction on **N** that might be considered, in this case denoted Set-Ind$_{\mathbf{N}}$, since $\mathcal{P}(\mathbf{N}) = (\mathbf{N} \to \{0, 1\})$ is being called the class of all subsets of **N**. The next level is Class-Ind$_{\mathbf{N}}$, with the axiom

$$(\forall X \subseteq \mathbf{N})[0 \in X \wedge \forall n(n \in X \Rightarrow n' \in X) \Rightarrow \forall n(n \in X)].$$

By axiom I, this implies the scheme

$$\phi[0,\ldots] \wedge \forall n(\phi[n,\ldots] \Rightarrow \phi[n',\ldots]) \Rightarrow \forall n\phi[n,\ldots]$$

for all bounded predicative formulas ϕ. The third level to consider is Full-Ind$_{\mathbf{N}}$, which takes this scheme for *all* formulas $\phi[n,\ldots]$ of $\mathcal{L}$(W).

6. The axiom V(iv) for definition by primitive recursion is the weaker of two such axioms that might be considered. In place of it, we might take an operator Rc with axiom

$$\begin{aligned} f = Rc(g,h) \wedge g \in (X \to Y) \wedge h \in (\mathbf{N} \times X \times Y \to X) \\ \Rightarrow f \in (\mathbf{N} \times X \to Y) \\ \wedge\ (\forall n)(\forall x \in X)[f(0,x) = g(x) \wedge\ f(n',x) = h(n,x,f(n,x))]. \end{aligned}$$

The difference is that $Rc_{\mathbf{N}}$ provides only for recursion on $\mathbf{N}$-valued functions. Using $Rc_{\mathbf{N}}$ and explicit definition by the λ-operator, we can define all primitive recursive functions and relations on $\mathbf{N}$, in particular $+, \cdot, <$, etc., and establish closure under bounded quantification and the *bounded minimum operator* $(\mu m \leq n) f(m) = 0$.

7. The symbol "μ" in the search operator axiom is used to suggest the *unbounded minimum operator* $\mu!$. But that is easily defined from the search operator μ and bounded minimum by

$$\begin{aligned} \mu!(a) &= \begin{cases} (\mu m \leq \mu(a))(m \in a) & \text{if } \mu(a) \in a \\ 0 & \text{otherwise} \end{cases} \\ &= D(a(\mu(a)), 0, (\mu m \leq \mu(a))(a(m) = 0), 0). \end{aligned}$$

8. Note that $\exists n(n \in a) \Leftrightarrow \mu(a) \in a$. By induction, we can associate with each arithmetical formula $\phi[n_1,\ldots,n_k]$ a term $t_\phi[n_1,\ldots,n_k]$ built up by explicit definition, $Rc_{\mathbf{N}}$ and μ, with

$$\begin{aligned} \forall n_1,\ldots,n_k\{(t_\phi[n_1,\ldots,n_k] = 0 \vee t_\phi[n_1,\ldots,n_k] = 1) \\ \wedge\ t_\phi[n_1,\ldots,n_k] = 0 \Leftrightarrow \phi[n_1,\ldots,n_k]\}. \end{aligned}$$

In particular, for each arithmetical $\phi[n,\ldots]$ it follows that $\exists a \in \mathcal{P}(\mathbf{N})$ $\forall n(n \in a \Leftrightarrow \phi[n,\ldots])$. Hence Set-Ind$_{\mathbf{N}}$ implies the scheme of induction for all arithmetical formulas. Thus the system PA may be considered to be a subsystem of W, with the variables of $\mathcal{L}$(PA) taken to range over $\mathbf{N}$.

9. The system obtained from W by adding Class-Ind$_{\mathbf{N}}$ is stronger than PA. This may be seen in various ways; one of these is to prove Con(PA) by first showing the principle TI(ϵ_0) of transfinite induction up to ϵ_0. Let $\prec$ be a standard primitive recursive ordering in

N of order-type ϵ_0, and let $I(n)$ express that every set $a \in \mathcal{P}(\mathbf{N})$ which has a nonempty intersection with $\{m : m \prec n\}$ has a $\prec$-least element. Since the sets in $\mathcal{P}(\mathbf{N})$ are closed under numerical quantification, standard arguments (cf., for example, Feferman 1977, p. 946) show that $\forall n(I(n) \Rightarrow I(\omega^n))$. The property *I(n)* defines a class $X = \{n \in \mathbf{N} | I(n)\}$. Then Class-Ind$_{\mathbf{N}}$ allows one to prove $\forall n(\omega_n \in X)$ where $\omega_0 = 0$ and $\omega_{n+1} = \omega^{\omega_n}$. Hence, finally we obtain $\forall n I(n)$, which is TI(ϵ_0) (for sets).

10. Weyl's system $\mathrm{K}^{(\beta)}$ may be interpreted in W together with an axiom Rc for the full recursion operator Rc (6 above), since by taking $Y = \mathcal{P}(\mathbf{N})$ for the value class we can treat iteration of set-valued functions and the other forms of recursion principles stated by Weyl. Thus W + Rc cannot have an arithmetical interpretation. Moreover, we can convert this into a formal proof that W + Rc is stronger than PA.

11. If we weaken the search axiom to give only a partial operator, as $\mu_0 \in \mathcal{P}(\mathbf{N}) \stackrel{\sim}{\rightarrow} \mathbf{N}$, we can explicitly define all partial recursive functions in the thus weakened theory W_0. Moreover W_0 has a model in which $(\mathbf{N} \stackrel{\sim}{\rightarrow} \mathbf{N})$ consists of all partial recursive functions; for this we take the domain of individuals of the model to be ω and let $x(y) \simeq \{x\}(y)$, from ordinary recursion theory.

12. For a definability model of W with full μ we must go to Kleene's notion of recursion in ${}^2E$ (Kleene 1959). Again the domain of individuals of the model is ω, but now we take $x(y) \simeq \{x\}({}^2E, y)$. The total operator μ (or more specifically $\mu!$) is recursively defined from ${}^2E$. This model also satisfies Full-Ind$_{\mathbf{N}}$ and Rc. In this model, $(\mathbf{N} \rightarrow \mathbf{N})$ consists of the *hyperarithmetic functions* and $(\mathbf{N} \stackrel{\sim}{\rightarrow} \mathbf{N})$ of the Π^1_1 partial functions from $\mathbf{N} \rightarrow \mathbf{N}$.

In contrast to 12, there is no obvious definability model of *W* in which $(\mathbf{N} \rightarrow \mathbf{N})$ consists of the arithmetical functions. Nevertheless we have the following:

Metatheorem. *W is a conservative extension of PA.*

This result is due jointly to Gerhard Jäger and the author and is established by a proof-theoretical argument [in Feferman and Jäger (1993) and (1996)].[33] The argument is much harder than that indicated above for conservativity of $\mathrm{K}^{(\alpha)}$ over PA, so no effort will be made to indicate the details here, except to say that we first embed W into a subtheory of the system

[33] I first demonstrated that the theory W′, which differs from W in using only total types $(S \rightarrow T)$ instead of $(S \stackrel{\sim}{\rightarrow} T)$, is conservative over PA by a kind of reduction to the theory Res-VT + (μ) of Feferman (1985); it was shown there that the latter theory is conservative over PA.

T_1 of Feferman (1975), and work in that framework. In the joint work with Jäger we also give proof-theoretical information about the strengthening of W considered above, in particular showing that W + (Full-Ind$_{\mathbf{N}}$) + Rc is a predicative theory in the precise sense of Feferman and Schütte.

9 Scientifically Applicable Mathematics in W

In Feferman (1985), an outline was given for a direct development of much of classical and modern analysis in the restricted theory of variable types Res-VT + (μ). Since this latter theory is contained in W, the development may be carried out just as well in W, and in fact is even simpler there due to the greater flexibility of W. I shall not repeat this development here, but refer the reader to the paper mentioned, or to the last section of Feferman (1987) [reproduced as chapter 12 in this volume], where there are also some indications given. Of course, we may begin (as Weyl did) by constructing the nonnegative rationals $\mathbf{Q}^+$ as the class $\mathbf{N} \times (\mathbf{N} - \{0\})$ of pairs $\langle m, n\rangle$ representing m/n with the usual equivalence relation $=_{\mathbf{Q}^+}$, and then $\mathbf{Q}$ as the class of pairs $\mathbf{Q}^+ \times \mathbf{Q}^+$ of pairs (r, s) representing $r - s$, with the usual equivalence relation $=_{\mathbf{Q}}$. Next $\mathbf{R}$ is introduced as the class of Dedekind sections in $\mathcal{P}(\mathbf{Q})$ or alternatively as the class of Cauchy sequences in $(\mathbf{N} \to \mathbf{Q})$, again with a suitable equivalence relation $=_{\mathbf{R}}$. Then the class $F(\mathbf{R})$ of real-valued functions on $\mathbf{R}$ is the class

$$\{f \in (\mathbf{R} \to \mathbf{R}) | \forall x \in \mathbf{R}\, \forall y \in \mathbf{R}(x =_{\mathbf{R}} y \Rightarrow f(x) =_{\mathbf{R}} f(y))\}.$$

Various spaces S of real-valued functions are then contained as subclasses in F, and operators on such spaces belong to $(S \to \mathbf{R})$.

The development of nineteenth-century analysis is carried out in a very straightforward way, making systematic use of Cauchy completeness rather than the l.u.b. principle, and sequential compactness rather than the Heine-Borel theorem. Moving on to twentieth-century analysis, the theory of Lebesgue measurable sets and functions can be formalized in a rather direct way in W via sequential approximations by "nice" sets and functions; however, the operation of outer measure in general and the existence of nonmeasurable sets cannot be established, as these require impredicative constructions. Moving on to functional analysis, again the "positive" theory can be developed in a straightforward way, at least for separable Banach and Hilbert spaces, and can be applied to various L_p spaces as principal examples. Among the general results which can be obtained are forms of the Riesz Representation Theorem, the Hahn-Banach Theorem, the Uniform Boundedness Theorem, and the Open Mapping Theorem. In this way, the part of Weyl's program that was left open at the end of *Das Kontinuum* can now be realized via W.

Indeed, it has been shown by Brown (1987) in pursuit of Friedman's and Simpson's "Reverse Mathematics" program that much of the above portion of functional analysis can already be formalized in a weak second-order subtheory WKL_0^+ of W, which is conservative over PRA (Primitive Recursive Arithmetic). It would be interesting to find a subtheory of W with the same properties as WKL_0^+ together with the advantages offered by the flexible type theory of W.

In view of these developments, I conjectured at the end of Feferman (1987) [chapter 12 in this volume] that all scientifically applicable mathematics can be formalized in (a subtheory of) W, and hence does not require the assumption of impredicative set theory or of uncountable cardinal numbers for its eventual justification.[34] However, this conjecture has been questioned because of certain uses of nonseparable spaces in applied mathematics and physics. I asked a colleague of mine who works in applied mathematics about the latter and was told that while nonseparable spaces such as L^∞ are common, natural results for which the issue of (non-)separability arises are really somewhat rare. The question to what extent analysis on nonseparable spaces can be treated in such systems as W has not been touched at all. As a test case (scientific applications aside) one might start with the nonseparable Hilbert space of almost periodic functions (cf. for example, Akhiezer and Glazman 1961, §13). A principal result here is the theorem of H. Bohr that each such function is the uniform limit of "polynomials"

$$\sum_{k=0}^{n} a_k e^{i\lambda_k t} .$$

Incidentally, an elegant proof of this was given by Weyl using the theory of completely continuous operators (cf. *op. cit.*, §57). It would be interesting to see whether this can be carried out in W or, at least, a predicative theory [the work of Margenstern (1978) on a constructive treatment of almost-periodic functions may be relevant]. In any case one should go on to investigate the role of nonseparable spaces in applied mathematics and mathematical physics to see to what extent they are essential and, if so, whether they can be treated in W or similar theories. But, taking it for granted that such uses are exceptional, it is fair to say that Weyl's program is now substantially vindicated for scientifically applicable mathematics by means of systems like W. [This matter is discussed further in the following postscript and in the following chapter 14.]

Postscript [Added 23 May 1996]

Two sources of possible counterexamples to my conjecture discussed just above—that all scientifically applicable mathematics can be formalized in W—have been brought to my attention. The first concerns the use of nonseparable spaces to deal with systems with an infinite number of degrees

[34] Of course in W itself, the class $\mathcal{P}(\mathbf{N})$ is provably not equivalent to $\mathbf{N}$; thus internally (in W) it has higher cardinality than $\mathbf{N}$, and we obtain successively "higher" cardinals in W by iterating the power-set operation.

of freedom in quantum mechanics and statistical mechanics, while the second concerns the appearance of nonmeasurable sets in a proposed hidden-variable theory for quantum mechanics. Both raise prima facie problems for my conjecture because (as explained above) separability seems to be an essential hypothesis for the formalization of applicable functional analysis in W, and because it is consistent with W, all sets are measurable (in the measure spaces that are usually dealt with).

As to the first, it is stated in Emch (1972, p. 103) that it is necessary to deal with nonseparable spaces in the C*-algebra approach to systems with infinitely many degrees of freedom, at least in his global theory. The matter is taken up more specifically by Streater and Wightman (1978, p. 87).[35] They say that one might call on nonseparable Hilbert spaces in two cases in quantum mechanics. The first is to model a Bose field as a system composed of an infinity of oscillators via an infinite tensor product of Hilbert spaces (of dimension greater than one), which is necessarily nonseparable. Of this, Streater and Wightman say:

> [O]ne might think that such an infinite tensor product is the natural state space. However, it is characteristic of field theory that some of its observables involve all the oscillators at once and it turns out that such observables can be naturally defined only on vectors belonging to a tiny separable subset. . . . It is the subspace spanned by such a subset which is the natural state space rather than the whole infinite tensor product itself. Thus, while it may be a matter of convenience to regard the state space as part of an infinite tensor product, it is not necessary.

The second case where nonseparable Hilbert spaces appear (taken up *loc. cit.*) is in statistical mechanics

> when one passes to the limit in which the conventional box containing the system becomes arbitrarily large, the density being maintained constant. . . . It seems to us that such phenomena are consequences of considering systems in which every state contains an actual infinity of physical particles and provide no argument for analogous phenomena in relativistic quantum field theory.

Streater and Wightman leave the subject of nonseparability at that. On the other hand, Emch (1972, pp. 30–31) has argued that it is necessary to go beyond standard accounts (implicitly confined to separable spaces) in order to avoid "inescapable paradoxes" in connection with passage to the thermodynamic limit. In a (private) discussion of this, Geoffrey Hellman has stressed that there are, in general, some rather important theoretical

[35] Their discussion *loc. cit.* continues unmodified from the first edition (1964) of that monograph.

advantages—avoiding contradictions and gaining theoretical understanding of physical phenomena—that go well beyond mere "convenience." This, of course, is independent of the question whether those advantages accrue to Emch's algebraic approach.

The possible use of nonmeasurable sets in quantum theory is presented in Pitowsky (1989). These appear in a kind of generalized hidden variable theory which allows for nonadditive "probability" assignments to disjoint sets. However, Pitowsky states in his introduction (*op. cit.*, p. 10) that he does not endorse this theory but is merely concerned to lay it out as a reasonable possibility, demonstrated by a proof of its consistency (*op. cit.*, p. 166). On the other hand, this result depends crucially on his rejection of one of two constraints on hidden variable theories due to Kochen and Specker, namely, the one having to do with negation. In his critical notice of Pitowsky's work, Malament (1992, pp. 317ff.), has argued that Pitowsky's stated reasons for rejecting the condition do not make sense—indeed, that Pitowsky is himself committed to the condition—and for this reason his strategy for getting around the Kochen-Specker theorem "does not work." Malament has also questioned the interest of *any* hidden variable theory that denies the (finite) additivity of "probability" assignments.

Evidently, both sources of possible counterexamples to my conjecture depend on controversial theoretical models. But even if one were to accept these on a theoretical level, there would remain the practical question of applying those models. Here, some comments of my colleague George Papanicolaou are highly apropos:

> The general trend in mathematical physics and other applications (control theory, optimization, stochastic processes) is to make the minimal assumptions needed so that the objects one works with are measurable and the Hilbert spaces separable. This usually takes the form of some topological condition, that is, some continuity property. If, for example, we have a bounded measurable function of two variables and consider its minimum over one variable while fixing the other then the minimum as a function of the other variable will not in general be measurable. But if the function is lower semicontinuous then we are OK. In stochastic processes a similar problem arises and we impose stochastic continuity to avoid pathologies. I can only think of how to avoid the situations you describe, at the moment. I cannot think of a compelling physical situation where nonmeasurability or nonseparability are essential. (Private communication)[36]

[36]I wish to thank Robert Geroch, Geoffrey Hellman, David Malament, and George Papanicolaou for helpful discussions concerning these questions.

14

Why a Little Bit Goes a Long Way: Logical Foundations of Scientifically Applicable Mathematics

1 Introduction

Does science justify any part of mathematics and, if so, what part? These questions are related to the so-called indispensability arguments propounded, among others, by Quine and Putnam. The general idea of the arguments has been formulated (for critical assessment) by Penelope Maddy in a recent article as follows:

> We have good reason to believe our best scientific theories, and mathematical entities are indispensable to those theories, so we have good reason to believe in mathematical entities. Mathematics is thus on an ontological par with natural science. Furthermore, the evidence that confirms scientific theories also confirms the required mathematics, so mathematics and science are on an epistemological par as well. (Maddy 1992, p. 78)

If one accepts the indispensability arguments, there still remain two critical questions:

Q1. Just which mathematical entities are indispensable to current scientific theories?

Q2. Just what principles concerning those entities are needed for the required mathematics?

"Why a little bit goes a long way: Logical foundations of scientifically applicable mathematics" was originally published in *PSA 1992* Vol. 2 (1993) pp. 442–455 (Feferman 1993a)and is here reprinted with the kind permission of the publisher, the Philosophy of Science Association, East Lansing, MI. There are some minor corrections and additions.

The material of this chapter was first presented in a symposium, "Is foundational work in mathematics relevant to the philosophy of science?" at the meeting of the Philosophy of Science Association, 1 Nov. 1992.

Here we consider answers of an underlying character to these questions, that is, from the point of view of the foundations of mathematics.[1] In this respect, both Quine and Putnam were led to accept set-theoretical notions and principles to some significant extent or other. However, neither one relied on any detailed examination of just what is needed for scientifically applicable mathematics in arriving at their positions. Nor did they seem to consider whether any of the alternative foundational schemes actively developed during this century—namely, those of predicativism, constructivism, and finitism—ought to be preferred on philosophical grounds, particularly when natural science is given such primacy. On the face of it, scientific realism is at odds with the strong form of platonic realism required to justify set theory through its assumption of the independent existence of abstract entities (such as sets of sets of sets . . . of unbounded infinite cardinality).

The failure to consider other foundational approaches no doubt stems from the common impression that—whatever their philosophical merits—these schemes are simply inadequate to meet the needs of everyday mathematics by being too restrictive and too foreign to practice. This impression needs to be corrected: there has been considerable logical work in recent years which has established in some detail the unexpected mathematical reach of each of these programs. Moreover, one result of the work in question is that surprisingly meager (in the proof-theoretical sense) predicatively justified systems suffice for the direct formalization of almost all, if not all, scientifically applicable mathematics. It is my main purpose here to describe this result (with the background leading to it) and, in its light, to reexamine the indispensability arguments.

In considering what mathematics is actually used in science it suffices to restrict attention to physics since, among all the sciences, that subject makes the heaviest use of mathematics and there is hardly any branch of mathematics that has some scientific application which is not applied there. It would be foolish to claim detailed knowledge of the vast body of mathematics that has been employed in mathematical physics. However, in general terms one can say that it makes primary use of mathematical analysis on Euclidean, complex, and Riemannian spaces, and of functional analysis on various Hilbert and Banach spaces. Any logical foundation for scientifically applicable mathematics should, at a minimum, cover all of nineteenth-century mathematical analysis of (piecewise) continuous functions on the former kind of spaces and should then go on to cover the theory of (Lebesgue) measurable functions and basic parts of twentieth-century functional analysis on the latter spaces.

We begin in the next section with a brief review of the set-theoretical foundations of analysis, followed in section 3 by a discussion of the reasons for rejecting that framework on philosophical grounds. Section 4 then

[1] Geoffrey Hellman's contribution to the symposium mentioned in the source footnote above (Hellman 1993a) contains a discussion which is closely related to ours at a number of points.

traces the historical development of predicative foundations of analysis. This leads in section 5 to the description of a formal theory W of variable finite types with the following properties: (i) W is proof-theoretically reducible to the system PA of Peano Arithmetic (the starting system for predicativity), and (ii) almost all, if not all, of scientifically applicable mathematics, as described above, can be formalized directly in W. By way of comparison, section 6 is devoted to a brief description of results concerning the mathematical reach of constructivist and finitist foundations. We return in section 7 to a discussion of the significance of these results for the indispensability arguments, and conclude in section 8 with some more speculative philosophical remarks.

2 Set-Theoretical Foundations of Analysis

The real number system or continuum $\mathbf{R}$ is the basic system for analysis; from it we define directly the complex numbers $\mathbf{C}$ and various Euclidean and non-Euclidean (Riemannian) spaces which figure in the applications. Then in functional analysis one makes use of certain spaces of functions satisfying continuity, measurability, or integrability conditions (for example, the L^p spaces). Concretely, in the case of functions of one variable, these are subspaces of the set $\mathbf{R} \overset{\sim}{\rightarrow} \mathbf{R}$ of all partial functions from $\mathbf{R}$ to $\mathbf{R}$. Abstractly, given any spaces X and Y of a certain kind, one will form subspaces of the set $X \overset{\sim}{\rightarrow} Y$ of all partial functions from X to Y; functionals then act on one such space to another.

The reals are characterized set-theoretically as the unique ordered field satisfying the l.u.b. axiom or Dedekind's continuity axiom. A specific realization of these axioms may be constructed set-theoretically as the set D of all lower Dedekind sections in the rationals $\mathbf{Q}$. Let $\mathcal{P}(X)$ be the set of all subsets of X; then $D \subseteq \mathcal{P}(\mathbf{Q})$.

Moreover, $D \sim \mathcal{P}(\mathbf{Q}) \sim \mathcal{P}(\mathbf{N})$ where $X \sim Y$ is the relation of equinumerosity (or set-theoretic equivalence) and $\mathbf{N} = \{0, 1, 2, \ldots\}$ is the set of natural numbers. Hence the cardinality of the reals is $2^{\aleph_0}$. Alternatively, following Cantor, the reals may be represented as Cauchy sequences of rationals, which are members of $\mathbf{N} \rightarrow \mathbf{Q}$, under a suitable equivalence relation.

In set theory, functions are defined as sets of ordered pairs satisfying the many-one condition and pairs are in turn defined by a set construction. With the preceding, this leads to a representation of the real numbers in the cumulative hierarchy $\langle V_n(\mathbf{N})\rangle_{n\in\omega}$ over $\mathbf{N}$, where

(1) $V_0(\mathbf{N}) = \mathbf{N},\ V_{n+1}(\mathbf{N}) = V_n(\mathbf{N}) \cup \mathcal{P}(V_n(\mathbf{N}))$, and $V_\omega(\mathbf{N}) = \bigcup_{n\in\omega} V_n(\mathbf{N})$.

By the above, $\mathbf{R}$ is located in $V_\omega(\mathbf{N})$ essentially at the level $V_1(\mathbf{N})$, $\mathbf{R} \overset{\sim}{\rightarrow} \mathbf{R}$ at the level $V_2(\mathbf{N})$, etc.; for any $X, Y \in V_\omega(\mathbf{N})$, $X \overset{\sim}{\rightarrow} Y$ is a subset of $\mathcal{P}(X \times Y)$ and hence is also in $V_\omega(\mathbf{N})$.

An alternative to the representation in the cumulative hierarchy over **N** is that in the pure cumulative hierarchy defined by

(2) $V_0 = \phi$, $V_{\alpha+1} = V_\alpha \cup \mathcal{P}(V_\alpha)$, and $V_\lambda = \bigcup_{\alpha<\lambda} V_\alpha$

(for λ a limit ordinal).

Using the von Neumann representation of ordinals, $\omega = \{0, 1, 2 \ldots\} \in V_{\omega+1}$ and thus $V_{\omega+\omega}$ includes $V_\omega(\mathbf{N})$ when we identify **N** with ω. $V_{\omega+\omega}$ is the "standard" model of the system ZC of Zermelo Set Theory with the Axiom of Choice. Set theorists now commonly accept the much stronger system ZFC of Zermelo-Fraenkel with Choice, whose standard model is V_κ for the first strongly inaccessible ordinal κ. Indeed, most working set theorists go far beyond ZFC in accepting various axioms of "large" transfinite cardinals beyond κ. One of the first to urge the plausibility of such extensions of ZFC was Gödel (1947/1964). In contrast, Quine's acceptance of some part of set theory is moderated by his version of the indispensability argument:

> So much of mathematics as is wanted for use in empirical science is for me on a par with the rest of science. Transfinite ramifications are on the same footing insofar as they come of a simplificatory rounding out, but anything further is on a par with uninterpreted systems. (Quine 1984, p. 788)

And, further:

> I recognize indenumerable infinities only because they are forced on me by the simplest known systematizations of more welcome matters. Magnitudes in excess of such demands, for example, Beth_ω [the cardinal number of $V_\omega(\mathbf{N})$ and of $V_{\omega+\omega}$] or inaccessible numbers, I look upon only as mathematical recreation and without ontological rights. (Quine 1986, p. 400)

It seems from these quotations that Quine would readily accept Zermelo set theory because the power-set operation $\mathcal{P}$ applied to **N** leads to **R**, and then its application once more to $\mathbf{R} \to \mathbf{R}$, etc.; "simplificatory rounding out" thus suggests acceptance of the Power-Set Axiom in general.

3 What's Wrong with Set-Theoretical Foundations?

Philosophically, set theory—even in its "moderate" form given by Zermelo's axioms—requires for its justification a strong form of platonic realism. This is not without its defenders, most notably Gödel (1944 and 1947/1964) (cf. also Maddy 1990). For its critics, however, the following are highly problematic features of this philosophy:

(i) abstract entities are assumed to exist independently of any means of human definition or construction;

(ii) classical reasoning (leading to nonconstructive existence results) is admitted, since the statements of set theory are supposed to be about such an independently existing reality and thus have a determinate truth value (true or false);

(iii) completed infinite totalities and, in particular, the totality of all subsets of any infinite set are assumed to exist;

(iv) in consequence of (iii) and the Axiom of Separation, impredicative definitions of sets are routinely admitted;

(v) the Axiom of Choice is assumed in order to carry through the Cantorian theory of transfinite cardinals.

The question of admissibility of impredicative definitions is discussed in the next section. The Axiom of Choice has been the focus of much specific criticism, since choice sets are not in general definable; however, it is entirely in keeping with the other assumptions (and it would not be coherent to accept (i)–(iv) and reject the Axiom of Choice).

Quite surprisingly, Gödel at one time sided with the critics: in an unpublished lecture he delivered to a meeting of the Mathematical Association of America in 1933, after explaining the use of systems like ZFC for the axiomatic foundation of "all of mathematics," Gödel raised pointed objections to its features (ii), (iv), and (v) and went on to say

> The result of the preceding discussion is that our axioms [of set theory], if interpreted as meaningful statements, necessarily presuppose a kind of Platonism, which cannot satisfy any critical mind and which does not even produce the conviction that they are consistent. (Gödel 1933, p. 19)[2]

In the present context for assessing the indispensability argument(s), one should also note the ontological and epistemological anomalies in accepting set-theoretical foundations for mathematics, first of all because highly infinitary abstract objects are put on an ontological par with physical objects, and second because there is no observational knowledge of abstract objects (either directly or indirectly).[3]

[2]The complete text for Gödel's 1933 lecture, entitled "The present situation in the founcations of mathematics," was found in his *Nachlass*; it is reproduced in Gödel (1995), pp. 45–53. My introduction to that text is reproduced as chapter 8 in the present volume.

[3]Maddy (1990) does attempt to advance a modified form of realism which would root our knowledge of sets in everyday physical experience. In my view, this is convincing only for the most elementary parts of set theory, if at all.

4 The Development of Predicative Foundations for Analysis

Poincaré and Russell

In seeking ways to avoid the set-theoretical paradoxes while pursuing the logicist program, Russell introduced the term *predicative* for properties which determine legitimate classes, but he had no settled criterion for telling which those are. Poincaré saw the root of the paradoxes in the existence of a vicious circle, namely, in the use of definitions which purport to single out a member of a totality by reference to that totality. Such apparent definitions are called *impredicative.* There can be no objection to impredicative definitions when the totality in question is regarded as having a clear and determinate extent. For example, if one accepts the totality $\mathbf{N} = \{0, 1, 2, \dots\}$ as definite, there can be no objection to definitions singling out some natural number as the least n satisfying a property $\varphi(n)$ when φ refers, by quantification, to the totality of objects in $\mathbf{N}$. Poincaré regarded the natural number sequence and the principle of induction on it as an irreducible basis of our mathematical intuition, and he argued that the attempts of the logicists such as Frege and Russell to derive these from purely logical principles are fundamentally misguided and question-begging.[4] On the other hand, Poincaré thought all other mathematical notions should be introduced by proper definitions; in particular, sets are to be defined only by reference to prior defined sets and notions, not by reference to any presumed totality of sets. The latter would thus constitute impredicative definitions; Poincaré banned their use under his *vicious circle principle.* (For a useful recent analysis of Poincaré's philosophy of mathematics, see Folina (1992).)

Russell adopted Poincaré's proscription of impredicative definitions in setting up the Ramified Theory of Types (RTT) for the *Principia Mathematica* (Whitehead and Russell 1910–1913). However, he did not follow Poincaré in taking the natural number system for granted, but aimed to construct that "logically" within RTT. Each set variable of RTT is ranked, and the Comprehension Axiom schema for existence of sets $\{x : \varphi(x)\}$ of a given rank is restricted to φ with all bound variables of a smaller rank; membership is restricted to successive ranks (the basic syntactic feature of typed theories of sets). But Russell then found that he was faced with a multiplicity of notions of natural number and could not even derive the

[4] Parsons (1983, 1992) has argued for the impredicativity of the induction principle on $\mathbf{N}$. Nelson (1986) (in evident agreement with this view) has developed an axiom system for what he calls predicative arithmetic which drastically restricts the use of the induction principle. (The computational content of Nelson's system is related to the feasibly computable functions in the sense of computer science.) [In contrast, Feferman and Hellman (1995) provide predicative foundations of full arithmetic on the basis of rather weak and intuitively evident axioms for finite sets. So one sees that it is reasonable to consider notions of predicativity relative to various basic conceptions; the one dealt with in the present text has come to be called *predicativity given the natural numbers.*]

simplest closure principles by induction using ranked formulas. Similarly, he was faced with an unworkable theory of the continuum, since one could only deal with real numbers of different ranks, for which the l.u.b. axiom would not hold in any one rank. For purely pragmatic reasons then, Russell introduced the so-called Axiom of Reducibility, which asserts that every set is coextensive with a set of lowest rank. This, in effect, completely compromised the predicative/impredicative distinction; Russell recognized the objections to this "axiom," but thought it could somehow be justified and in any case saw no alternative. Later Ramsey pointed out that if one dropped rankings but maintained the restriction of membership to successive types, thus yielding the Simple Theory of Types (STT), no obvious set-theoretical paradoxes would arise and some of the above problems would be avoided. STT allows impredicative definitions $\{x : \varphi(x)\}$ of sets of objects x in any given type n, with $\{x : \varphi(x)\}$ in the next type $n+1$. Contrary to Poincaré's view, impredicativity is thus not an essential ingredient of the paradoxes.

As with set theory, STT requires a platonistic philosophy for its justification. Its "standard" model is given by

$$S_{n+1} = \mathcal{P}(S_n) \text{ for } n < \omega,$$

where S_0 is a basic set of individuals. In order to derive arithmetic in STT one must assume that S_0 is infinite; hence it is taken for granted that one accepts completed infinite totalities, and that for any totality S one also has the totality $\mathcal{P}(S)$ of all subsets of S. Impredicativity already appears at type 1, since in forming the type 1 set $\{x : \varphi(x)\}$ for x of type 0, with bound variables in φ of type 1, one is implicitly assuming the existence of the totality of sets of type 1.

Weyl's Approach

The first steps, after Russell's aborted attempt, to see what part of analysis could be developed in strictly predicative terms were made by Weyl in his monograph *Das Kontinuum* (1918). Like Poincaré, Weyl accepted the natural numbers and the associated principles of proof by induction and definition by recursion as basic. This simplified his task in comparison with Russell's (modified) logicist program. The main question was how to develop a workable theory of real numbers, via an unramified yet predicatively acceptable theory of sets of natural numbers. His answer in essence was to restrict to arithmetical definitions $\{n : A(n)\}$ of such sets, that is, where A contains no bound set variables but may contain free set variables and free or bound numerical variables. Relative arithmetic definitions determine functions $F(X) = \{n : A(n, X)\}$ under which the arithmetically definable sets are closed. Weyl's further principles have been analyzed in modern terms in Feferman (1988a) [chapter 13 of this volume] and a certain ambiguity was revealed, leading to two possible formal systems for his work. The first of these turns out to be a conservative extension of Peano Arithmetic PA, and it suffices for all his applications.

Using a standard representation of the rational numbers in terms of the natural numbers, Weyl defined the reals as lower Dedekind sections in **Q**. And, though he was blocked from inferring the l.u.b. axiom for sets of reals because that requires (impredicative) quantification over "all" reals, he could obtain the l.u.b. for *sequences* of reals, since this only requires quantification over **N**: for a sequence of lower sections $\langle X_n \rangle_{n \in \mathbf{N}}$ in **Q**, we have $\bigcup_{n \in \mathbf{N}} X_n = \{x : (\exists n \in \mathbf{N})(x \in X_n)\}$ as the l.u.b. With continuous functions from reals to reals treated via approximating functions from **Q** to **Q**, Weyl was then able to sketch in his system a straightforward reconstruction of the whole of nineteenth-century analysis of (piecewise) continuous functions of a real variable. He also suggested the possibility of extension to more general classes of functions, including the Lebesgue measurable functions, but gave no indications how this would be carried out.

Modern Developments

Weyl did no more work with his system after 1918,[5] and his program lay dormant until the 1950s when the logical study of predicativity was taken up in a variety of ways by Lorenzen, Kleene, Kreisel, Grzegorczyck, Wang, Spector, and others (the development is traced in Feferman (1964) part I). In particular, Kreisel proposed a characterization of predicativity in terms of a certain "autonomous" transfinite progression of ramified second-order systems whose exact extent was determined independently by Schütte and myself in 1964 to be given by a certain recursively described ordinal Γ_0 (cf. *op. cit.* for references). Peano Arithmetic PA is contained in the base system for this sequence of theories.

In order to develop analysis in a workable but still predicatively acceptable way, it was necessary to follow Weyl's lead in setting up suitable unramified systems which would be justified (at least indirectly) by reduction to the autonomous progression of ramified theories. This was achieved in my 1964 paper in a second-order unramified system, followed by some improved versions in later papers. However, higher types are called for in order to have greater ease of development of analysis. For this, Feferman (1977) and, independently, Takeuti (1978) introduced certain systems of finite type which are conservative over PA by proof-theoretic arguments and, a fortiori, certainly predicatively justifiable. These permitted a direct development of nineteenth-century analysis and the beginnings of twentieth-century analysis, but the latter called for an even more flexible and expressive formalism.

5 A Flexible System W of Variable Finite Types for the Modern Development of Weyl's Program

The Formal System

As indicated in section 2 above, in order to represent the notions of modern analysis directly and develop analysis flexibly, it is necessary to have not

[5] For reasons explained in Feferman (1988a) [cf. chapter 13].

only the set **R** of real numbers but also for any set S, its definable subsets $X = \{x \in S : \varphi(x)\}$, and for any sets X and Y also the sets $X \times Y$ and $X \xrightarrow{\sim} Y$, where the latter is understood to be the set of all partial functions from X to Y. Now if functions were defined as many-one relations as in set theory, $X \xrightarrow{\sim} Y$ would simply be the set of all many-one $Z \subseteq X \times Y$, and this would call on $\mathcal{P}(X \times Y)$ for its definition; that route would, in effect, take us back to Zermelo's set theory. The crucial first step toward a flexible, predicatively justifiable system is to treat functions and classes as conceptually independent basic notions, that is, with neither explained in terms of the other. A system taking this lead was worked out successively in Feferman (1975, 1985, and 1988a). We follow the last of these papers [reproduced in chapter 13] in describing the system W (so designated in honor of Weyl) presented there in detail.

The language of W is two-sorted, with *individual variables a, b, c, . . . , x, y, z* ranging over a universe containing numbers, sets, functions, functionals, etc., and closed under pairing, while the variables *A, B, C, . . . , X, Y, Z* range over *classes* or (variable) *types.* Note the terminological shift here: *sets* will be reserved for subclasses of a given class S which have a characteristic function. There are constants for specific individuals and functions (indicated below); there are also binary operations $\langle x, y\rangle$ and $x(y)$, where the latter is interpreted as the value of x at y when x is a partial function and y is in its domain, otherwise undefined. *Individual terms s,t, . . .* are built up from individual variables by use of the given constants and operations by closure under the general process of explicit definition, including function definition $(\lambda x \in S).t(x)$. *Class terms S, T, . . .* are built up from class variables and the class constant **N** by closure under the operations $S \times T, S \xrightarrow{\sim} T$ and $\{x \in S | \varphi(x)\}$ where φ is a *bounded predicative formula* (explained below). Formulas of W in general are built up from atomic formulas $s = t, t\downarrow$ and $t \in S$ by the propositional operations and quantification with respect to both individual and class variables; $t \downarrow$ is used to express that t is defined. The bounded predicative formulas admit no quantification over class variables, and individual quantifiers are restricted as $(\forall x \in S)\varphi$ and $(\exists x \in S)\varphi$. By definition, $(S \to T) = \{z \in (S \xrightarrow{\sim} T) | (\forall x \in S) zx\downarrow\}$.

The individual and function constants are 0 and Sc (for *successor* on **N**), and μ (for the *nonconstructive least number* or *search operator*), P_1 and P_2 (for *projections* of pairing), D (for *definition by cases*), and $Rc_{\mathbf{N}}$ (for *definition by recursion* on **N**). The *function axioms* of W follow these intended meanings together with usual axioms for explicit definition. Taking $1 = Sc(0)$, we write $\mathcal{P}(S) = (S \to \{0, 1\})$ for *the class of subsets* of S regarded as the class of characteristic functions on S. Then for $a \in \mathcal{P}(S)$ we write $x \in a$ for $a(x) = 0$. In particular, the axiom for μ is simply

$$(\mu) \quad \mu \in \mathcal{P}(\mathbf{N}) \to \mathbf{N} \wedge \forall a \in \mathcal{P}(\mathbf{N})[(\exists n)(n \in a) \Rightarrow \mu(a) \in a]$$

The *class axioms* of W are the evident ones for $S \times T, S \xrightarrow{\sim} T$ and $\{x \in S | \varphi(x)\}$.

Induction on **N** is taken as so-called *set induction*:

$$(\text{Ind}_{\text{Set}}) \quad \forall a \in \mathcal{P}(\mathbf{N})[0 \in a \wedge \forall n(n \in a \Rightarrow Sc(n) \in a) \Rightarrow \forall n \in \mathbf{N}(n \in a)].$$

A stronger natural principle to consider is class induction:

$$(\text{Ind}_{\text{Class}}) \quad \forall X \subseteq \mathbf{N}[0 \in X \wedge \forall n(n \in X \Rightarrow Sc(n) \in X) \Rightarrow \mathbf{N} \subseteq X].$$

This is still predicative, but it leads us beyond PA, so is not included in W.

Using the μ operator one shows in W that sets are closed under numerical quantification, since $\exists n(n \in a) \Leftrightarrow a(\mu(a)) = 0$. It follows that every arithmetical formula defines a subset of **N**; then by Ind_{Set}, we obtain the induction scheme for arithmetical formulas. The logic of W is assumed to be that of classical two-sorted predicate calculus. Hence the system PA of classical first-order arithmetic is contained in W. The main metatheoretical result about W is the following:

Main Theorem. *W is proof-theoretically reducible to PA and is a conservative extension of PA.*

The proof of this appears in Feferman and Jäger (1993 and 1996). It follows from this theorem that W rests on entirely predicative grounds, though it has much of the conceptual richness and flexibility of systems like Zermelo set theory.

Analysis in W

Space does not begin to permit the demonstration that all (or practically all) of the necessary nineteenth-and twentieth-century analysis needed for scientific applications can be carried out directly in W; indications are given in Feferman (1985 and 1988a) [cf. chapter 13]. To begin with, the class **R** of real numbers may be introduced as the class of Dedekind sections in $\mathcal{P}(\mathbf{Q})$ or alternatively as the class of Cauchy sequences in $(\mathbf{N} \to \mathbf{Q})$. By closure under numerical quantification, we obtain closure under l.u.b. of bounded sequences, just as in Weyl's system. Now instead of dealing with (partial) functions of a real variable by some reduction or other to second-order functions, we treat them directly as members of $\mathbf{R} \xrightarrow{\sim} \mathbf{R}$. Then the classes of continuous functions, measurable functions, etc., are obtained as suitable subclasses of $\mathbf{R} \xrightarrow{\sim} \mathbf{R}$.[6] Given a function space S, functionals on

[6]The notion of Lebesgue outer measure cannot be defined in W since it makes essential use of the g.l.b. applied to sets, not sequences. However, the measure of measurable sets can be defined as the g.l.b. of measures of a sequence of approximating open covers. Incidentally, W does not prove the existence of nonmeasurable sets of reals in the sense of Lebesgue; that is, it is consistent with W to assume that all sets of reals are measurable.

that space are simply members of $S \xrightarrow{\sim} \mathbf{R}$. In this way, basic concepts and examples from functional analysis are readily represented in W. In that subject I have verified (in unpublished notes) that such results as the Riesz Representation Theorem, Hahn-Banach Theorem, Uniform Boundedness Theorem, and Open Mapping Theorem for separable Banach and Hilbert spaces are derivable in W—and that, finally, one can obtain the principal results of the spectral theory of bounded self-adjoint linear operators on a separable Hilbert space. Extension of this work to the case of unbounded self-adjoint operators has been carried out in a preliminary way, using an approach to their spectral theory via limits of bounded operators.

While there are clearly parts of theoretical analysis that cannot be carried out in W because they make essential use of the l.u.b. axiom applied to sets rather than sequences, or because they make essential use of transfinite ordinals or cardinals, or because they deal with nonseparable spaces, *the working hypothesis that all of scientifically applicable analysis can be developed in W has been verified in its core parts.* What remains to be done is to examine results closer to the margin to see whether this hypothesis indeed holds in full generality.*

6 Comparisons with the Reverse Mathematics Program and Constructivist and Finitist Foundations of Analysis

The Reverse Mathematics (R.M.) program was originated and initially developed by H. Friedman (1975) and subsequently pursued in detail mainly by S. Simpson and his students (cf. Simpson 1987 and 1988). The main question addressed in that program is, Which set-existence principles are necessary to establish the (known) propositions of ordinary non-set-theoretical mathematics? To fix matters, though, only results which can be formulated (or reformulated) in the language of second-order arithmetic have been considered. The pattern of the work is to find for each mathematical theorem τ (which can thus be expressed) a set-existence principle σ such that $\sigma \Rightarrow \tau$ is provable in a system based on principles weaker than σ; the "reverse" part comes in showing that $\tau \Rightarrow \sigma$ is derivable (in the same system), so that σ is exactly necessary for τ. A great number of results from analysis (as well as in algebra and logic) have been examined successfully in the R.M. program.[7] Moreover, it has been found that five principles σ of increasing strength come up repeatedly in this process: RCA_0 (Recursive Comprehension Axiom), WKL_0 (Weak König's Lemma), ACA_0 (Arithmetical Comprehension Axiom), ATR_0 (Arithmetical Transfinite Recursion), and Π^1_1-CA_0 (Π^1_1-Comprehension Axiom); the [restricting]

*[In this respect see, for example, the postscript to chapter 13 in this volume.]

[7] Simpson is currently preparing a book on subsystems of second-order arithmetic in which many of these results will be presented in full. [See now Simpson (1998).]

subscript "0" indicates that Ind_{Set} is the only form of induction on **N** used with each set-existence principle. ATR_0 is proof-theoretically equivalent to the full progression of predicative systems referred to at the end of section 4, while $\Pi^1_1\text{-CA}_0$ is the first patently impredicative system beyond that; we have the full l.u.b. principle for sets of reals provable in $\Pi^1_1\text{-CA}_0$. Going back down, ACA_0 is contained in our system W and is also conservative over PA. On the other hand, both RCA_0 and WKL_0 are conservative over the much weaker system PRA of Primitive recursive Arithmetic. All of the results shown to be equivalent to RCA_0, WKL_0, or ACA_0 are thus provable in W.

There are two main differences of the R.M. program from that described in section 5 with W. The first is the R.M. restriction to statements in the language of second-order arithmetic: this requires considerable coding once one moves beyond the nineteenth-century analysis of continuous functions (and even the representation of the latter in second-order terms is less than natural). On the other hand, the equivalences established in R.M. of results τ from ordinary mathematics with one of RCA_0, WKL_0, or ACA_0, are evidently sharper than the fact that such τ are consequence of the principles of W. In any case, the work done on RCA_0, WKL_0, or ACA_0 in analysis corroborates fully the development of all of nineteenth-century analysis and substantial tracts of twentieth-century functional analysis within W as described in the preceding section.

Further evidence comes from the development of constructive analysis in the hands of the Bishop school (cf. Bishop and Bridges 1985). Bishop and his followers found constructive substitutes for considerable portions of classical analysis. In general, with each classical theorem τ for which this is successful is associated a constructive theorem τ^* such that LPO implies $\tau^* \Rightarrow \tau$ where LPO is the "Limited Principle of Omniscience" $\forall n(f(n)) = 0 \vee \exists n(f(n) \neq 0)$ for all $f : \mathbf{N} \to \mathbf{N}$. Evidently LPO is a special consequence of the Law of Excluded Middle (LEM). It was shown in Feferman (1979) how all of Bishop's development of constructive analysis (except for his theory of Borel sets which Bishop had shown to be dispensable) could be formalized in a constructive theory of variable finite types which is proof-theoretically reducible to HA (Heyting Arithmetic). Indirectly, then, all of the classical analysis for which constructive substitutes were found in the Bishop school are accounted for by principles reducible to PA (=HA + LEM). This again acts to corroborate the work described in section 5 above.[8]

A word, finally, about finitist foundations of analysis. Initial efforts in this direction were made by Goodstine (1961), by considering which results of analysis can be accounted for on the basis of PRA. That is a quantifier-free system generally acknowledged to represent part (if not all)

[8]Naturally, one may expect that many results τ of classical analysis are not constructively provable as given; cf., for example, Pour-El and Richards (1989) and Hellman (1993 and 1993a). That doesn't imply that τ has *no* constructive substitute in Bishop's sense.

of finitist constructions and arguments in number theory. One of the impressive results of the R.M. program is to show how much of nineteenth- and twentieth-century analysis can be established in WKL_0; since that is proof-theoretically reducible to PRA (cf. Sieg 1985), an exceptional amount of analysis is already accounted for on finitistically justifiable grounds.[9] In view of this it would be worthwhile setting up a subsystem of W conservative over PRA in which that same part of analysis can be formalized more directly.

7 Significance for the Indispensability Arguments

The questions Q1 and Q2 raised in section 1 take the indispensability arguments for granted. Answers given in the past to these questions have been extremely broad, on the order of the following: mathematical analysis is indispensable to science, the real numbers and functions and sets of reals are the basic objects of analysis, set theory provides our best account of the real number continuum and of functions and sets in general, so the entities and principles of set theory are justified by science. This sweeping passage leaves undetermined just which of those entities and principles are thereby justified, except perhaps to say that the farther reaches of set theory are evidently unnecessary for science and so may be disregarded.

The work described in the preceding two sections allows one to tell quite different and much more specific stories in response to these questions. In all of the indicated formal systems one can speak *within* the language of these systems about arbitrary real numbers, functions of real numbers, sets of real numbers, etc. Only the existence principles (closure conditions) concerning these objects are much more restricted than in the case of systems of set theory like Zermelo's. To be specific, let us concentrate on the system W. Acceptance of W and the entities with which one can deal in its language does not commit one to a platonistic ontology of those entities, though the platonist is free to understand W in those terms. By the fact of the proof-theoretical reduction of W to PA, the only ontology it commits one to is that which justifies acceptance of PA. But even there, the answer to Q1 and thence to Q2 is underdetermined. One view of PA is that it is about the natural numbers as independently existing abstract objects; that is again a platonistic view, albeit an extremely moderate one. Another view is that PA is about the mental conception of the structure of natural numbers, which is of such clarity that statements concerning these numbers have a determinate truth value and their properties can be established in an indisputable intersubjective way; this is more or less the predicativistic view. Or one can make use of the fact that PA is reducible to HA to justify it on the basis of a more constructive ontology. In all these cases except the

[9]Cf., for example, Feferman (1977) and (1993) [chapter 10 in this volume]) for surveys of work in this direction, as well as Simpson (1987).

fully platonistic point of view concerning W, it is treated in an instrumental way, its entities outside the natural numbers are regarded as "theoretical" entities, and the justification for its use lies in whatever justification we give to the use of PA; but even there we do not arrive at a unique ontology.

My conclusion from all this is that even if one accepts the indispensability arguments, practically nothing philosophically definitive can be said of the entities which are then supposed to have the same status—ontologically and epistemologically—as the entities of natural science. That being the case, what do the indispensability arguments amount to? As far as I'm concerned, they are completely vitiated. This does not mean, however, that questions Q1 and Q2 lose their interest. Rather that is retained if one regards them instead from a phenomenological point of view, and here the kind of logical work described in sections 5 and 6 already has much to tell us. This work needs, of course, to be continued in order to make it fully conclusive, and here it will be necessary to investigate questions at the margin, for example, the possible essential use in physical applications of such objects as nonmeasurable sets or nonseparable spaces, which are not accounted for in systems like W [cf. the postscript to chapter 13].

8 Final Remarks

As a complement to the preceding discussion one should mention Maddy's critical examination (1992) of the indispensability arguments, whose conclusion (*op. cit.* p. 289) is that they "do not provide a satisfactory approach to the ontology or the epistemology of mathematics," for two reasons, quite different from those given here. The first is that "fundamental mathematized science is 'idealized' (that is, literally false)," for example, "the analysis of water waves by assuming the water to be infinitely deep or the treatment of matter as continuous in fluid dynamics or the treatment of energy as a continuously varying quantity" (*op. cit.* p. 281), and hence that the mathematics involved cannot be regarded as true, other than "true in the model" (my quote marks). The second is that "[scientific] indispensability cannot account for mathematics as it is actually done."

The first of these objections puts me in mind of the oft referred to article by Wigner, "The unreasonable effectiveness of mathematics in natural science" (1960). I agree with Maddy completely about the extent to which mathematized science depends on highly idealized models. What is remarkable, then, is not the unreasonable effectiveness of mathematics so much as the *unreasonable effectiveness of (mathematized) natural science.*

In her second objection, Maddy has particularly in mind the mathematics carried on by current set theorists, but it is not unjustly applied to the bulk of pure mathematics as it is actually practised. Here one meets a different kind of indispensability argument, stemming from Gödel, for the need of "higher" set theory to account for that body of mathematical work. But, again, appearances are deceiving, and logical results from recent years have much to tell us about just what is needed to carry on all but the patently set-theoretical reaches of modern mathematics. It is not claimed that predicative systems account for that, but it has been established that systems far weaker than Zermelo set theory and even than second-order arithmetic suffice for the bulk of mathematical practice.

Like most scientists, philosophers of science could simply take mathematics for granted and not concern themselves with its foundations, as being irrelevant to their main concerns. But, as Hellman (1993a) has emphasized in his introduction, debates like those discussed here as to realism versus (for example) instrumentalism, and as to the indispensability of highly theoretical concepts and principles, are equally central to the philosophy of science. Whether the kind of logical results described here will be more directly relevant to those debates remains to be seen. But as long as science takes the real number system for granted, its philosophers must eventually engage the basic foundational question of modern mathematics: "What are the real numbers, really?"

Symbols

Note: This is a comprehensive reference list of symbols used at one point or another in the book. A broad division is made between mathematical symbols and metamathematical symbols, but also included under the former are symbols for set-theoretical principles and axiomatic systems for sets and classes. The symbolism in the individual chapters followed the original sources and thus there are occasional discrepancies; these are noted below as alternatives. The reader can rely on context to deal with cases of actual conflict of notation.

Mathematical symbols

Number systems

$\mathbf{N}$	The (set of) natural numbers
$\mathbf{Q}$	The (set of) rational numbers
$\mathbf{Z}$	The (set of) integers
$\mathbf{R}$	The (set of) real numbers
$\mathbf{C}$	The (set of) complex numbers
$\mathcal{D}$ or $(\mathcal{D}(\mathbf{Q}))$	Dedekind sections in $\mathbf{Q}$

Sets and classes

$\in$	Membership relation
$x \in S$; $x \notin S$	x is an element (or member) of S; x is not an element of S
$\{x : P(x)\}$ (or $\{x \mid P(x)\}$)	The set of all x such that $P(x)$
$\{x \in S : P(x)\}$ (or $\{x \in S \mid P(x)\}$)	The set of all x in S such that $P(x)$
$\{a_0, \ldots, a_n\}$	The set consisting of $a_0, \ldots, a_n$
$\{a_0, \ldots, a_n, \ldots\}$	The set consisting of $a_0, \ldots, a_n, \ldots$
$\emptyset$	Empty set
V (or V)	Universal class
$X \subseteq Y$; $X \nsubseteq Y$	X is included (or contained) in Y; X is not contained in Y

$X \sim Y; X \not\sim Y$	X and Y are equinumerous; X and Y are not equinumerous
$f : X \to Y$	f is a function (or map) from X to Y
$\langle X_n \rangle_{n \in N}$	Sequence of X_n for $n \in \mathbf{N}$
$X \cup Y$	Union of X and Y
$X \cap Y$	Intersection of X and Y
$X - Y$	Difference of X and Y (or complement of Y in X)
$\bigcup X[P(X)]$	Union of all X such that $P(X)$
$\bigcup X_n[n \in \mathbf{N}]$ (or $\bigcup_{n \in \mathbf{N}} X_n$)	Union of the sequence $\langle X_n \rangle_{n \in \mathbf{N}}$
$\bigcap X[P(X)]$	Intersection of all X such that $P(X)$
$\bigcap X_n[n \in \mathbf{N}]$ (or $\bigcap_{n \in \mathbf{N}} X_n$)	Intersection of the sequence $\langle X_n \rangle_{n \in \mathbf{N}}$
$X + Y$	Disjoint union of X and Y
$\mathrm{X} \times \mathrm{Y}$	Cartesian product of X and Y
X^Y (or $Y \to X$)	Cartesian power of X by Y, i.e. $\{f \mid f : Y \to X\}$
$\mathcal{P}(X)$	Power set (or set of all subsets) of X, i.e. $\{Y \mid Y \subseteq X\}$
$[X]^k$	Set of all k element subsets of X

Cardinal and ordinals

card (X) (or $\overline{\overline{X}}$)	Cardinal number of X
$m, n, \ldots$ (or $\kappa, \ldots$)	Variables for cardinal numbers
$m + n$	*Card* $(X + Y)$ when $m = card(X)$ and $n = card(Y)$
$m \times n$	*Card*$(X \times Y)$ when $m = card(X)$ and $n = card(Y)$
m^n	*Card*(X^Y) when $m = card(X)$ and $n = card(Y)$
$m < n$	m is smaller than (or less than) n
m^+	The least n such that $m < n$
$\aleph_0$	*Card*$(\mathbf{N})$
$\mathbf{c}$ (or $2^{\aleph_0}$)	*Card*$(\mathbf{R})$ (same as *card*$(\mathcal{P}(\mathbf{N}))$, and as *card*$(\{0,1\}^{\mathbf{N}})$)
$\alpha, \beta, \ldots$	Variables for ordinal numbers
$\alpha + \beta$	Ordinal sum of α and β
$\alpha \cdot \beta$	Ordinal product of α and β
α^β	Ordinal power of α to the β
ω	Least infinite ordinal
ϵ_0	Cantor epsilon-number (= least α with $\omega^\alpha = \alpha$)

Ω (or ω_1)	Least uncountable ordinal
$\aleph_\alpha$	α-th transfinite cardinal
$\aleph_{\alpha+1}$	$(\aleph_a)^+$, i.e. least cardinal greater than $\aleph_\alpha$
$\aleph_\lambda$	λ-th transfinite cardinal, for λ a limit ordinal

Set-theoretical hierarchies

V_α (or V_α, C_α)	α-th level of the cumulative hierarchy
V (or V)	Union of all V_α, for α an arbitrary ordinal
$V_\alpha(\mathbf{N})$	α-th level of the cumulative hierarchy over **N**
L_α (or L_α)	α-th level of the constructible hierarchy
L (or L)	Union of all L_α

Classical and set-theoretical principles and hypotheses

LEM	Law of Excluded Middle (or Tertium non Datur)
AC	Axiom of Choice
WO	Well-ordering Principle
CH	(Cantor's) Continuum Hypothesis, $2^{\aleph_0} = \aleph_1$
GCH	Generalized Continuum Hypothesis, $2^{\aleph_\alpha} = \aleph_{\alpha+1}$
V=L	Axiom of Constructibility

Axiomatic systems of sets and classes

Z	Zermelo Set Theory
ZC	Z + AC
ZF	Zermelo-Fraenkel Set Theory
ZFC	ZF + AC
BG	Bernays-Gödel theory of sets and classes

Metamathematical symbols

Logical symbols

$\neg$	Negation ("not")
$\wedge$	Conjunction ("and")
$\vee$	Disjunction ("or")
$\rightarrow$ (or $\Rightarrow$, or $\supset$)	Conditional, or implication ("implies," or "if...then")
$\leftrightarrow$ (or $\Leftrightarrow$)	Biconditional ("if and only if")
$\perp$	Absurdity, or falsity
$\forall$; $\forall x$ (or (x))	Universal quantifier; "for all x"

$\exists$; $\exists x$	Existential quantifier; "for some x", or "there exists x"
$\forall x \in U$ (or $\forall x^U$)	"for all x in U"
$\exists x \in U$ (or $\exists x^U$)	"for some x in U"
Qx	Either $\forall x$ or $\exists x$

Formal systems—general symbols

S, T,...	Variables for formal systems
L_S (or L(S), or $\mathcal{L}$(S))	The language of S
PC (or PC_1)	Classical first-order predicate calculus
IPC	Intuitionistic first-order predicate calculus
$S^{(i)}$	S with logic restricted to IPC
Ax_S	Axioms of S
$Rules_S$	Rules of S
$Thms_S$	The (set of) theorems of S
s,t,...	Variables for terms of S
$\phi, \psi, \ldots$ (or A, B,...)	Variables for (well-formed) formulas of S
Φ (or $\mathcal{F}$)	A set of formulas of S
S + {A} (or S + A)	S with A added as an axiom
S + Φ	S with the set Φ of formulas added as an axiom scheme
$S \vdash \phi$	ϕ is a theorem of S (or, S proves ϕ)
$\Gamma \vdash \phi$	A formal sequent, where Γ is a finite sequence of formulas
$S \leq T$	S is proof-theoretically reducible to T
f: $S \leq T$ for Φ	S is proof-theoretically reducible to T by f, conservatively for Φ
$Proofs_S(m, n)$	Number-theoretical definition of the proof relation for S
Con_S (or Con(S))	Number-theoretical expression of the consistency of S
$\ulcorner A \urcorner$	Gödel number of A (or numeral of the same)

Classifications of arithmetical and analytical formulas

QF	Quantifier-free formulas
Σ^0_n	Arithmetical (first-order) formulas with n alternating quantifiers, beginning with $\exists$ (with a QF matrix)

Π^0_n	Arithmetical (first-order) formulas with n alternating quantifiers, beginning with $\forall$ (with a QF matrix)
Π^0_∞ (or Arith)	The class of all arithmetical formulas (formulas of first-order number theory)
Σ^1_n	Analytical (second-order) formulas with n alternating set or function quantifiers, beginning with $\exists$ (with an arithmetical matrix)
Π^1_n	Analytical (second-order) formulas with n alternating set or function quantifiers, beginning with $\forall$ (with an arithmetical matrix)
Δ^0_n	Assumed equivalence of formlas ϕ, ψ with ϕ in Σ^0_n and ψ in Π^0_n
Δ^1_n	Assumed equivalence of formulas ϕ, ψ, with ϕ in Σ^1_n and ψ in Π^1_n
$\Delta_{\mathcal{F}}$	Assumed equivalence of formulas ϕ, ψ with ϕ in $\mathcal{F}$ and $\neg\psi$ in $\mathcal{F}$

First-order number-theoretical principles and systems

IA	Induction axiom scheme for the natural numbers
IR	Induction rule for the natural numbers
Φ-IA (or $\mathcal{F}$-IA)	Induction axiom scheme restricted to formulas in Φ (or $\mathcal{F}$)
TI (α)	Transfinite induction scheme up to α (in a recursive notation system for ordinals)
PRA	Primitive Recursive Arithmetic
PA	Peano Arithmetic
HA	Heyting Arithmetic
ID_α	System for α-times iterated positive arithmetical definitions
$\mathrm{ID}_{<\alpha}$	Union of the ID_β for $\beta < \alpha$
$\mathrm{ID}^{(i)}_{<\alpha}$	Intuitionistic $\mathrm{ID}_{<\alpha}$

Second-order number-theoretical principles and systems

PA_2	Peano Arithmetic with IA for all second-order formulas

Ind (or I_1)	Second-order formulation of induction as a single axiom
$S\restriction$ (or Res-S, or S_0)	Restricted induction form of S, with IA replaced by Ind
CA	Comprehension Axiom scheme
Φ-CA (or $\mathcal{F}$-CA)	CA for formulas restricted to be in Φ (in $\mathcal{F}$)
RCA	CA for formulas assumed in Δ_1^0
RCA_0	Restricted induction system based on RCA
KL	König's Lemma
$2^{<\omega}$	Full binary tree (finite sequences from $\{0,1\}$)
WKL	Weak König's Lemma (KL for $2^{<\omega}$)
WKL_0	Restricted induction system based on RCA and WKL
ACA (or Π_∞^0-CA, or Π_1^0-CA)	Arith-CA scheme (equivalent to both Π_∞^0-CA and Π_1^0-CA schemes)
ACA_0	Restricted induction system based on ACA
$(\Pi_1^0\text{-CA})_\alpha$	System based on α times iterated Π_1^0-CA
$(\Pi_1^0\text{-CA})_{<\alpha}$ (or $(ACA)_{<\alpha}$)	Union of the systems $(\Pi_1^0\text{-CA})_\beta$ for $\beta < \alpha$ (when $\alpha = \omega \cdot \alpha$)
Δ_1^1-CA	System based on Δ_1^1-CA
Π_1^1-CA	System based on Π_1^1-CA
$(\Pi_1^1\text{-CA})\restriction$ (or $(\Pi_1^1\text{-CA})_0$)	Restricted induction system based on Π_1^1-CA
$(\Pi_1^1\text{-CA})_\alpha$	System based on α times iterated Π_1^1-CA
$(\Pi_1^1\text{-CA})_{<\alpha}$	Union of the systems $(\Pi_1^1\text{-CA})_\beta$ for $\beta < \alpha$
Δ_2^1-CA	System based on Δ_2^1-CA
BI	Bar Induction
ATR	Arithmetical Transfinite Recursion
ATR_0	Restricted induction system based on ATR
AC	Second order choice scheme, $\forall x \exists Y \phi(x, Y) \rightarrow \exists X \forall x \phi(x, X_x)$

Finite type principles and systems

STT	Simple Type Theory
RTT	Ramified Type Theory
R_α	System of RTT up to level α
$\sigma, \tau, \ldots$	Finite type symbols
$\sigma \to \tau$	Type of functions from type σ to type τ
L^ω	Finite type language closed under function types
$K_{\sigma,\tau}$	Typed combinators for constant functions
$S_{\rho,\sigma,\tau}$	Typed substitution combinators
R_σ	Typed recursors for recursion on **N** (with values of type σ)
PR^ω (or T)	Quantifier free system of primitive recursive functionals of finite type (Gödel's system T)
HA^ω	Extension of PR^ω by intuitionistic quantificational logic in L^ω
PA^ω	Extension of PR^ω by classical quantificational logic in L^ω (equivalently, HA^ω + LEM)
ϕ^D	D-interpretation form of ϕ ("*Dialectica*" functional interpretation)
ϕ_D	Quantifier-free matrix of ϕ^D
ϕ^-	Negative (or "double-negation") translation *quad* of ϕ
AC	Axiom of Choice scheme in L^ω
M′	Extension of Markov's Principle to L^ω
IP′	Independence of Premise scheme in L^ω
$\widetilde{HA}^\omega$	A special D-interpretable extension of HA^ω
$\widetilde{PA}^\omega$	A special $(-, D)$-interpretable extension of PA^ω
BR	Bar Recursion scheme of functional definition in L^ω

Applicative theories with variable types

$s, t, \ldots$	Type-free terms
$t[x]$	t with some free occurrences of x
$t\downarrow$	t is "defined" (has a value)
LPT	Logic of Partial Terms
$s(t)$ (or st)	Application of the partial function s to t
$s \simeq t$	$s = t$ if either $s\downarrow$ or $t\downarrow$
$\lambda x.t[x]$	Partial function whose value at x is given by $t[x]$

$(\lambda x \in S)t[x]$	Partial function whose value at x in S is given by $t[x]$
$\langle s,t\rangle$ (or (s,t))	Ordered pair of s and t
P_1	Projection operation on first term of a pair
P_2	Projection operation on second term of a pair
D	Definition by cases operation
sc	Successor operation
t'	Successor of t, $sc(t)$
$Rc_{\mathbf{N}}$ (or $r_{\mathbf{N}}$)	Recursor on **N** with values in **N**
μ	Unbounded (non-constructive) minimum operator
$E_{\mathbf{N}}$ (or ${}^2E$)	Existential quantification over **N** as an operator
$f : X \xrightarrow{\sim} Y$	f is a partial function from X to Y
$X \xrightarrow{\sim} Y$	Class of all partial functions from X to Y, i.e. $\{f \mid f : X \xrightarrow{\sim} Y\}$
VT	Applicative theory of variable types (or classes)
VT_μ	VT with axiom for μ
$\mathrm{Ind}_{\mathrm{Set}}$ (or Set-Ind)	Induction axiom restricted to sets (given by characteristic functions)
$\mathrm{Ind}_{\mathrm{Class}}$ (or Class-Ind)	Induction axiom restricted to classes
$\mathrm{VT}_\mu \restriction$	VT_μ with $\mathrm{Ind}_{\mathrm{Set}}$
W	System superseding $\mathrm{VT}_\mu \restriction$
T_0	System of operations and classes ("Explicit Mathematics")
EM_0	Base subsystem of T_0

Miscellaneous symbols (by chapter, in order of appearance)

$M \equiv M'$	The structures M and M' satisfy the same sentences (ch. 2)
PC_2	Second-order logic for validity in the standard sense (ch. 2)
PA_2	Peano's Axioms with 2nd-order Ind axiom, in PC_2 (ch. 2)
$\mathcal{M} \equiv_{\mathrm{L}} \mathcal{M}$	$\mathcal{M}$ and $\mathcal{M}'$ satisfy the same sentences in L (chs. 4, 5)

CS	Formal theory of choice sequences (chs. 4, 5)
α	A choice sequence α (chs. 4, 5)
$\bar{\alpha}(n)$	Sequence of the first n terms of α (chs. 4, 5)
$\{e\}(n)$	Value of e-th partial recursive function at n (ch. 5)
CT	Church's Thesis (ch. 5)
CST	Constructive Set Theory (ch. 5)
IZF	ZF in intuitionistic logic (i.e. $\mathrm{ZF}^{(i)}$) (ch. 5)
ML	Martin-Löf constructive theory of types (ch. 5)
Γ_0	Proof-theoretic ordinal bound to predicative systems (ch. 5)
PM	Gödel's version of STT over number theory (ch. 7)
$(a, \in \cap a)$	Structure on set of sets a with $\in$-relation restricted to a (ch. 7)
$A \mapsto \neg A$	Operation of negation (ch. 9)
$A, B \mapsto A \wedge B$	Operation of conjunction (ch. 9)
$\mathcal{M}$	An informal body of mathematics (ch. 10)
$\mathcal{F}$	A general foundational framework (ch. 10)
H. P.	Hilbert's Program (ch. 10)
$\mathcal{C}$	Continuous functionals of finite type (ch. 11)
$\mathcal{RC}$	Recursively continuous functionals of finite type (ch. 11)
Tr_1	Truth predicate for arithmetical sentences (ch. 12)
$Tr_1(\mathrm{PA})$	Extension of PA by axioms for Tr_1 (ch. 12)
$Inacc(\alpha)$	α is an inaccessible cardinal (ch. 12)
RH	Riemann Hypothesis (ch. 12)
$meas(X)$	Measure of a Lebesgue measurable set X (ch. 12)
$meas^\star(X)$	Outer measure of X (ch. 12)
LPO	Limited Principle of Omniscience, i.e. LEM for Π^0_1 formulas (ch. 12)
$\mathrm{X}_i^{(j)}, \mathrm{Y}_i^{(j)}, \ldots$	Variables of type i, ramified level j in RTT (ch. 13)
$\mathrm{X}^{(j)}, \mathrm{Y}^{(j)}, \ldots$	Variables of type 1, ramified level j in a 2nd-order ramified system (ch. 13)
$\Phi(x)$	Set-valued function of x, given by $\mathrm{R}_\Phi(u, x)$ in *Das Kontinuum* (ch. 13)

$\Phi(x, X)$	Set-valued function of x, X given by $\mathrm{R}_\Phi(u\|x, X)$ in *Das Kontinuum* (ch. 13)
Φ^n	n-th iterate of Φ (ch. 13)
$\mathrm{K}^{(\alpha)}$	First formalization of the system of *Das Kontinuum* (ch. 13)
$\mathrm{K}^{(\beta)}$	Second formalization of the system of *Das Kontinuum* (ch. 13)
$\phi[n, a, f]$	Formula of $\mathrm{K}^{(\alpha)}$ with free number variables n, set variables a, and function variables f (ch. 13)
L^p	Space of functions f with $\|f\|^p$ Lebesgue integrable (ch. 14)
R. M.	Reverse Mathematics program (ch. 14)

References

Aberth, O. (1980) *Computable Analysis* (McGraw-Hill, New York).

Ackermann, W. (1924) Begründung des "tertium non datur" mittels der Hilbertschen Theorie der Widerspruchsfreiheit. *Mathematische Annalen* 93, 1–36.

Aczel, P. (1977) An introduction to inductive definitions. In Barwise (1977), 739–782.

———(1988) *Lectures on Non-well-founded Sets.* CSLI Lecture Notes 14 (Center for the Study of Language and Information, Stanford).

Akhiezer, N. I., and Glazman, I. M. (1961) *Theory of Linear Operators in Hilbert Space*, Vol. 1 (F. Ungar, New York).

Avigad, J., and Feferman, S. (1998) Gödel's functional ("Dialectica") interpretation. In S. Buss (ed.), *Handbook of Proof Theory* (North-Holland, Amsterdam), 337–405.

Ax, J., and Kochen, S. (1965) Diophantine problems over local fields I, II. *American Journal of Mathematics* 87, 605–648.

Barendregt, H. P. (1977) The type free lambda calculus. In Barwise (1977), 1091–1132.

———(1984) *The Lambda Calculus, Its Syntax and Semantics* (2nd edn.) (North Holland, Amsterdam).

Barwise, J. (ed.) (1977) *Handbook of Mathematical Logic* (North-Holland, Amsterdam).

Barwise, J., and Etchemendy, J. (1987) *The Liar: An Essay on Truth and Circularity* (Oxford University Press, Oxford).

Barwise, J., and Schlipf, J. (1975) On recursively saturated models of arithmetic. In *Model Theory and Algebra*, Lecture Notes in Mathematics 498, 42–55.

———(1976) An introduction to recursively saturated and resplendent models. *The Journal of Symbolic Logic* 41, 531–536.

Beeson, M. (1977) Principles of continuous choice and continuity of functions in formal systems for constructive mathematics. *Annals of Mathematical Logic* 12, 249–322.

———(1980) Extensionality and choice in constructive mathematics. *Pacific Journal of Mathematics* 88, 1–28.

———(1981) Formalizing constructive mathematics: why and how? In F. Richman (ed.), *Constructive Mathematics*, Lecture Notes in Mathematics 873, 146–190.

———(1982) Problematic principles in constructive mathematics. In D. van Dalen, D. Lascar, and T. J. Smiley (eds.), *Logic Colloquium '80* (North-Holland, Amsterdam), 11–55.

———(1985) *Foundations of Constructive Mathematics: Metamathematical Studies.* (Springer-Verlag, Berlin).

———(1987) Some theories conservative over intuitionistic arithmetic. In W. Sieg (ed.), *Logic and Computation.* Contemporary Mathematics 106 (American Mathematical Society, Providence), 1–16.

Bernays, P. (1967) Hilbert, David. In P. Edwards (1967), Vol. 3, 496–504

Bishop, E. (1967) *Foundations of Constructive Analysis* (McGraw-Hill, New York).

Bishop, E., and Bridges, D. (1985) *Constructive Analysis* (Springer-Verlag, Berlin).

Bishop, E., and Cheng, H. (1972) *Constructive Measure Theory.* A. M. S. Memoirs 116 (American Mathematical Society, Providence).

Bridges, D. S. (1979) *Constructive Functional Analysis.* Research Notes in Mathematics 28 (Pitman, London).

Brouwer, L. E. J. (1975) *Collected Works*, Vol. I. *Philosophy and Foundations of Mathematics.* A. Heyting (ed.) (North-Holland, Amsterdam).

Browder, F. E. (ed.) (1976) *Mathematical Developments Arising from Hilbert Problems.* Proceedings of Symposia in Pure Mathematics 28 (American Mathematical Society, Providence).

Brown, D. K. (1987) *Functional Analysis in Weak Subsystems of Second Order Arithmetic.* Ph. D. Dissertation, Pennsylvania State University.

Buchholz, W. (1987) An independence result for $\prod_1^1$-CA + BI. *Annals of Pure and Applied Logic* 33, 131–155.

Buchholz, W., and Schütte, K. (1988) *Proof Theory of Impredicative Subsystems of Analysis* (Bibliopolis, Naples).

Buchholz, W., Feferman, S., Pohlers, W., and Sieg, W. (1981) *Iterated Inductive Definitions and Subsystems of Analysis: Recent Proof-Theoretical Studies.* Lecture Notes in Mathematics 897.

Buss, S. (1986) *Bounded Arithmetic* (Bibliopolis, Naples).

Cherlin, G. (1976) *Model Theoretic Algebra: Selected Topics.* Lecture Notes in Mathematics 521.

Chevalley, C., and Weil, A. (1957) Hermann Weyl (1885–1955). *Enseignement Mathématique* 3, 157–187; reprinted in Weyl (1968) Vol. IV, 655–685.

Chihara, C. (1973) *Ontology and the Vicious-Circle Principle* (Cornell University Press, Ithaca).

Christian, C. (1980) Leben und Wirken Kurt Gödels. *Monatshefte für Mathematik*, 89, 261–273.

Church, A. (1934) The Richard paradox. *American Mathematical Monthly* 41, 356–361.

Cohen, P. J. (1963) The independence of the continuum hypothesis I. *Proceedings of the National Academy of Sciences USA* 50, 1143–1148.

———(1964) The independence of the continuum hypothesis II. *Proceedings of the National Academy of Sciences USA* 51, 105–110.

———(1966) *Set Theory and the Continuum Hypothesis* (W. A. Benjamin, New York).

———(1969) Decision procedures for real and p-adic fields. *Communications on Pure and Applied Mathematics* 22, 131–151.

———(1971) Comments on the foundations of set theory. In D. S. Scott (ed.), *Axiomatic Set Theory, Part I* (American Mathematical Society, Providence), 9–15.

Cook, S. A., and Urquhart, A. (1993) Functional interpretations of feasibly constructive arithmetic. *Annals of Pure and Applied Logic* 63, 103–200.

Cousot, P. (1990) Methods and logics for proving programs. In J. van Leeuwen (ed.), *Handbook of Theoretical Computer Science*, Vol. B (Elsevier Science Publishers, Amsterdam), 843–993.

Cutland, N. (1980) *Computability. An Introduction to Recursive Function Theory* (Cambridge University Press, Cambridge).

Davis, M. (ed.) (1965) *The Undecidable: Basic Papers on Undecidable Propositions, Unsolvable Problems and Computable Functions.* (Raven Press, Hewlett).

———(1977) Unsolvable problems. In Barwise (1977), 567–594.

———(1982) Why Gödel didn't have Church's thesis. *Information and Control* 54, 3–24.

Davis, M., Matiyasevich, Y., and Robinson, J. (1976) Hilbert's tenth problem. Diophantine equations: positive aspects of a negative solution. In Browder (1976), 323–378.

Davis, P. J., and Hersh, R. (1981) *The Mathematical Experience* (Birkhäuser, Boston).

Dawson, J. W., Jr. (1983) The published work of Kurt Gödel: An annotated bibliography. *Notre Dame Journal of Formal Logic* 24, 255–284; *addenda and corrigenda, ibid.* 25, 283–287.

———(1984) Discussion on the foundations of mathematics. *History and Philosophy of Logic* 5, 111–129.

———(1984a) Kurt Gödel in sharper focus. *The Mathematical Intelligencer* 6, no. 4, 9–17.

———(1985) The reception of Gödel's incompleteness theorems. *PSA 1984*, Vol. 2 (Philosophy of Science Association, East Lansing).

———(1985a) Completing the Gödel-Zermelo correspondence. *Historia Mathematica* 12, 66–70.

————(1996) *Logical Dilemmas: The Life and Work of Kurt Gödel* (A. K. Peters, Wellesley).

de Bruijn, N. G. (1970) The mathematical language AUTOMATH, its usage and some of its extensions. In M. Laudet et al. (eds.), *Symposium on Automatic Demonstration V*, Lecture Notes in Mathematics 125, 29–61.

————(1980) A survey of the project AUTOMATH. In J. P. Seldin and J. R. Hindley (eds.), *To H. B. Curry: Essays on Combinatory Logic, Lambda Calculus and Formalism* (Academic Press, London), 579–606.

Delzell, C. (1996) Kreisel's unwinding of Artin's proof–Part I. In P. Odifreddi (ed.), *Kreiseliana. About and around Georg Kreisel* (A. K. Peters, Wellesley), 113–246.

Devlin, K. (1983) *Mathematics: The New Golden Age* (Penguin Books, London).

Dieudonné, J. (1982) Mathématiques vides et mathématiques significatives. In *Penser les Mathématiques* (Editions du Seuil, Paris), 15–38.

Drake, F. R. (1974) *Set Theory: An Introduction to Large Cardinals.* (North-Holland, Amsterdam).

Dreben, B., and Denton, J. (1966) A supplement to Herbrand. *The Journal of Symbolic Logic* 31, 393–398.

Dreben, B., Andrews, P., and Åanderaa, S. (1963) False lemmas in Herbrand. *Bulletin of the American Mathematical Society* 69, 699–706.

Dummett, M. (1977) *Elements of Intuitionism* (Clarendon Press, Oxford).

Dyson, F. (1983) Unfashionable pursuits. *The Mathematical Intelligencer* 5, 47–54.

Edwards, P. (ed.) (1967) *The Encyclopedia of Philosophy* (Macmillan and the Free Press, New York).

Eklof, P. (1973) Lefschetz' principle and local functors. *Proceedings of the American Mathematical Society* 37, 333–339.

Emch, G. G. (1972) *Algebraic Methods in Statistical Mechanics and Quantum Field Theory* (Wiley, New York).

Feferman, S. (1960) Arithmetization of metamathematics in a general setting. *Fundamenta Mathematicae* 49, 35–92.

————(1962) Transfinite recursive progressions of axiomatic theories. *The Journal of Symbolic Logic* 27, 259–316.

————(1964) Systems of predicative analysis. *The Journal of Symbolic Logic* 29, 1–30.

————(1965) Some applications of the notions of forcing and generic sets. *Fundamenta Mathematicae* 56, 325–345.

————(1968) Systems of predicative analysis, II. Representation of ordinals. *The Journal of Symbolic Logic* 33, 193–220.

————(1968a) Autonomous transfinite progressions and the extent of predicative mathematics. In B. Van Rootselaar and J. F. Staal (eds.), *Logic, Methodology and Philosophy of Science* III (North-Holland, Amsterdam), 121–135.

————(1969) Set-theoretical foundations of category theory (with an appendix by G. Kreisel). In S. MacLane (ed.), *Reports of the Midwest Category Seminar* III, Lecture Notes in Mathematics 106, 201–247.

————(1970) Formal theories for transfinite iterations of generalized inductive definitions and some subsystems of analysis. In Kino et al. (1970), 303–326.

————(1971) Ordinals and functionals in proof theory. In *Actes du Congrés International des Mathématiciens (Nice, 1970)*, Vol. 1 (Gauthier-Villars, Paris), 220–233.

————(1972) Infinitary properties, local functors, and systems of ordinal functions. In W. Hodges (ed.), *Conference in Mathematical Logic–London '70*, Lecture Notes in Mathematics 255, 63–97

————(1975) A language and axioms for explicit mathematics. In J. N. Crossley (ed.), *Algebra and Logic*, Lecture Notes in Mathematics 450, 87–139.

————(1977) Theories of finite type related to mathematical practice. In Barwise (1977), 913–972.

————(1977a) Inductive schemata and recursively continuous functionals. In R. O. Gandy and M. Hyland (eds.), *Logic Colloquium '76* (North-Holland, Amsterdam), 373–392.

————(1977b) Categorical foundations and foundations of category theory. In R. E. Butts and J. Hintikka (eds.), *Logic, Foundations of Mathematics and Computability Theory* (Reidel, Dordrecht), 149–169.

————(1979) Constructive theories of functions and classes. In M. Boffa, D. van Dalen, and K. McAloon (eds.), *Logic Colloquium '78* (North-Holland, Amsterdam), 159–224.

————(1979a) A more perspicuous formal system for predicativity. In K. Lorenz (ed.), *Konstruktionen versus Positionen* (Walter de Gruyter, Berlin), 87–139.

————(1979b) What does logic have to tell us about mathematical proofs? *The Mathematical Intelligencer* 2, 20–24. (Chapter 9 in this volume.)

————(1981) The logic of mathematical discovery vs. the logical structure of mathematics. In P. D. Asquith and I. Hacking (eds.), *PSA 1978* Vol. 2 (Philosophy of Science Association, East Lansing), 309–327. (Chapter 3 in this volume.)

————(1982) Inductively presented systems and the formalization of metamathematics. In D. van Dalen, D. Lascar, and T. J. Smiley (eds.), *Logic Colloquium '80* (North-Holland, Amsterdam), 95–128.

————(1982a) Monotone inductive definitions. In A. S. Troelstra and D. van Dalen (eds.), *The L. E. J. Brouwer Centenary Symposium* (North-Holland, Amsterdam), 77–89.

———(1984) Toward useful type free theories I. *The Journal of Symbolic Logic* 49, 75–111.

———(1984a) Kurt Gödel: Conviction and Caution. *Philosophia Naturalis* 21, 546–562. (Chapter 7 in this volume.)

———(1984b) Foundational ways. In W. Jäger, J. Moser, and R. Remmert (eds.), *Perspectives in Mathematics* (Birkhäuser, Basel), 147–158. (Chapter 4 in this volume.)

———(1985) Working foundations. *Synthese* 62, 229–254.

———(1985a) A theory of variable finite types. *Revista Colombiana de Matemáticas* 19, 95–105.

———(1986) Gödel's life and work. In Gödel (1986), 1–36. (Chapter 6 in this volume.)

———(1987) Proof theory: A personal report. Appendix to Takeuti (1987), 447–485.

———(1987a) Infinity in mathematics: Is Cantor necessary? In G. Toraldo di Francia (ed.), *L'infinito nella scienzia* (*Infinity in Science*) (Istituto della Enciclopedia Italiana, Rome), 151–209. (Divided between chapters 2 and 12 in this volume.)

———(1988) Hilbert's program relativized: Proof-theoretical and foundational reductions. *The Journal of Symbolic Logic*, 53, 364–384.

———(1988a) Weyl vindicated: *Das Kontinuum* 70 years later. In *Temi e prospettive della logica e della filosofia della scienza contemporanee*, Vol. I (CLUEB, Bologna), 59–93. (Chapter 13 in this volume.)

———(1988b) Turing in the land of $O(z)$. In R. Herken (ed.), *The Universal Turing Machine. A Halfcentury Survey*, (Oxford University Press, Oxford), 113–147.

———(1989) Finitary inductively presented logics. In R. Ferro et al. (eds.), *Logic Colloquium '88* (North-Holland, Amsterdam), 191–220.

———(1990) Polymorphic typed lambda-calculi in a type-free axiomatic framework. In W. Sieg (ed.), *Logic and Computation*, Contemporary Mathematics 106, 101–136.

———(1990a) Milking the *Dialectica* interpretation. Unpublished notes for a lecture at the Conference on Proof Theory, Arithmetic and Complexity, University of California, San Diego, 25–27 June.

———(1991) Reflecting on incompleteness. *The Journal of Symbolic Logic* 56, 1–49.

———(1992) Logics for termination and correctness of functional programs. In Y. Moschovakis (ed.), *Logic for Computer Science*, MSRI Publications 21 (Springer-Verlag, New York), 95–127.

———(1993) What rests on what? The proof-theoretic analysis of mathematics. In J. Czermak (ed.), *Philosophy of Mathematics*, Part I, Proceedings of the 15th International Wittgenstein Symposium (Verlag Hölder-Pichler-Tempsky, Vienna), 147–171. (Chapter 10 in this volume.)

———(1993a) Why a little bit goes a long way: Logical foundations of scientifically applicable mathematics. In *PSA 1992*, Vol. 2 (Philosophy of Science Association, East Lansing), 442–455. (Chapter 14 in this volume.)

———(1993b) Working foundations '91. In G. Corsi, M. L. Dalla Chiara, and G. C. Ghirardi (eds.), *Bridging the Gap: Philosophy, Mathematics and Physics*, Boston Studies in the Philosophy of Science, Vol. 140 (Kluwer, Dordrecht), 99–124. (Chapter 5 in this volume.)

———(1993c) Gödel's *Dialectica* interpretation and its two-way stretch. In G. Gottlob, A. Leitsch, and D. Mundici (eds.), *Computational Logic and Proof Theory*, Lecture Notes in Computer Science 713, 23–40. (Chapter 11 in this volume.)

———(1996) Gödel's program for new axioms: Why, where, how and what? In P. Hájek (ed.), *Gödel '96*, Lecture Notes in Logic 6, 3–22.

———(1996a) Kreisel's "unwinding" program. In P. Odifreddi (ed.), *Kreiseliana. About and around Georg Kreisel* (A. K. Peters, Wellesley), 247–273.

———(1998) Does mathematics need new axioms? *American Mathematical Monthly* (to appear).

Feferman, S., and Hellman, G. (1995) Predicative foundations of arithmetic. *Journal of Philosophical Logic* 24, 1–17.

Feferman, S., and Jäger, G. (1983) Choice principles, the bar rule and autonomously iterated comprehension schemes in analysis. *The Journal of Symbolic Logic* 48, 63–70.

———(1993) Systems of explicit mathematics with non-constructive μ-operator, I. *Annals of Pure and Applied Logic* 65, 243–263.

———(1996) Systems of explicit mathematics with non-constructive μ-operator, II. *Annals of Pure and Applied Logic* 79, 37–52.

Feferman, S., and Sieg, W. (1981) Iterated inductive definitions and subsystems of analysis. In Buchholz et al. (1981), 16–77.

———(1981a) Proof-theoretic equivalences between classical and constructive theories. In Buchholz et al. (1981), 78–142.

Feigl, H. (1969) The Wiener Kreis in America. In D. Fleming and B. Bailyn (eds.), *The Intellectual Migration: Europe and America, 1930–1960* (Harvard University Press, Cambridge, MA).

Fenstad, J. E. (1980) *General Recursion Theory, an Axiomatic Approach* (Springer Verlag, Berlin).

Ferreira, F. (1988) *Polynomial Time Computable Arithmetic and Conservative Extensions.* Ph. D. Thesis, Pennsylvania State University.

———(1994) A feasible theory for analysis. *The Journal of Symbolic Logic* 59, 1001–1011.

Fitting, M. (1981) *Fundamentals of Generalized Recursion Theory* (North-Holland, Amsterdam).

Folina, J. (1992) *Poincaré and the Philosophy of Mathematics* (Macmillan, London).

Fraenkel, A. (1922) Der Begriff 'definit' und die Unabhängigkeit des Auswahlaxioms. *Sitzungsberichte der Preussischen Akademie der Wissenschaften, Physikalisch-mathematische Klasse*, 253–257; English translation in van Heijenoort (1967), 284–289.

Fraenkel, A., Bar-Hillel, Y., and Levy, A. (1973) *Foundations of Set Theory* (North-Holland, Amsterdam).

Frege, G. (1879) *Begriffsschrift, eine der arithmetischen nachgebildete Formelsprache des reinen Denkens* (Nebert, Halle); English translation in van Heijenoort (1967), 1–82.

Freiling, C. (1986) Axioms of symmetry: Throwing darts at the real line. *The Journal of Symbolic Logic*, 51, 190–200.

Friedman, H. (1970) Iterated inductive definitions and Σ^1_2-AC. In Kino et al. (1970), 435–442.

————(1971) Algorithmic procedures, generalized Turing algorithms and elementary recursion theories. In R. O. Gandy and C. E. M. Yates (eds.), *Logic Colloquium '69* (North-Holland, Amsterdam), 361–390.

————(1975) Some systems of second order arithmetic and their use. In *Proceedings of the International Congress of Mathematicians, Vancouver 1974*, Vol. I, 235–242.

————(1976) Systems of second order arithmetic with restricted induction I, II (Abstracts) *The Journal of Symbolic Logic* 41, 557–559.

————(1977) Set-theoretic foundations for constructive analysis. *Annals of Mathematics* 105, 1–28.

————(1980) A strong conservative extension of Peano arithmetic. In Barwise et al. (eds.), *The Kleene Symposium* (North-Holland, Amsterdam), 113–122.

————(1986) Necessary uses of abstract set theory in finite mathematics. *Advances in Mathematics* 60, 92–122.

————(1998) Finite functions and the necessary use of large cardinals. *Annals of Mathematics* (2) (to appear).

Friedman, H., McAloon, K., and Simpson, S. G. (1982) A finite combinatorial principle which is equivalent to the 1-consistency of predicative analysis. In G. Metakides (ed.), *Patras Logic Symposion* (North-Holland, Amsterdam), 197–230.

Friedman, H., Simpson, S. G., and Smith, R. (1983) Countable algebra and set existence axioms. *Annals of Pure and Applied Logic* 25, 141–181.

Gentzen, G. (1935) Untersuchungen über das logische Schließen. *Mathematische Zeitschrift* 39, 176–210, 405–431; English translation in Gentzen (1969), 68–131.

————(1936) Die Widerspruchsfreiheit der reinen Zahlentheorie. *Mathematische Annalen* 112, 493–565; English translation in Gentzen (1969), 132–213.

————(1969) *The Collected Papers of Gerhard Gentzen.* Edited and translated into English by M. E. Szabo (North-Holland, Amsterdam).

Girard, J. Y. (1981) $\prod_2^1$-logic, Part I: Dilators. *Annals of Mathematical Logic* 21, 75–219.

————(1987) *Proof Theory and Logical Complexity* (Bibliopolis, Naples).

Gödel, K. (1929) *Über die Vollständigkeit des Logikkalküls.* Doctoral dissertation, University of Vienna; reprinted with an English translation in Gödel (1986), 60–101.

————(1930) Die Vollständigkeit der Axiome des logischen Funktionenkalküls.*Monatshefte für Mathematik und Physik* 37, 349–360; reprinted with an English translation in Gödel (1986), 102–123.

————(1931) Über formal unentscheidbare Sätze der *Principia Mathematica* und verwandter Systeme I. *Monatshefte für Mathematik und Physik* 38, 173–198; reprinted with an English translation in Gödel (1986), 144–195.

————(1931a) Diskussion zur Grundlegung der Mathematik. *Erkenntnis* 2, 147–151; reprinted with an English translation in Gödel (1986), 200–205.

————(1933) Zur intuitionistischen Arithmetik und Zahlentheorie. *Ergebnisse eines mathematische Kolloquiums* 4, 34–38; reprinted with an English translation in Gödel (1986), 286–295.

————(1933a) The present situation in the foundations of mathematics. In Gödel (1995), 45–53.

————(1934) *On Undecidable Propositions of Formal Mathematical Systems.* Mimeographed lecture notes taken by S. C. Kleene and J. B. Rosser; reprinted with revisions in Davis (1965), 39–74; reprinted in Gödel (1986), 346–371.

————(1938) The consistency of the axiom of choice and of the generalized continuum hypothesis. *Proceedings of the National Academy of Sciences USA* 24, 556–557; reprinted in Gödel (1990), 26–27.

————(1939) Consistency proof for the generalized continuum hypothesis. *Proceedings of the National Academy of Sciences USA* 25, 220–224; reprinted in Gödel (1990), 28–32.

————(1940) *The Consistency of the Axiom of Choice and the Generalized Continuum Hypothesis with the Axioms of Set Theory.* Annals of Mathematics Studies, Vol. 3 (Princeton University Press, Princeton); reprinted in Gödel (1990), 33–101.

————(1941) In what sense is intuitionistic logic constructive? In Gödel (1995), 189–200.

————(1944) Russell's mathematical logic. In P. A. Schilpp (ed.), *The Philosophy of Bertrand Russell*, Library of Living Philosophers, Vol. 5 (Northwestern University Press, Evanston); reprinted in Gödel (1990), 119–141.

———(1946) Remarks before the Princeton bicentennial conference on problems in mathematics; first published in Davis (1965), 84–88; reprinted in Gödel (1990), 150–153.

———(1947) What is Cantor's continuum problem? *American Mathematical Monthly* 54, 515–525; errata, 55, 151; reprinted in Gödel (1990), 176–187.

———(1958) Über eine bisher noch nicht benützte Erweiterung des finiten Standpunktes. *Dialectica* 12, 280–287; reprinted with an English translation in Gödel (1990), 240–251.

———(1964) What is Cantor's continuum problem? (revised version of 1947). In P. Benacerraf and H. Putnam (eds.), *Philosophy of Mathematics: Selected Readings* (Prentice-Hall, Englewood Cliffs, NJ); reprinted in Gödel (1990), 254–270.

———(1972) On an extension of finitary mathematics which has not yet been used. In Gödel (1990), 271–280.

———(1986) *Collected Works, Vol. I: Publications 1929–1936.* S. Feferman et al. (eds.) (Oxford University Press, Oxford).

———(1990) *Collected Works, Vol. II: Publications 1938–1974.* S. Feferman et al. (eds.) (Oxford University Press, Oxford).

———(1995) *Collected Works, Vol. III: Unpublished Essays and Lectures.* S. Feferman et al. (eds.) (Oxford University Press, Oxford).

Goldfarb, W. (1988) Poincaré against the logicists. In W. Aspray and P. Kitcher (eds.), *History and Philosophy of Modern Mathematics* (University of Minnesota Press, Minneapolis), 61–81.

Goldstine, H. H. (1972) *The Computer from Pascal to von Neumann* (Princeton University Press, Princeton).

Goodman, N. D. (1970) A theory of constructions equivalent to arithmetic. In Kino et al. (1970), 101–120.

———(1979) Mathematics as an objective science. *American Mathematical Monthly* 86, 540–551.

Goodstein, R. L. (1957) *Recursive Number Theory* (North-Holland, Amsterdam).

———(1961) *Recursive Analysis* (North-Holland, Amsterdam).

Grattan-Guinness, I. (1970) *The Development of the Foundations of Mathematical Analysis from Euler to Riemann* (MIT Press, Cambridge, MA).

———(1979) In memoriam Kurt Gödel: His 1931 correspondence with Zermelo on his incompleteness theorem. *Historia Mathematica* 6, 294–304.

———(ed.)(1980) *From the Calculus to Set-Theory, 1630–1910.* (Duckworth, London).

Gunter, C. (1992) *Semantics of Programming Languages* (MIT Press, Cambridge, MA).

Hacking, I. (1979) Imré Lakatos' philosophy of science. *British Journal of the Philosophy of Science* 30, 381–402.

Hahn, H. (1980) *Empiricism, Logic and Mathematics: Philosophical Papers* (Reidel, Dordrecht).

Hájek, P., and Pudlák, P. (1993) *Metamathematics of First-Order Arithmetic* (Springer-Verlag, Berlin).

Hallett, M. (1984) *Cantorian Set-Theory and the Limitation of Size* (Clarendon Press, Oxford).

————(1995) Hilbert and logic. *Québec Studies in the Philosophy of Science* 7, 135–187.

Harel, D. (1987) *Algorithmics. The Spirit of Computing.* (Addison-Wesley, Reading, MA).

Heims, S. (1980) *John von Neumann and Norbert Wiener: From Mathematics to the Technologies of Life and Death* (MIT Press, Cambridge, MA).

Hellman, G. (1993) Constructive mathematics and quantum mechanics: Unbounded operators and the spectral theorem. *Journal of Philosophical Logic* 22, 221–248.

————(1993a) On the scope and force of indispensability arguments. In *PSA 1992*, Vol. 2 (Philosophy of Science Association, East Lansing), 456–464.

Herbrand, J. (1930) *Recherches sur la théorie de la démonstration.* Doctoral Dissertation, University of Paris; English translation by B. Dreben and J. van Heijenoort in Herbrand (1971).

————(1931) Sur la non-contradiction de l'arithmétique. *Journal für die reine und angewandte Mathematik* 166, 1–8; English translation in van Heijenoort (1967), 618–628.

————(1968) *Écrits logiques.* J. van Heijenoort (ed.) (Presses Universitaires de France, Paris).

————(1971) *Logical Writings.* English translation of Herbrand (1968) by W. Goldfarb (Reidel, Dordrecht).

Hersh, R. (1978) Introducing Imré Lakatos. *The Mathematical Intelligencer* 1, 148–151.

————(1979) Some proposals for reviving the philosophy of mathematics. *Advances in Mathematics* 31, 31–50.

Hilbert, D. (1900) Mathematische Probleme. Vortrag, gehalten auf dem internationalen Mathematiker-Kongress zu Paris 1900. *Göttinger Nachrichten* 1900, 253–297.

————(1902) Mathematical problems. English translation of Hilbert (1900). *Bulletin of the American Mathematical Society* 8, 437–479; reprinted in Browder (1976), 1–34.

————(1904) Über die Grundlagen der Logik und der Arithmetik. In *Verhandlungen des dritten internationalen Mathematiker-Kongresses in Heidelberg vom 8. bis 13. August 1904* (Teubner, Leibzig), 174–185; English translation in van Heijenoort (1967), 129–138.

———(1918) Axiomatisches Denken. *Mathematische Annalen* 78, 405–415.

———(1926) Über das Unendliche. *Mathematische Annalen* 95, 161–190; English translation in van Heijenoort (1967), 367–392.

Hilbert, D., and Ackermann, W. (1928) *Grundzüge der theoreti-schen Logik* (Springer-Verlag, Berlin); English translation in Hilbert and Ackermann (1950).

———(1950) *Principles of Mathematical Logic.* English translation of Hilbert and Ackerman (1928) (Chelsea, New York).

Hilbert, D., and Bernays, P. (1934) *Grundlagen der Mathematik,* Vol. I (Springer-Verlag, Berlin) (2nd rev. edn. 1968).

———(1939) *Grundlagen der Mathematik,* Vol. II (Springer-Verlag, Berlin). (2nd rev. edn. 1970).

Hodges, A. (1983) *Alan Turing. The Enigma* (Simon and Schuster, New York).

Hofstadter, D. (1979) *Gödel, Escher, Bach: An Eternal Golden Braid* (Basic Books, New York).

Howard, W. (1968) Functional interpretation of bar induction by bar recursion. *Compositio Mathematica* 20, 107–124.

———(1973) Hereditarily majorizable functionals of finite type. Appendix to Troelstra (1973), 454–461.

Isaacson, D. (1987) Arithmetical truth and higher order concepts. In Paris Logic Group (eds.), *Logic Colloquium '85* (North-Holland, Amsterdam).

Jäger, G. (1986) *Theories of Admissible Sets. A Unifying Approach to Proof Theory* (Bibliopolis, Naples).

Jäger, G., and Pohlers, W. (1982) Eine beweistheoretische Untersuchung von (Δ^1_2-CA)+(BI) und verwandter Systeme. *Sitzungsberichte der Bayerischen Akademie der Wissenschaften* 1–28.

Jensen, R. (1969) On the consistency of a slight (?) modification of Quine's *New Foundations.* In D. Davidson and J. Hintikka (eds.), *Words and Objections: Essays on the Work of W. V. O. Quine* (Reidel, Dordrecht), 278–291.

Kanamori, A. (1994) *The Higher Infinite* (Springer-Verlag, Berlin).

Kanamori, A., and Magidor, M. (1978) The evolution of large cardinal axioms in set theory. In *Higher Set Theory,* Lecture Notes in Mathematics 669, 99–275.

Kino, A., Myhill, J., and Vesley, R. E. (eds.) (1970) *Intuitionism and Proof Theory* (North-Holland, Amsterdam).

Kleene, S. C. (1945) On the interpretation of intuitionistic number theory. *The Journal of Symbolic Logic,* 10, 109–124.

———(1952) *Introduction to Metamathematics* (North-Holland, Amsterdam).

———(1959) Countable functionals. In A. Heyting (ed.), *Constructivity in Mathematics* (North-Holland, Amsterdam), 81–100.

———(1976) The work of Kurt Gödel. *The Journal of Symbolic Logic* 41, 761–778; addendum, 43, 613.

———(1981) Origins of recursive function theory. *Annals of the History of Computing* 3, 52–67; corrections in Davis (1982), footnotes 10 and 12.

———(1987) Gödel's impression on students of logic in the 1930s. In P. Weingartner and L. Schmetterer (eds.), *Gödel Remembered* (Bibliopolis, Naples), 49–64.

———(1987a) Kurt Gödel. April 28, 1906–January 14, 1978. *Biographical Memoirs, National Academy of Sciences USA* 56, 134–178.

Kleene, S. C., and Vesley, R. E. (1965) *The Foundations of Intuitionistic Mathematics, Especially in Relation to Recursive Function Theory* (North-Holland, Amsterdam).

Kline, M. (1972) *Mathematical Thought from Ancient to Modern Times* (Oxford University Press, Oxford).

Kohlenbach, U. (1990) *Theorie der majorisierbaren und stetige Funktionale und ihre Anwendung bei der Extraktion von Schranken aus inkonstruktiven Beweisen.* Dissertation, University of Frankfurt.

———(1992) Pointwise hereditary majorization and some applications. *Archives of Mathematical Logic* 31, 227–241.

———(1992a) Effective bounds from ineffective proofs in analysis: An application of functional interpretation and majorization. *The Journal of Symbolic Logic* 57, 1239–1273.

———(1993) Effective moduli from ineffective uniqueness proofs. An unwinding of de La Vallée Poussin's proof for Chebycheff approximation. *Annals of Pure and Applied Logic* 64, 27–94.

———(1993a) New effective moduli of uniqueness and uniform a priori estimates for constants of strong unicity by logical analysis of known proofs in best approximation theory. *Numerical Functional Analysis and Optimization* 14, 581–606.

———(1994) Analyzing proofs in analysis. W. Hodges et al. (eds.), *Logic: From Foundations to Applications* (Clarendon Press, Oxford), 225–260.

Kreisel, G. (1951) On the interpretation of non-finitist proofs–Part I. *The Journal of Symbolic Logic* 16, 241–267.

———(1952) On the interpretation of non-finitist proofs–Part II. Interpretation of number theory. Applications. *The Journal of Symbolic Logic* 17, 43–58.

———(1957) Gödel's interpretation of Heyting's arithmetic. *Summaries of Talks Presented at the Summer Institute for Symbolic Logic, Cornell University 1957* (Institute for Defense Analyses, Princeton) (2nd edn. 1960).

———(1958) Mathematical significance of consistency proofs. *The Journal of Symbolic Logic* 23, 155–182.

———(1958a) Hilbert's programme. *Dialectica* 12, 346–372.

———(1959) Interpretation of analysis by means of constructive functionals of finite type. In A. Heyting (ed.), *Constructivity in Mathematics* (North-Holland, Amsterdam), 101–128.

———(1960) Ordinal logics and the characterization of informal concepts of proof. *Proceedings of the International Congress of Mathematicians, 14–21 August 1958* (Cambridge University Press, Cambridge), 289–299.

———(1962) Foundations of intuitionistic logic. In E. Nagel et al. (eds.), *Logic, Methodology and Philosophy of Science* (Stanford University Press, Stanford), 198–210.

———(ed.) (1963) *Reports of Seminar on the Foundations of Analysis, Stanford Summer 1963* (Stanford University, Stanford).

———(1967) Mathematical logic: What has it done for the philosophy of mathematics? In R. Schoenman (ed.), *Bertrand Russell, Philosopher of the Century* (Allen and Unwin, London), 201–272.

———(1968) A survey of proof theory. *The Journal of Symbolic Logic* 33, 321–388.

———(1970) Principles of proof and ordinals implicit in given concepts. In Kino et al. (1970), 489–516.

———(1971) Observations on popular discussions of foundations. In D. S. Scott (ed.), *Axiomatic Set Theory*, Part I (American Mathematical Society, Providence), 189–198.

———(1976) What have we learned from Hilbert's second problem? In Browder (1976), 93–130.

———(1977) Review of L. E. J. Brouwer *Collected Works*, Vol. I. *Bulletin of the American Mathematical Society* 83, 86–93.

———(1980) Kurt Gödel, 28 April 1906–14 January 1978. In *Biographical Memoirs of Fellows of the Royal Society* 26, 148–224; corrections 27, 697, and 28, 718.

———(1987) Gödel's excursions into intuitionistic logic. In P. Weingartner and L. Schmetterer (eds.), *Gödel Remembered* (Bibliopolis, Naples), 65–186.

Kreisel, G., and Troelstra, A. S. (1970) Formal systems for some branches of intuitionistic analysis. *Annals of Mathematical Logic* 1, 229–387.

Lakatos, I. (1976) *Proofs and Refutations: The Logic of Mathematical Discovery* (Cambridge University Press, Cambridge)

———(1978) *Mathematics, Science and Epistemology.* Philosophical Papers 2. J. Worrall and G. Currie (eds.) (Cambridge University Press, Cambridge)

Lavine, S. (1994) *Understanding the Infinite* (Harvard University Press, Cambridge, MA).

Luckhardt, H. (1996) Bounds extracted by Kreisel from ineffective proofs. In P. Odifreddi (ed.), *Kreiseliana. About and around Georg Kreisel* (A. K. Peters, Wellesley), 289–300.

MacLane, S. (1961) Locally small categories and the foundations of mathematics. In *Infinitistic Methods* (Pergamon, Oxford), 25–43.

————(1971) *Categories for the Working Mathematician.* (Springer-Verlag, Berlin).

————(1981) Mathematical models: A sketch for the philosophy of mathematics. *American Mathematical Monthly* 88, 462–472.

Maddy, P. (1990) *Realism in Mathematics* (Clarendon Press, Oxford).

————(1992) Indispensability and practice. *Journal of Philosophy* 59, 275–289.

Malament, D. (1992) Critical notice: Itamar Pitowsky's *Quantum Probability - Quantum Logic. Philosophy of Science* 59, 300–320.

Manin, Y. I. (1977) *Course in Mathematical Logic* (Springer-Verlag, Berlin).

Margenstern, M. (1978) On a variant of the constructivisation of the theory of almost periodic functions. *Zeitschrift für mathematische Logik und Grundlagen der Mathematik*, 24, 495–507.

Martin, D. A. (1976) Hilbert's first problem: The continuum hypothesis. In Browder (1976), 81–92.

Martin, R. L. (ed.) (1984) *Recent Essays on Truth and the Liar Paradox* (Oxford University Press, Oxford).

Martin-Löf, P. (1982) Constructive mathematics and computer programming. In L. J. Cohen et al. (eds.), *Logic, Methodology and the Philosophy of Science VI* (North-Holland, Amsterdam).

————(1984) *Intuitionistic Type Theory* (Bibliopolis, Naples).

Matiyasevich, Y. (1993) *Hilbert's Tenth Problem* (MIT Press, Cambridge, MA).

Menger, K. (1994) *Reminiscences of the Vienna Circle and the Mathematical Colloquium* (Kluwer, Dordrecht).

Mines, R., Richman, F., and Ruitenberg, W. (1988) *A Course in Constructive Algebra* (Springer-Verlag, Berlin).

Mints, G. (1976) What can be done in PRA? *Zapiski Nauchuyh Seminarov, LOMI*, 60, 93–102 (in Russian); English translation in *Journal of Soviet Mathematics* 14 (1980), 1487–1494.

Montgomery, D. (1963) Oswald Veblen. *Bulletin of the American Mathematical Society* 69, 26–36.

Mooij, J. J. A. (1966) *La philosophie de mathématique de Henri Poincaré* (Gauthier-Villars, Paris).

Moore, G. H. (1980) Beyond first-order logic: The historical interplay between mathematical logic and axiomatic set theory. *History and Philosophy of Logic* 1, 85–137.

————(1982) *Zermelo's Axiom of Choice: Its Origins, Development and Influence* (Springer-Verlag, Berlin).

Moschovakis, Y. N. (1977) On the basic notions in the theory of induction. In R. E. Butts and J. Hintikka (eds.), *Logic, Foundations of Mathematics, and Computability Theory* (Reidel, Dordrecht), 207–236.

———(1991) The formal language of recursion. *The Journal of Symbolic Logic* 54, 1216–1252.

Mosses, P. D. (1990) Denotational semantics. In J. van Leeuwen (ed.), *Handbook of Theoretical Computer Science, Vol. B: Formal Models and Semantics* (Elsevier, Amsterdam), 595–631.

Myhill, J. (1975) Constructive set theory. *The Journal of Symbolic Logic* 40, 347–383.

Nagel, E., and Newman, J. R. (1960) *Gödel's Proof* (New York University Press, New York).

Nelson, E. (1986) *Predicative Arithmetic* (Princeton University Press, Princeton).

Newman, M. H. A. (1957) Hermann Weyl. *Biographical Memoirs of the Fellows of the Royal Society* 3, 305–328; reproduced in shortened form in *Journal of the London Mathematical Society* 33, 500–511.

Paris, J., and Harrington, L. (1977) A mathematical incompleteness in Peano Arithmetic. In Barwise (1977), 1133–1142.

Parsons, C. (1970) On a number-theoretic choice scheme and its relation to induction. In Kino et al. (1970), 459–473.

———(1983) The impredicativity of induction. In L. S. Caumann et al. (eds.), *How Many Questions? Essays in Honor of S. Morgenbesser* (Hackett, Indianapolis), 132–154.

———(1992) The impredicativity of induction. Revised and expanded version of Parsons (1983). In M. Detlefsen (ed.), *Proof, Logic and Formalization* (Routledge, London), 139–161.

Pitowsky, I. (1989) *Quantum Probability – Quantum Logic.* Lecture Notes in Physics 321.

Pohlers, W. (1989) *Proof Theory: An introduction.* Lecture Notes in Mathematics 1407.

———(1991) Proof theory and ordinal analysis. *Archive for Mathematical Logic* 30, 311–376.

Poincaré, H. (1913) *Dernières penseés* (Flammarion, Paris); English translation in Poincaré (1963).

———(1963) *Mathematics and Science: Last essays.* English translation of Poincaré (1913) (Dover Press, New York).

Pólya, G. (1957) *How to Solve It* (2nd edn.) (Princeton University Press, Princeton).

———(1965) *Mathematical Discovery: On Understanding, Learning and Teaching Problem Solving*, Vols. 1 and 2 (Wiley, New York).

———(1968) *Mathematics and Plausible Reasoning*, Vols. 1 and 2, (2nd edn.) (Princeton University Press, Princeton).

————(1972) Eine Erinnerung an Hermann Weyl. *Mathematische Zeitschrift.* 126, 296–298.

Pour-El, M. B., and Richards, J. I. (1989) *Computability in Analysis and Physics* (Springer-Verlag, Berlin).

Prawitz, D. (1965) *Natural Deduction: A Proof-Theoretical Study* (Almquist and Wiksell, Stockholm).

————(1971) Ideas and results in proof theory. In J. E. Fenstad (ed.), *Proceedings of the Second Scandinavian Logic Symposium* (North-Holland, Amsterdam), 235–307.

Quine, W. V. O. (1963) Predicate-functor logic. In J. E. Fenstad (ed.), *Proceedings of the Second Scandinavian Logic Symposium* (North-Holland, Amsterdam).

————(1979) Kurt Gödel (1906–1978). *The Year Book 1978 of the American Philosophical Society*, 81–84.

————(1984) Review of Charles Parsons' *Mathematics in Philosophy. Journal of Philosophy* 81, 783–794.

————(1986) Reply to Charles Parsons. In H. Hahn and P. A. Schilpp (eds.), *The Philosophy of W. V. O. Quine, I* (Open Court, La Salle), 396–403.

Rabin, M. O. (1977) Decidable theories. In Barwise (1977), 595–629.

Rang, B., and Thomas, W. (1981) Zermelo's discovery of the "Russell paradox." *Historia Mathematica* 8, 15–22.

Rathjen, M. (1991) Proof-theoretic analysis of KPM. *Archive for Mathematical Logic* 30, 377–403.

————(1995) Recent advances in ordinal analysis: Π^1_2-CA and related systems. *Bulletin of Symbolic Logic* 1, 468–485.

Reid, C. (1970) *Hilbert* (Springer Verlag, Berlin).

Reinhardt, W. N. (1974) Remarks on reflection principles, large cardinals, and elementary embeddings. In T. Jech (ed.), *Axiomatic Set Theory,* Part II (American Mathematical Society, Providence), 189–206.

Robinson, A. (1966) *Non-standard Analysis* (North-Holland, Amsterdam) (rev. edn. 1974).

Robinson, J. (1996) *The Collected Works of Julia Robinson.* S. Feferman (ed.) (American Mathematical Society, Providence).

Rogers, H. (1967) *Theory of Recursive Functions and Effective Computability*
(McGraw-Hill, New York).

Russell, B. (1903) *The Principles of Mathematics* (Allen and Unwin, London).

————(1919) *Introduction to Mathematical Philosophy* (Allen and Unwin, London).

Sacks, G. (1990) *Higher Recursion Theory* (Springer-Verlag, Berlin).

Schütte, K. (1965) Eine Grenze für die Beweisbarkeit der transfiniten Induktion in der verzweigten Typenlogik. *Archiv für Mathematische Logik und Grundlagenforschung* 7, 45–60.

————(1977) *Proof Theory* (Springer-Verlag, Berlin).

Scott, D. S. (1961) Measurable cardinals and constructible sets. *Bulletin de l'académie Polonaise des Sciences* 9, 521–524.

————(1972) Continuous lattices. In F. W. Lawvere (ed.), *Toposes, Algebraic Geometry and Logic*, Lecture Notes in Mathematics 274, 97–136.

————(1976) Data types as lattices. *SIAM Journal of Computing* 5, 522–587.

Shankar, N. (1994) *Metamathematics, Machines, and Gödel's Proof* (Cambridge University Press, Cambridge, MA).

Shoenfield, J. (1954) A relative consistency proof. *The Journal of Symbolic Logic* 53, 338–348.

Sieg, W.(1984) Foundations for analysis and proof theory. *Synthese* 60, 159–200.

————(1985) Fragments of arithmetic. *Annals of Pure and Applied Logic* 28, 33–72.

————(1990) Relative consistency and accessible domains. *Synthese* 84, 259–297.

————(1991) Herbrand analyses. *Archive for Mathematical Logic* 30, 409–441.

Simpson, S. G. (1984) Which set existence axioms are needed to prove the Cauchy/Peano theorem of ordinary differential equations? *The Journal of Symbolic Logic* 49, 783–802.

————(1985) Nonprovability of certain combinatorial properties of finite trees. In L. A. Harrington et al (eds.), *Harvey Friedman's Research in the Founations of Mathemaitcs* (North-Holland, Amsterdam), 87–117.

————(1985a) Reverse mathematics. In A. Nerode and R. Shore (eds.), *Recursion Theory*, Proceedings of Symposia in Pure Mathematics XLII (American Mathematical Society, Providence), 461–471.

————(1987) Subsystems of Z_2 and reverse mathematics. Appendix to Takeuti (1987), 432–446.

————(1987a) Unprovable theorems and fast-growing functions. In S. G. Simpson (ed.), *Logic and Combinatorics*, Contemporary Mathematics 65 (American Mathematical Society, Providence), 359–394.

————(1988) Partial realizations of Hilbert's program. *The Journal of Symbolic Logic* 53, 349–363.

————(1998) *Subsystems of Second Order Arithmetic* (Springer-Verlag, Berlin).

Skolem, T. (1923) Begründung der elementaren Arithmetik durch die rekurrierende Denkweise ohne Anwendung scheinbarer Veränderlichen mit unendlichem Ausdehnungsbereich. *Skrifter utgit av Videnskapsselskapet i Kristiana*, I. *Matematisk-naturvidenskabelig klasse*, no. 6, 1–38; English translation in van Heijenoort (1967), 302–333.

———(1923a) Einige Bemerkungen zur axiomatischen Begründung der Mengenlehre. *Matematikerkongressen i Helsingfors 4–7 juli 1922, Den femte skandinaviska matematikerkongressen, Redogörelse* (Akademiska Bokhandeln, Helsinki); English translation in van Heijenoort (1967), 290–301.

Smorynski, C. (1977) The incompleteness theorems. In Barwise (1977), 821–865.

Smullyan, R. (1961) *Theory of Formal Systems* (Princeton University Press, Princeton).

———(1987) *Forever Undecided: A Puzzle Guide to Gödel* (Knopf, New York).

———(1992) *Gödel's Incompleteness Theorems* (Oxford University Press, Oxford).

Spector, C. (1962) Provably recursive functionals of analysis: A consistency proof of analysis by an extension of principles formulated in current intuitionistic mathematics. In J. C. E. Dekker (ed.), *Recursive Function Theory*, Proceedings of Symposia in Pure Mathematics 5 (American Mathematical Society, Providence), 1–27.

Stoy, J. E. (1979) *Denotational Semantics: The Scott-Strachey Approach to Programming Language Theory.* (MIT Press, Cambridge, MA).

Straus, E. G. (1982) Reminiscences. In G. Holton and Y. Elkana (eds.), *Albert Einstein: Historical and Cultural Perspectives. The Centennial Symposium in Jerusalem* (Princeton University Press, Princeton), 417–423.

Streater, R. F., and Wightman, A. S. (1978) *PCT, Spin and Statistics, and All That* (Benjamin/Cummings, New York), (2nd edn.).

Stroyan, K. (1977) Infinitesimal analysis of curves and surfaces. In Barwise (1977), 197–231.

Tait, W. W. (1981) Finitism. *Journal of Philosophy* 78, 524–546.

Takeuti, G. (1975) *Proof Theory.* (North-Holland, Amsterdam).

———(1978) A conservative extension of Peano arithmetic. *Two Applications of Logic to Mathematics*, Part II (Princeton University Press, Princeton, and Publications of the Mathematics Society of Japan 13), 73–137.

———(1987) *Proof Theory* (2nd edn.) (North-Holland, Amsterdam).

Tarski, A. (1944) The semantic conception of truth and the foundations of semantics. *Philosophy and Phenomenological Research* 4, 341–376.

———(1956) *Logic, Semantics, Metamathematics. Papers from 1923 to 1938.* J. H. Woodger (ed.) (Clarendon Press, Oxford).

Tarski, A., Mostowski, A., and Robinson, R. M. (1953) *Undecidable Theories* (North-Holland, Amsterdam).

Taussky-Todd, O. (1987) Remembrances of Kurt Gödel. In P. Weingartner and L. Schmetterer (eds.), *Gödel Remembered* (Bibliopolis, Naples), 29–41.

Troelstra, A. S. (ed.) (1973) *Metamathematical Investigations of Intuitionistic Arithmetic and Analysis*, Lecture Notes in Mathematics 344 (Springer-Verlag, Berlin).

———(1977) Aspects of constructive mathematics. In Barwise (1977), 973–1052.

———(1990) Introductory note to *1958* and *1972*. In Gödel (1990), 217–241.

Troelstra, A. S., and van Dalen, D. (1988) *Constructivism in Mathematics*, Vols. I and II (North-Holland, Amsterdam).

Turing, A. (1939) Systems of logic based on ordinals. *Proceedings of the London Mathematical Society* (2) 45, 161–228; reprinted in Davis (1965), 155–222.

Turner, R. (1991) *Constructive Foundations for Functional Languages* (McGraw-Hill, Maidenhead, U. K.).

Tutte, W. T. (1978) Colouring problems. *The Mathematical Intelligencer* 1, 72–75.

Tymoczko, T. (ed.) (1986) *New Directions in the Philosophy of Mathematics* (Birkhäuser, Boston).

Ulam, S. (1958) John von Neumann, 1903–1957. *Bulletin of the American Mathematical Society* 64, no. 3, part 2 (May supplement), 1–49.

———(1976) *Adventures of a Mathematician* (Scribner's, New York).

van Heijenoort, J. (ed.) (1967) *From Frege to Gödel: A Source Book in Mathematical Logic* (Harvard University Press, Cambridge, MA).

———(1986) *Selected Essays* (Bibliopolis, Naples).

van Stigt, W. P. (1990) *Brouwer's Intuitionism* (North-Holland, Amsterdam).

Visser, A. (1989) Semantics and the liar paradox. In D. Gabbay (ed.), *Handbook of Philosophical Logic* 4, (Reidel, Dordrecht) 617–706.

von Neumann, J. (1927) Zur Hilbertschen Beweistheorie. *Mathematische Zeitschrift* 26, 1–46.

———(1929) Über die Widerspruchsfreiheitsfrage in der axiomatischen Mengenlehre. *Journal für die reine und angewandte Mathematik* 160, 227–241.

———(1966) *Theory of Self-Reproducing Automata*. A. W. Burks (ed.) (University of Illinois Press, Urbana).

Wang, H. (1974) *From Mathematics to Philosophy* (Humanities Press, New York).

———(1978) Kurt Gödel's intellectual development. *The Mathematical Intelligencer* 1, 182–184.

———(1981) Some facts about Kurt Gödel. *The Journal of Symbolic Logic* 46, 653–659.

Weyl, H. (1910) Über die Definitionen der mathematischen Grundbegriffe. *Mathematisch-naturwissenschaftliche Blätter* 7, 93–95 and 109–113; reprinted in Weyl (1968), Vol. I, 298–304.

———(1918) *Das Kontinuum. Kritische Untersuchungen über die Grundlagen der Analysis* (Veit, Leipzig).

———(1918a) *Raum, Zeit, Materie* (Springer-Verlag, Berlin).

———(1919) Der circulus vitiosus in der heutigen Begründung der Analysis. *Jahresbericht der Deutschen Mathematiker Vereinigung* 28, 85–92; reprinted in Weyl (1968), 43–50.

———(1946) Mathematics and logic. A brief survey serving as a preface to a review of *The Philosophy of Bertrand Russell. American Mathematical Monthly* 53, 2–13; reprinted in Weyl (1968), vol. IV, 268–279.

———(1949) *Philosophy of Mathematics and Natural Science* (Princeton University Press, Princeton).

———(1968) *Gesammelte Abhandlungen*, Vols. I–IV. K. Chandrasekharan (ed.) (Springer-Verlag, Berlin).

———(1977) *Il Continuo. Indagini critiche sui fondamenti dell 'Analisi.* Italian translation of Weyl (1918) by A. B. Veit Riccioli (Bibliopolis, Naples).

———(1985) Axiomatic vs. constructive procedures in mathematics (lecture text). T. Tonietti (ed.), *The Mathematical Intelligencer* 7, 10–17 and 38.

———(1987) *The Continuum. A Critical Examination of the Foundations of Analysis.* English translation of Weyl (1918), by S. Pollard and T. Bole (Thomas Jefferson Press, University Press of America, Latham, distributors).

Whitehead, A. N., and Russell, B. (1910–1913) *Principia Mathematica*, Vols. 1–3 (Cambridge University Press, Cambridge).

Wigner, E. (1960) The unreasonable effectiveness of mathematics in natural sciences. *Communications in Pure and Applied Mathematics* 13, 1–14.

Winskel, G. (1993) *The Formal Semantics of Programming Languages* (MIT Press, Cambridge, MA).

Wittgenstein, L. (1922) *Tractatus Logico-Philosophicus.* English translation of *Logisch-philosophische Abhandlung*, with corrected German text (Harcourt Brace, New York).

Zemanek, H. (1978) Oskar Morgenstern (1902–1977) – Kurt Gödel (1906–1978). *Elektronische Rechenanlagen* 20, 209–211.

Zermelo, E. (1904) Beweis, daß jede Menge wohlgeordnet werden kann. (Aus einem an Herrn Hilbert gerichteten Briefe.) *Mathematische Annalen* 59, 514–516; English translation in van Heijenoort (1967), 139–141.

———(1908) Untersuchungen über die Grundlagen der Mengenlehre, I. *Mathematische Annalen* 65, 261–281; English translation in van Heijenoort (1967), 199–215.

———(1908a) Neuer Beweis für die Möglichkeit einer Wohlordnung. *Mathematische Annalen* 65, 107–128; English translation in van Heijenoort (1967), 183–198.

———(1930) Über Grenzzahlen und Mengenbereiche: neue Untersuchungen über die Grundlagen der Mengenlehre. *Fundamenta Mathematicae* 16, 29–47.

Index